Studies in Fuzziness and Soft Computing

Volume 385

The series "Studies in Fuzziness and Soft Computing" contains publications on various topics in the area of soft computing, which include fuzzy sets, rough sets, neural networks, evolutionary computation, probabilistic and evidential reasoning, multi-valued logic, and related fields. The publications within "Studies in Fuzziness and Soft Computing" are primarily monographs and edited volumes. They cover significant recent developments in the field, both of a foundational and applicable character. An important feature of the series is its short publication time and world-wide distribution. This permits a rapid and broad dissemination of research results.

Indexed by ISI, DBLP and Ulrichs, SCOPUS, Zentralblatt Math, GeoRef, Current Mathematical Publications, IngentaConnect, MetaPress and Springerlink. The books of the series are submitted for indexing to Web of Science.

More information about this series at http://www.springer.com/series/2941

Amarpreet Kaur · Janusz Kacprzyk ·
Amit Kumar

Fuzzy Transportation and Transshipment Problems

Springer

Amarpreet Kaur
Irving K. Barber School of Arts
and Sciences
The University of British Columbia
Kelowna, BC, Canada

Janusz Kacprzyk
Systems Research Institute
Polish Academy of Sciences
Warsaw, Poland

Amit Kumar
School of Mathematics
Thapar Institute of Engineering
and Technology
Patiala, Punjab, India

ISSN 1434-9922 ISSN 1860-0808 (electronic)
Studies in Fuzziness and Soft Computing
ISBN 978-3-030-26678-3 ISBN 978-3-030-26676-9 (eBook)
https://doi.org/10.1007/978-3-030-26676-9

This Springer imprint is published by the registered company Springer Nature Switzerland AG
The registered company address is: Gewerbestrasse 11, 6330 Cham, Switzerland

Contents

Chapter 1
Introduction

In today's highly competitive market, the pressure on organizations to find better ways to send the products to the customers in a cost-effective manner, becomes more challenging. Transportation models [33] provide a powerful framework to meet this challenge.

To find the optimal solution of transportation problems it is assumed that the direct route between a source node and a destination node is a minimum-cost route. However, in real life transportation problems there may exist some nodes, called intermediate nodes, at which the product may be stored in case of excess of the available product and later on the product may be supplied from these intermediate nodes to the destinations. Such types of transportation problems are known as transshipment problems.

Also, to find the optimal solution of transportation problems it is assumed that same type of conveyances are used to transport the product from sources to destinations. However, in real life problems different types of conveyances are used for transporting the product from sources to destinations e.g., in many industrial problems, a homogeneous product is delivered from a source to a destination by different conveyances such as trucks, cargo flights, trains, ships etc. For such a case, the transportation problem turns into the solid transportation problem.

In conventional transportation problems, transshipment problems and solid transportation problems it is assumed that decision maker is sure about the precise values of transportation cost, availability and demand of the product. In the real world applications all the parameters of the transportation problems may not be known precisely due to uncontrollable factors.

To quantitatively deal with imprecise information, the concepts and techniques of probability could be employed. However, probability distributions require either a priori predictable regularity or a posteriori frequency distribution to construct. Moreover, the premise that imprecision can be equated with randomness is still questionable.

A. Kaur et al., *Fuzzy Transportation and Transshipment Problems*, Studies in Fuzziness and Soft Computing 385,
https://doi.org/10.1007/978-3-030-26676-9_1

As an alternative, uncertain values can be represented by fuzzy sets, more specifically by their membership functions in the classic sens of Zadeh [71]. The main advantages of methodologies based on fuzzy theory are that they do not require prior predictable regularities or posterior frequency distributions and they can deal with imprecise input information containing feelings and emotions quantified based on the decision maker's subjective judgment. Due to the same reason several authors have represented some or all the parameters of transportation problems, transshipment problems and solid transportation problems by fuzzy numbers and proposed different methods for solving these problems.

The existing methods for solving fuzzy transportation problems, depending upon the fuzziness of decision variables, can be broadly divided into two groups. In the first group, the decision variables are assumed as real numbers i.e., in an uncertain environment, a crisp decision is to be made to meet some decision criteria. In the second group, decision variables are assumed as fuzzy numbers. Since, on assuming the decision variables as fuzzy numbers instead of a single optimal solution, a set of possible optimal solutions are obtained and an appropriate solution can be chosen from the set of possible solutions i.e., fuzzy optimal solution provides a range of flexibility to the decision maker. So, in the literature the use of fuzzy numbers is advocated, cf. Allahviranloo et al. [5], Allahviranloo et al. [2, 3], Allahviranloo et al. [4], Avineri and Prashket [7], Baykasoglu and Gocken [9], Buckley and Feuring [10], Buckley et al. [12], Buckley and Jowers [11], Dehgan et al. [20], Ghazanfari et al. [29], Hashemi et al. [32], Jowers et al. [39], Kacprzyk and Fedrizzi [41], Kacprzyk and Orlovsky [40, 42], Kaur and Kumar [43], Kumar et al. [46], Kumar et al. [47], Lai and Hwang [48], Lotfi et al. [55], Maleki et al. [56], Rommelfanger [60], Tanaka and Asai [68, 69], Tanaka et al. [70], Zimmermann [72–74], to just name a few.

1.1 A Brief Literature Review

In this section, a brief review of the main works done in the area of single-objective fuzzy transportation problems, fuzzy transshipment problems and fuzzy solid transportation problems are presented.

Oheigeartaigh [57] proposed an algorithm to find the crisp optimal solution of such fuzzy transportation problems in which the availabilities and demands are represented by triangular fuzzy numbers. Chanas et al. [15] presented a fuzzy linear programming model to find the crisp optimal solution of fuzzy transportation problems with crisp cost coefficients, fuzzy availability and fuzzy demand values. Ishii et al. [34] considered a fuzzy version of the transportation problem by introducing two kinds of membership functions which characterize fuzzy supply and fuzzy demands to determine a crisp optimal flow that maximize the smallest value of all membership functions under the constraints that the total transportation cost must not exceed a certain upper limit.

Tada et al. [67] generalized the existing fuzzy transportation problem due to Ishii et al. [34] into an integer fuzzy transportation problem by adding an additional integral constraint of flow and proposed a method for solving integer fuzzy transportation problems. Chanas et al. [16] formulated the fuzzy transportation problems in three different situations and proposed a method for finding the crisp optimal solution of fuzzy transportation problems. Tada and Ishii [66] pointed out that in the existing fuzzy transportation problem [67] the budget constraint is not considered due to which there may exist more than one crisp optimal solution with the same objective value and considered another generalized problem by adding the budget constraint in the existing fuzzy transportation problem [67] and modified the existing method [67] for solving these types of transportation problems. Chanas and Kuchta [13] proposed a method to find the crisp optimal solution of such fuzzy transportation problems in which all the cost parameters are represented by LR type fuzzy numbers.

Jimenez and Verdegay [38] pointed out that there is no solution method to solve the parametric solid transportation problem and proposed a genetic algorithm based solution method to find crisp optimal solution of such fuzzy solid transportation problems in which all the parameters except cost parameters are represented by trapezoidal fuzzy numbers. Chanas and Kuchta [14] proposed a method to find the integer crisp optimal solution of such fuzzy transportation problems in which availability and demand are represented by fuzzy numbers. Jimenez and Verdegay [36] proposed the methods for finding the crisp optimal solution of two kinds of fuzzy solid transportation problem i.e., the availabilities, demands and conveyance capacities are interval numbers and fuzzy numbers, respectively. Jimenez and Verdegay [37] designed an evolutionary algorithm based parametric approach to find the crisp optimal solution of such fuzzy solid transportation problems in which all the parameters except cost parameters are represented by trapezoidal fuzzy numbers. Kikuchi [44] represented all the parameters of transportation problem by triangular fuzzy numbers and used the fuzzy linear programming approach to find the set of values such that the smallest membership grade among them is maximized. Ahlatcioglu et al. [1] proposed an algorithm for finding the crisp optimal solution of fuzzy transportation problem by converting all the fuzzy parameters into intervals. Saad and Abbas [61] discussed the solution algorithm for solving the transportation problem in fuzzy environment.

Liu and Kao [53] proposed a method, based on Zadeh's extension principle, to find the fuzzy optimal solution of such fuzzy transportation problems in which all the parameters are represented by trapezoidal fuzzy numbers. Liu and Kao [54] proposed a method for find the crisp optimal solution of such fuzzy transshipment problems in which the cost parameters are represented by trapezoidal fuzzy numbers. Chiang [17] proposed a method to find the crisp optimal solution of fuzzy transportation problems with fuzzy demand and fuzzy availability. Liang et al. [50] proposed an interactive possibilistic linear programming approach for finding the crisp optimal solution of transportation planning decision problems with imprecise cost, availabilities and demand. Gani and Samuel [23] proposed an algorithm for finding the fuzzy optimal solution of such fuzzy transportation problems in which availabilities and demands are represented by triangular fuzzy numbers. Liu and Kao [54] extended the existing method proposed in their paper [53] to find a fuzzy optimal solution of such fuzzy solid transportation problems in which all the parameters are represented by trapezoidal fuzzy numbers. Gupta and Mehlawat [30] proposed a method to find the crisp optimal solution of such fuzzy transportation problems in which the availability and demand are represented by (λ, ρ) interval-valued fuzzy numbers, and used it to solve a practical example of selecting a new type of coal for a steel manufacturing unit.

Smimou [63] proposed a method to find the crisp optimal solution of a fuzzy transshipment model with fuzzy costs and transshipment model with fuzzy supply and fuzzy demand. Gani, Samuel and Balasubramanian [24] proposed an algorithm for finding the fuzzy initial basic feasible solution of such fuzzy transportation problems in which the cost, availabilities and demands are represented by the triangular fuzzy numbers.

Das and Baruah [18] proposed the use of Vogel's approximation method to find a fuzzy initial basic feasible solution and a modified method to find the fuzzy optimal solution of such fuzzy transportation problems in which all the parameters are represented by triangular fuzzy numbers. Ghatee and Hashemi [26] proposed a method to find the fuzzy optimal solution of such balanced fuzzy transshipment problems in which all the parameters are represented by LR type fuzzy numbers. Ghatee and Hashemi [27] and Ghatee et al. [28] applied the existing Ghatee and Hashemi's [26] method for solving real life problems. Li et al. [49] proposed a method, based on goal programming, to find a crisp optimal solution of such fuzzy transportation problems in which cost parameters are represented by fuzzy numbers. Jana and Roy [35] proposed a method for solving fuzzy linear programming problems with fuzzy variables and used it to find the fuzzy optimal solution of such fuzzy transportation problems in which availability and demand are represented by triangular fuzzy numbers. Lin [51] used a genetic algorithm for finding the crisp optimal solution of fuzzy transportation problems with fuzzy coefficients. Stephen Dinagar and Palanivel [65] proposed a fuzzy modified distribution method to find the fuzzy optimal solution of such fuzzy transportation problems in which all the parameters are represented by trapezoidal fuzzy numbers. Stephen Dinagar and Palanivel [64] proposed fuzzy Vogel's approximation method for finding an initial fuzzy solution of such fuzzy transportation problems in which all the parameters are represented by trapezoidal fuzzy numbers and proposed a fuzzy modified distribution method to find the fuzzy optimal solution from the initial fuzzy solution obtained.

Pandian and Natarajan [58] proposed a new method, a so called fuzzy zero point method, for finding a fuzzy optimal solution of such fuzzy transportation problems in which all the parameters are represented by trapezoidal fuzzy numbers. In their another paper, Pandian and Natarajan [59] proposed a new method based on fuzzy zero point method for finding a more-or-less fuzzy optimal solution for such fuzzy transportation problems with mixed constraints in which all the parameters are represented by trapezoidal fuzzy numbers. De and Yadav [19] modified the existing method due to Kikuchi [44] by using the trapezoidal fuzzy numbers instead of the triangular fuzzy numbers.

On the other extreme, that is in the context of using nature inspired, to be more specific evolutionary computation based, methods, Lin [52] pointed out that differential evolution has received increasing attention owing to its simplicity and effectiveness in solving numerical optimization problems and introduced a differential evolution algorithm to find a crisp optimal solution of fuzzy transportation problems with fuzzy coefficients. Guzel [31] proposed a method to find the crisp optimal solution of such fuzzy transportation problems in which all the parameters are represented by triangular fuzzy numbers.

Dutta and Murthy [21] proposed a method to find the crisp optimal solution of such fuzzy transportation problems with additional impurity restrictions in which cost parameters are represented by fuzzy numbers. Basirzadeh [8] used classical algorithms for finding a fuzzy optimal solution of fully fuzzy transportation problems by transforming the fuzzy parameters into the crisp parameters. Samuel and Venkatachalapathy [62] proposed a new method, a modified Vogel's approximation method, for finding a fuzzy optimal solution of fully fuzzy transportation problems. Gani et al. [25] used Arsham and Khan's simplex algorithm [6] to find a fuzzy optimal solution of such fuzzy transportation problems in which all the parameters are represented by trapezoidal fuzzy numbers. Kumar and Murugesan [45] proposed a modified revised simplex method to find the fuzzy optimal solution of such fuzzy transportation problems in which the availability and demand are represented by the triangular fuzzy numbers. Gani et al. [22] proposed a method to find the fuzzy optimal solution of such fuzzy transshipment problems with mixed constraints in which all the parameters are represented by the triangular fuzzy numbers.

After reviewing the literature, it can be concluded that very few methods are proposed in the literature for solving such fuzzy transportation problems, fuzzy transshipment problems and fuzzy solid transportation problems in which at least one parameter of each type (at least one of the cost parameters, at least one of the availability parameters, at east one of the demand parameters) and all the decision variables are represented by the fuzzy numbers.

In the literature (c.f., for instance: Allahviranloo et al. [3], Allahviranloo et al. [4], Buckley et al. [12], Hashemi et al. [32], Dehghan et al. [20], Lotfi et al. [55]; see also Kaur and Kumar's [43] book) such linear programming problems in which at least one parameter of each type as well as all the decision variables are represented by fuzzy numbers are usually called *fully fuzzy linear programming problems*.

Therefore, using the same line of reasoning, such fuzzy transportation problems, fuzzy transshipment problems and fuzzy solid transportation problems in which at least one of the cost parameters, at least one of the availability parameters, at least one of the demand parameters and all the decision variables are represented by fuzzy numbers may be termed *fully fuzzy transportation problems*, *fully fuzzy transshipment problems* and *fully fuzzy solid transportation problems*.

In this book we will present a novel approach to the formulation and solution of such fuzzy transportation and transshipment problems, pointing out some limitations of the existing formulations and approaches, indicating some possible, conceptually and algorithmically attractive solutions to alleviate or avoid these shortcomings and limitations of all the existing methods. To be more specific, we will do this by proposing these new conceptual and algorithmic solutions for finding the fuzzy optimal solutions of: the single-objective fully fuzzy transportation problems, the fully fuzzy transshipment problems and the fully fuzzy solid transportation problems.

Moreover, based on the novel concepts and solutions proposed by combining the concept of a fully fuzzy solid transportation problem and a fully fuzzy transshipment problem, new class of problems, a *fully fuzzy solid transshipment problem*, is proposed with its fuzzy linear programming formulation and methods for finding its fuzzy optimal solution.

1.2 Outline of the Book

Chapter 2 constitutes a brief introduction to fuzzy logic, fuzzy relations, fuzzy number and fuzzy arithmetic, and related concepts. Basic concepts and problems of fuzzy optimization are in turn discussed in Chap. 4 with our attention limited to general aspects, and not to the fuzzy transportation and transshipment problems which will be discussed in mode detail in next chapters.

In Chap. 4 we proceed to the very fuzzy transportation problem and, as a point of departure, we consider the following existing and well known methods: Basirzadeh's [8], Gani et al. [25], Liu and Kao [53], Pandian and Natarajan's [58, 59] which are meant to find the fuzzy optimal solution of a fully fuzzy transportation problem. We analyze the pros and cons of all these methods and, to overcome some of their shortcomings and limitations, we propose two new methods for solving fully fuzzy transportation problems. First, we point out some advantages, from both an analytic and algorithmic points of view, of the new proposed methods over the existing methods mentioned above. Then, we show how an existing real life fully fuzzy transportation problem can be solved by using our new methods.

In Chap. 5 we further consider the new methods for the solution of the fully fuzzy transportation problem presented in Chap. 4, and point out again some of their limitations. Then, we propose some new methods for solving the fully fuzzy transportation problems by modifying the methods, proposed in Chap. 4. The advantages of the proposed methods over the methods proposed in Chap. 4 are discussed. To illustrate the proposed methods, a realistic fully fuzzy transportation problem is solved.

In Chap. 6 we are concerned with the approach of Ghatee and Hashemi [26] which is presumably the only method for finding a fuzzy optimal solution of the fully fuzzy transshipment problem reported in the literature. We analyze this method in much detail, point out some of its limitations, and propose some new means to alleviate, or even overcome these limitations. We propose two new methods for solving the fully fuzzy transshipment problems. The advantages of the proposed methods over the existing method mentioned above [26], but also over the new methods proposed in previous chapters, are discussed. For illustration we solve an example of a fully fuzzy transshipment problem, as well as a real life fully fuzzy transshipment problem.

In Chap. 7 we consider first Liu and Kao's [53] method for finding the fuzzy optimal solution of the fully fuzzy solid transportation problems. This seems to be the only method existing in the literature that is developed for this purpose. In this chapter, we first analyze some limitations and shortcomings of existing methods, and—to alleviate or even overcome them—we propose two new methods for solving the fully fuzzy solid transportation problem. The advantages of the proposed methods over the existing method due to Liu and Kao [53, 54], and then over the new method proposed in Chaps. 4 and 5, are discussed. To illustrate the new methods proposed, an example of the fully fuzzy solid transportation problem is solved, followed by the solution of a real life problem of this class.

Chapter 8 deals with the fully fuzzy transshipment problems which are obtained by introducing some intermediate nodes in the fully fuzzy transportation problems. Then, the fully fuzzy solid transportation problems are obtained by introducing the additional conveyances in the fully fuzzy transportation problems. However, in real life problems both the intermediate nodes and additional conveyances are simultaneously used so that, by combining the concept of the fully fuzzy solid transportation problems and the fully fuzzy transshipment problems, a new type of problems, termed *fully fuzzy solid transshipment problems* are introduced, followed by their corresponding fuzzy linear programming formulation. Then, two new methods for finding its fuzzy optimal solution, are proposed. The advantages of the proposed methods over the methods introduced in the previous chapters and over the well known existing method by Ghatee and Hashemi [26] are briefly presented. And, again, for illustration, we first solve a fully fuzzy solid transshipment problem using the new method proposed.

In Chap. 9 we briefly summarize the new methods proposed in the previous chapters, present some concluding remarks, and some possible and more promising future research directions.

References

1. M. Ahlatcioglu, M. Sivri, N. Guzel, Transportation of the fuzzy amounts using the fuzzy cost. J. Marmara Pure Appl. Sci. **18**, 141–157 (2002)
2. T. Allahviranloo, K.H. Shamsolkotabi, N.A. Kiani, L. Alizadeh, Fuzzy integer linear programming problems. Int. J. Contemp. Math. Sci. **2**, 167–181 (2007)
3. T. Allahviranloo, F.H. Lotfi, M.K. Kiasary, N.A. Kiani, L. Alizadeh, Solving fully fuzzy linear programming problem by the ranking function. Appl. Math. Sci. **2**, 19–32 (2008)
4. T. Allahviranloo, N. Mikaeilvand, F.H. Lotfi, M.K. Kiasari, Fully fuzzy linear programming problem with positive or negative core. Far East J. Appl. Math. **33**, 337–350 (2008)
5. T. Allahviranloo, F.H. Lotfi, L. Alizadeh, N. Kiani, Degeneracy in fuzzy linear programming problems. J. Fuzzy Math. **17**, 389–402 (2009)
6. H. Arsham, A.B. Kahn, A simplex-type algorithm for general transportation problems: an alternative to stepping-stone. J. Oper. Res. Soc. **40**, 581–590 (1989)
7. E. Avineri, J. Prashker, A. Ceder, Transportation projects selection process using fuzzy sets theory. Fuzzy Sets Syst. **116**, 35–47 (2000)
8. H. Basirzadeh, An approach for solving fuzzy transportation problem. Appl. Math. Sci. **5**, 1549–1566 (2011)
9. A. Baykasoglu, T. Gocken, A direct solution approach to fuzzy mathematical programs with fuzzy decision variables. Expert Syst. Appl. **39**, 1972–1978 (2012)
10. J.J. Buckley, T. Feuring, Evolutionary algorithm solution to fuzzy problems: fuzzy linear programming. Fuzzy Sets Syst. **109**, 35–53 (2000)
11. J.J. Buckley, L.J. Jowers, *Simulating Continuous Fuzzy Systems* (Springer-Verlag, Berlin, 2006)

12. J.J. Buckley, T. Feuring, Y. Hayashi, Multi-objective fully fuzzified linear programming. Int. J. Uncertain. Fuzz. Knowl.-Based Syst. **9**, 605–621 (2001)
13. S. Chanas, D. Kuchta, A concept of the optimal solution of the transportation problem with fuzzy cost coefficients. Fuzzy Sets Syst. **82**, 299–305 (1996)
14. S. Chanas, D. Kuchta, Fuzzy integer transportation problem. Fuzzy Sets Syst. **98**, 291–298 (1998)
15. S. Chanas, W. Kolodziejczyk, A.A. Machaj, A fuzzy approach to the transportation problem. Fuzzy Sets Syst. **13**, 211–221 (1984)
16. S. Chanas, M. Delgado, J.L. Verdegay, M.A. Vila, Interval and fuzzy extension of classical transportation problems. Transp. Plan. Technol. **17**, 203–218 (1993)
17. J. Chiang, The optimal solution of the transportation problem with fuzzy demand and fuzzy product. J. Inf. Sci. Eng. **21**(2), 439–451 (2005)
18. M.K. Das, H.K. Baruah, Solution of the transportation problem in fuzzified form. J. Fuzzy Math. **15**, 79–95 (2007)
19. P.K. De, B. Yadav, Approach to defuzzify the trapezoidal fuzzy number in transportation problem. Int. J. Comput. Cogn. **8**, 64–67 (2010)
20. M. Dehghan, B. Hashemi, M. Ghatee, Computational methods for solving fully fuzzy linear systems. Appl. Math. Comput. **179**, 328–343 (2006)
21. D. Dutta, A.S. Murthy, Fuzzy transportation problem with additional restrictions. ARPN J. Eng. Appl. Sci. **5**, 36–40 (2010)
22. A.N. Gani, R. Baskaran, S.N. Mohamed Assarudeen, Mixed constraint fuzzy transshipment problem. Appl. Math. Sci. **6**, 2385–2394 (2012)
23. A.N. Gani, A.E. Samuel, Transportation problem in fuzzy environment. Bull. Pure Appl. Sci. **25E**, 415–420 (2006)
24. A.N. Gani, A.E. Samuel, R. Balasubramanian, A new algorithm for solving a fuzzy transportation problem. Adv. Fuzzy Sets Syst. **2**, 301–310 (2007)
25. A.N. Gani, A.E. Samuel, D. Anuradha, Simplex type algorithm for solving fuzzy transportation problem. Tamsui Oxf. J. Inf. Math. Sci. **27**, 89–98 (2011)
26. M. Ghatee, S.M. Hashemi, Ranking function-based solutions of fully fuzzified minimal cost flow problem. Inf. Sci. **177**, 4271–4294 (2007)
27. M. Ghatee, S.M. Hashemi, Generalized minimal cost flow problem in fuzzy nature: an application in bus network planning problem. Appl. Math. Model. **32**, 2490–2508 (2008)
28. M. Ghatee, S.M. Hashemi, M. Zarepisheh, E. Khorram, Preemptive priority based algorithms for fuzzy minimal cost flow problem: an application in hazardous materials transportation. Comput. Ind. Eng. **57**, 341–354 (2009)
29. M. Ghazanfari, A. Yousefli, M.S. Jabal Ameli, A. Bozorgi-Amiri, A new approach to solve time-cost trade-off problem with fuzzy decision variables. Int. J. Adv. Manuf. Technol. **42**, 408–414 (2009)
30. P. Gupta, M.K. Mehlawat, An algorithm for a fuzzy transportation problem to select a new type of coal for a steel manufacturing unit. TOP **15**, 114–137 (2007)
31. N. Guzel, Fuzzy transportation problem with the fuzzy amounts and the fuzzy costs. World Appl. Sci. J. **8**, 543–549 (2010)
32. S.M. Hashemi, M. Modarres, E. Nasrabadi, M.M. Nasrabadi, Fully fuzzified linear programming, solution and duality. J. Intell. Fuzzy Syst. **17**, 253–261 (2006)
33. F.L. Hitchcock, The distribution of a product from several sources to numerous localities. J. Math. Phys. **20**, 224–230 (1941)
34. H. Ishii, M. Tada, T. Nishida, Fuzzy transportation problem. J. Jpn. Soc. Fuzzy Theory Syst. **2**, 79–84 (1990)
35. B. Jana, T.K. Roy, Fuzzy linear programming with fuzzy variables and its application in capacitated transportation model. J. Fuzzy Math. **17**, 1001–1016 (2009)
36. F. Jimenez, J.L. Verdegay, Obtaining fuzzy solutions to the fuzzy solid transportation problem with genetic algorithms, in *Proceedings of FUZZ?IEEE?97 - Sixth IEEE International Conference on Fuzzy Systems*, vol. III (IEEE Press, 1997), pp. 1657–1663

37. F. Jimenez, J.-L. Verdegay, Uncertain solid transportation problems. Fuzzy Sets Syst. **100**(1–2), 45–57 (1998)
38. F. Jimenez, J.L. Verdegay, An evolutionary algorithm for interval solid transportation problems. Evol. Comput. **7**(1), 103–107 (1999)
39. L.J. Jowers, J.J. Buckley, K.D. Reilly, Simulating continuous fuzzy systems. Inf. Sci. **177**, 436–448 (2007)
40. J. Kacprzyk, S.A. Orlovski, Fuzzy optimization and mathematical programming: a brief introduction and survey, in *Optimization Models Using Fuzzy Sets and Possibility Theory*, ed. by J. Kacprzyk, S.A. Orlovski (Reidel, Dordrecht/Boston/Lancaster, 1987), pp. 50–72
41. J. Kacprzyk, M. Fedrizzi (eds.), *Fuzzy Regression Analysis* (Omnitech Press, Warsaw and Physica-Verlag, Heidelberg, 1992)
42. J. Kacprzyk, S.A. Orlovski (eds.), *Optimization Models Using Fuzzy Sets and Possibility Theory* (Reidel, Dordrecht/Boston/Lancaster, 1987)
43. J. Kaur, A. Kumar, *An Introduction to Fuzzy Linear Programming Problems: Theory Methods and Applications* (Springer, Heidelberg and New York, 2016)
44. S. Kikuchi, A method to defuzzify the fuzzy number: transportation problem application. Fuzzy Sets Syst. **116**, 3–9 (2000)
45. B.R. Kumar, S. Murugesan, On fuzzy transportation problem using triangular fuzzy numbers with modified revised simplex method. Int. J. Eng. Sci. Technol. **4**, 285–294 (2012)
46. E.V. Kumar, S.K. Chaturvedi, A.W. Deshpande, Failure probability estimation using fuzzy fault tree analysis (FFTA) with PDM data in process plants. Int. J. Perform. Eng. **4**, 271–284 (2008)
47. A. Kumar, J. Kaur, P. Singh, A new method for solving fully fuzzy linear programming problems. Appl. Math. Model. **35**, 817–823 (2011)
48. Y.J. Lai, C.L. Hwang, *Fuzzy Mathematical Programming: Methods and Applications* (Springer-Verlag, New York, 1992)
49. L. Li, Z. Huang, Q. Da, J. Hu, A new method based on goal programming for solving transportation problem with fuzzy cost. in *Proceedings International Symposiums on Information Processing* (2008), pp. 3–8
50. T.F. Liang, C.S. Chiu, H.W. Cheng, Using possibilistic linear programming for fuzzy transportation planning decisions. Hsiuping J. **11**, 93–112 (2005)
51. F.T. Lin, Solving the transportation problem with fuzzy coefficients using genetic algorithms. in *Proceedings IEEE International Conference on Fuzzy Systems* (2009), pp. 1468–1473
52. F.T. Lin, Using differential evolution for the transportation problem with fuzzy coefficients, in *Proceedings International Conference on Technologies and Applications of Artificial Intelligence* (2010), pp. 299–304
53. S.T. Liu, C. Kao, Solving fuzzy transportation problems based on extension principle. Eur. J. Oper. Res. **153**, 661–674 (2004)
54. S.T. Liu, C. Kao, Network flow problems with fuzzy arc lengths. IEEE Trans. Syst. Man Cybern. Part B Cybern. **34**, 765–769 (2004)
55. F.H. Lotfi, T. Allahviranloo, M.A. Jondabeha, L. Alizadeh, Solving a fully fuzzy linear programming using lexicography method and fuzzy approximate solution. Appl. Math. Model. **33**, 3151–3156 (2009)
56. H.R. Maleki, M. Tata, M. Mashinchi, Linear programming with fuzzy variables. Fuzzy Sets Syst. **109**, 21–33 (2000)
57. M. Oheigeartaigh, A fuzzy transportation algorithm. Fuzzy Sets Syst. **8**, 235–243 (1982)
58. P. Pandian, G. Natarajan, A new algorithm for finding a fuzzy optimal solution for fuzzy transportation problems. Appl. Math. Sci. **4**, 79–90 (2010)
59. P. Pandian, G. Natarajan, An optimal more-for-less solution to fuzzy transportation problems with mixed constraints. Appl. Math. Sci. **4**, 1405–1415 (2010)
60. H. Rommelfanger, Interactive decision making in fuzzy linear optimization problems. Eur. J. Oper. Res. **41**, 210–217 (1989)
61. O.M. Saad, S.A. Abbas, A parametric study on transportation problem under fuzzy environment. J Fuzzy Math **11**, 115–124 (2003)

62. A.E. Samuel, M. Venkatachalapathy, Modified vogel's approximation method for fuzzy transportation problems. Appl. Math. Sci. **5**, 1367–1372 (2011)
63. K. Smimou, A fuzzy transshipment model for allocating foreign currencies. Int. J. Oper. Res. **2**, 284–307 (2007)
64. D. Stephen Dinagar, K. Palanivel, On trapezoidal membership functions in solving transportation problem under fuzzy environment. Int. J. Comput. Phys. Sci. **1**, 1–12 (2009)
65. D. Stephen Dinagar, K. Palanivel, The transportation problem in fuzzy environment. Int. J. Algorithms Comput. Math. **2**, 65–71 (2009)
66. M. Tada, H. Ishii, An integer fuzzy transportation problem. Comput. Math. Appl. **31**, 71–87 (1996)
67. M. Tada, H. Ishii, T. Nishida, Fuzzy transportation problem with integral flow. Math. Jpn. **35**, 335–341 (1990)
68. H. Tanaka, K. Asai, Fuzzy solution in fuzzy linear programming problems. IEEE Trans. Syst. Man Cybern. SMC-14 325–328 (1984)
69. H. Tanaka, K. Asai, Fuzzy linear programming problems with fuzzy numbers. Fuzzy Sets Syst. **13**, 1–10 (1984)
70. H. Tanaka, P. Guo, H.J. Zimmermann, Possibility distributions of fuzzy decision variables obtained from possibilistic linear programming problems. Fuzzy Sets Syst. **113**, 323–332 (2000)
71. L.A. Zadeh, Fuzzy sets. Inf. Control **8**, 338–353 (1965)
72. H.J. Zimmermann, Fuzzy programming and linear programming with several objective functions. Fuzzy Sets Syst. **1**, 45–55 (1978)
73. H.J. Zimmermann, *Fuzzy Set Decision Making and Expert Systems* (Kluwer, Dordrecht, 1987)
74. H.J. Zimmermann, *Fuzzy Set Theory and Its Applications* (Kluwer, Dordrecht, 2001)

Chapter 2
A Brief Introduction to Fuzzy Sets

The purpose of this chapter is to briefly expose a novice reader to basic elements of the theory of fuzzy sets and fuzzy systems. Our presentation will intuitive and constructive, and limited mainly to the elements needed for pour further analysis. Our perspective will be in the "pure" fuzzy setting, and possibility theory (which is related to fuzzy sets theory) will not be discussed; the interested reader is referred to, e.g., Dubois and Prade [4]. Basically, we will outline the idea of a fuzzy set, basic properties of fuzzy sets, operations on fuzzy sets, some extensions of the basic concept of a fuzzy set, fuzzy relations and their compositions, linguistic variables, fuzzy conditional statements, and the compositional rule of inference, the extension principle, and fuzzy arithmetic (notably with based on the LR fuzzy numbers). We will also briefly present Bellman and Zadeh's [3] general approach to decision making in a fuzzy environment which is a point of departure for virtually all fuzzy decision making, optimization, control, etc. and is therefore important for our purposes.

2.1 Basic Definitions and Properties of Fuzzy Sets

Fuzzy sets theory introduced by Zadeh in 1965 [14], is a simple yet very powerful, and effective and efficient means, or calculus, to represent and handle imprecise information exemplified by "*tall* buildings," "*large* numbers," etc.

The main purpose of a (conventional) set in mathematics can be viewed as to formally characterize some concept (or property), for instance the concept of "integer numbers which are greater than or equal 3 and less than or equal 10" which can be uniquely represented just by showing all integer numbers that satisfy this condition, that is given by the following set: $\{x \in I : 3 \leq x \leq 10\} = \{3, 4, 5, 6, 7, 8, 9, 10\}$ where I is the set of integers which is here the *universe of discourse* (universe,

A. Kaur et al., *Fuzzy Transportation and Transshipment Problems*, Studies in Fuzziness and Soft Computing 385,
https://doi.org/10.1007/978-3-030-26676-9_2

universal set, referential, reference set, etc.) that contains all those elements which are relevant for the particular concept.

A conventional set, say A, may be equated with its *characteristic function* defined as

$$\varphi_A : X \longrightarrow \{0, 1\} \quad (2.1)$$

which associates with each element x of a universe of discourse $X = \{x\}$ a number $\varphi(x) \in \{0, 1\}$ such that: $\varphi_A(x) = 0$ means that $x \in X$ does not belong to the set A, and $\varphi_A(x) = 1$ means that x belongs to the set A.

Therefore, for the set verbally defined as "integer numbers which are greater than or equal 3 and less than or equal 10", its equivalent set $A = \{3, 4, 5, 6, 7, 8, 9, 10\}$, listing all the respective integer numbers, may be represented by its characteristic function

$$\varphi_A(x) = \begin{cases} 1 \text{ for } x \in \{3, 4, 5, 6, 7, 8, 9, 10\} \\ 0 \text{ otherwise} \end{cases}$$

Notice that in a conventional set there is a clear-cut differentiation between elements belonging to the set and not, i.e. the transition from the belongingness to non-belongingness is clear-cut and abrupt.

However, it is very difficult to try to formalize by means of a conventional set vague concepts which commonly occur in everyday human discourse as, e.g., "integer numbers which are *more or less* equal to 6", because an abrupt and clear-cut differentiation between the elements belonging and not belonging to the set is artificial here.

Therefore, Zadeh [14] proposed the idea of a *fuzzy set* as a class of objects with unsharp boundaries, i.e. in which the transition from the belongingness to non-belongingness is not abrupt, i.e. in which elements may belong to *partial degrees*, from the full belongingness to the full non-belongingness through all intermediate values.

This can be formally done by the replacement of the characteristic function $\varphi : X \longrightarrow \{0, 1\}$ is by a *membership function* defined as

$$\mu_A : X \longrightarrow [0, 1] \quad (2.2)$$

such that $\mu_A(x) \in [0, 1]$ is the degree to which an element $x \in X$ belongs to the fuzzy set A: from $\mu_A(x) = 0$ for the full non-belongingness to $\mu_A(x) = 1$ for the full belongingness, through all intermediate $(0 < \mu_A(x) < 1)$ values.

For instance, for the fuzzy set "integer numbers which are *more or less* 6", we can consider $x = 6$ to certainly belong to this fuzzy set so that $\mu_A(6) = 1$, $x = 5$ and $x = 7$ belong to this set "almost surely" so that $\mu_A(5)$ and $\mu_A(7)$ are very close to 1, and the more a number differs from 6, the less its $\mu_A(.)$, and the numbers below 1 and above 10 do not belong to this set, so that their $\mu_A(.) = 0$. This may be sketched as in Fig. 2.1 (depicted in a continuous form to be more illustrative).

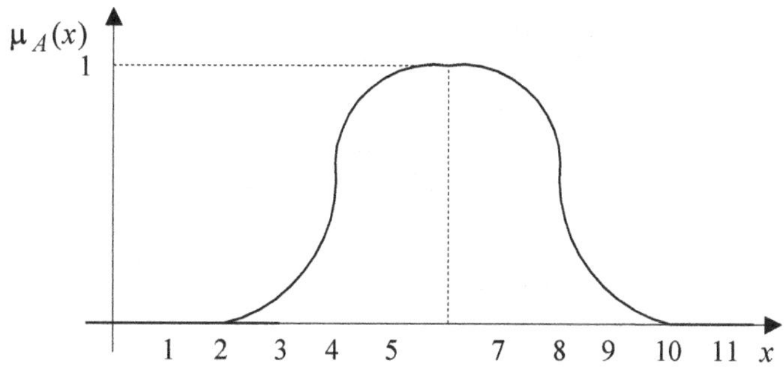

Fig. 2.1 Membership function of a fuzzy set "integer numbers which are *more or less* 6"

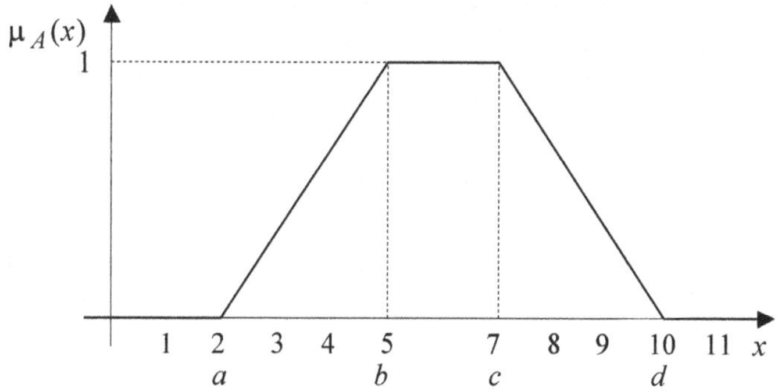

Fig. 2.2 Piecewise linear membership function of a fuzzy set "integer numbers which are *more or less* 6"

In practice the membership function is usually assumed to be piecewise linear as shown in Fig. 2.2 (for "integer numbers which are *more or less* 6" as in Fig. 2.1) in which the four numbers are only needed: $a, b, c,$ and d as, e.g., $a = 2, b = 5, c = 7,$ and $d = 10$ in Fig. 2.2.

Notice that the particular form of a membership function is *subjective* as opposed to an *objective* form of a characteristic function which is obvious because the concepts represented are subjective.

A *fuzzy set* A in a universe of discourse $X = \{x\}$, written A in X, is often defined, for convenience, as a set of pairs

$$A = \{(\mu_A(x), x)\} \tag{2.3}$$

where $\mu_A : X \longrightarrow [0, 1]$ is the *membership function* of A and $\mu_A(x) \in [0, 1]$ is the *grade of membership* of an element $x \in X$ in a fuzzy set A.

For practical reasons, it is often assumed (also here) that all the universes of discourse are finite as, e.g., $X = \{x_1, \ldots, x_n\}$. In such a case the pair $\{(\mu_A(x), x)\}$ will be denoted by "$\mu_A(x)/x$" which is called a *fuzzy singleton*.

Then, the fuzzy set A in X will be written as

$$A = \{(\mu_A(x), x)\} = \{\mu_A(x)/x\} = \\ = \mu_A(x_1)/x_1 + \cdots + \mu_A(x_n)/x_n = \sum_{i=1}^{n} \mu_A(x_i)/x_i \quad (2.4)$$

where "+" and "$\sum$" are meant in the set-theoretic sense, and, by convention, the pairs "0)/x" are omitted.

This basic concept of a (so-called *type 1*) fuzzy set presented above can be extended in various ways, for instance—to just list a few—to the L-fuzzy sets (Goguen [5]) in which the set of values of $\mu_A(.)$ is a lattice, a *type 2 fuzzy set* (Zadeh [14], Mendel [11]) in which the values of grades of membership are fuzzy sets themselves, an *intuitionistic fuzzy set* (Atanassov [1]) in which the grades of membership and non-membership do not need to sum up to 1, etc. These extensions will be employed here.

A fuzzy set A is said to be *empty*, written $A = \emptyset$, if and only if

$$\mu_A(x) = 0; \forall x \in X \quad (2.5)$$

Two fuzzy sets A and B in the same universe of discourse X are said to be *equal*, written $A = B$, if and only if

$$\mu_A(x) = \mu_B(x); \forall x \in X \quad (2.6)$$

Example 1 If $X = \{1, 2, 3\}$ and

$$A = 0.1/1 + 0.5/2 + 1/3 \\ B = 0.2/1 + 0.5/2 + 1/3 \\ C = 0.1/1 + 0.5/2 + 1/3$$

then $A = C$ but $A \neq C$ and $B \neq C$.

It is easy to see that this classic definition of the equality of two fuzzy sets by (2.6) is rigid and clear-cut, contradicting in a sense our intuition that the equality of fuzzy sets should be "softer," and not abrupt, i.e. should rather be to some degree, from 0 to 1. Some definitions of the equality to a degree can be found in, e.g., Kacprzyk [6], Dubois and Prade [4], Klir and Yuan [10].

A fuzzy set A defined in X is said to be *contained in* or a *subset of* a fuzzy set B in X, written $A \subseteq B$, if and only if

$$\mu_A(x) \leq \mu_B(x); \forall x \in X \quad (2.7)$$

Example 2 Suppose that $X = \{1, 2, 3\}$ and

$$A = 0.1/1 + 0.5/2 + 1/4$$
$$B = 0.1/1 + 0.4/2 + 0.9/3$$
$$C = 0.1/1 + 0.6/2 + 1/3$$

then only $B \subseteq A$.

Similarly as for the equality, the classic definition of the containment of two fuzzy sets by (2.7) is rigid and clear-cut, contradicting in a sense our intuition that the containment fuzzy sets should be "softer," and not abrupt, i.e. should rather be to some degree, from 0 to 1. Some definitions of the containment to a degree can be found in, e.g., Kacprzyk [6], Dubois and Prade [4], Klir and Yuan [10].

A fuzzy set A defined in X is said to be *normal* if and only if

$$\max_{x \in X} \mu_A(x) = 1 \tag{2.8}$$

and otherwise the fuzzy set is said to be *subnormal.*

There are now some important concepts of non-fuzzy sets associated with a fuzzy set.

The *support* of a fuzzy set A in X, written supp$A$, is the following (non-fuzzy) set

$$\text{supp}A = \{x \in X : \mu_A(x) > 0\} \tag{2.9}$$

and, evidently, $\emptyset \subseteq \text{supp}A \subseteq X$.

The *α-cut*, or *α-level set,* of a fuzzy set A in X, written A_α, is defined as the following (non-fuzzy) set

$$A_\alpha = \{x \in X : \mu_A(x) \geq \alpha\}; \forall \alpha \in (0, 1] \tag{2.10}$$

and if "$\geq$" in (2.10) is replaced by "$>$," then it stands for the *strong α-cut*, or *strong α-level set,* of a fuzzy set A in X. In principle, the α-cuts given by (2.10) will be used if not otherwise specified.

Example 3 If $X = \{1, 2, 3, 4\}$ and $A = 0.1/1 + 0.5/2 + 0.8/3 + 1/4$, then the following α-cuts are obtained

$$A_{0.1} = \{1, 2, 3, 4\} \; A_{0.5} = \{2, 3, 4\} \; A_{0.8} = \{3, 4\} \; A_1 = \{4\}$$

The α-cuts have many interesting and relevant properties, and among them one can mention the following

$$\alpha_1 \leq \alpha_2 \Longleftrightarrow A_{\alpha_1} \subseteq A_{\alpha_2} \tag{2.11}$$

The α-cuts play an extremely relevant role in both formal analyses and applications as they make it possible to uniquely replace a fuzzy set by a sequence of non-fuzzy sets (cf. Dubois and Prade [4], Klir and Yuan [10], Bede [2], or any book on fuzzy sets).

The following well known theorem, called the *representation theorem*, is very relevant both in theoretical analyses and applications.

Theorem 1 *Each fuzzy set A in X can be represented as*

$$A = \sum_{\alpha \in (0,1]} \alpha A_\alpha \tag{2.12}$$

where A_α is an α-cut of A defined as (2.11), "$\sum$" is in the set-theoretic sense, and αA_α denotes the fuzzy set whose degrees of membership are

$$\mu_{\alpha A_\alpha}(x) = \begin{cases} \alpha \text{ for } x \in A_\alpha \\ 0 \text{ otherwise} \end{cases} \tag{2.13}$$

Example 4 Let $X = \{1, 2, \ldots, 10\}$, and $A = 0.1/2 + 0.3/3 + 0.6/4 + 0.8/5 + 1/6 + 0.7/7 + 0.4/8 + 0.2/9$.

Then:

$$\begin{aligned} A = \sum_{\alpha \in (0,1]} \alpha A_\alpha = \\ &= 0.1(1/2 + 1/3 + 1/4 + 1/5 + 1/6 + 1/7 + 1/8 + 1/9) + \\ &\quad + 0.3(1/3 + 1/4 + 1/5 + 1/6 + 1/7 + 1/8 + 1/9) + \\ &\quad + 0.6(1/4 + 1/5 + 1/6 + 1/7) + 0.7(1/5 + 1/6 + 1/7) + \\ &\quad + 0.8(1/5 + 1/6) + 1(1/6) = \\ &= 0.1/2 + 0.3/3 + 0.6/4 + 0.8/5 + 1/6 + 0.7/7 + 0.4/8 + 0.2/9 \end{aligned}$$

Notice that the very essence of the representation theorem is that each fuzzy set can be uniquely represented by a set of its α-cuts, i.e. by a family of non-fuzzy sets.

An important issue, both in theory and application, is to be able to define the *cardinality* of a fuzzy set, i.e. to define how many elements it contains. Unfortunately, this is a difficult problem, and the definitions proposed have been criticized. The below one is presumably the most widely used.

A *nonfuzzy cardinality* of a fuzzy set $A = \mu_A(x_1)/x_1 + \cdots + \mu_A(x_n)/x_n$, the so-called *sigma-count*, denoted $\sum\text{Count}(A)$, is defined as (cf. Zadeh [13])

$$\sum\text{Count}(A) = \sum_{i=1}^{n} \mu_A(x_i) \tag{2.14}$$

Example 5 If $A = 1/x_1 + 0.8/x_2 + 0.6/x_3 + 0.2/x_4 + 0/x_5$, then

$$\sum \text{Count}(A) = 1 + 0.8 + 0.6 + 0.2 = 2.6$$

The $\sum$Count is very simple, and is hence widely used. However, an immediate objection may be that the set is fuzzy but its cardinality is not. A solution in this respect, a "fuzzy cardinality," was proposed by Zadeh [13], but its more complicated than a nonfuzzy cardinality defined by (2.14) and will not be discussed here. We refer the reader to Wygralak's [12] book.

Of a crucial importance, for theory and applications, is a *distance* between two fuzzy sets, notably the normalized one. Suppose that there are two fuzzy sets, A and B, both defined in $X = \{x_1, \dots, x_n\}$. Then, the following two basic (normalized) distances are:

- the *normalized linear* (Hamming) *distance* between A and B in X defined as

$$l(A, B) = \frac{1}{n} \sum_{i=1}^{n} \mid \mu_A(x_i) - \mu_B(x_i) \mid \tag{2.15}$$

- the *normalized quadratic* (Euclidean) *distance* between A and B in X defined as

$$q(A, B) = \sqrt{\frac{1}{n} \sum_{i=1}^{n} [\mu_A(x_i) - \mu_B(x_i)]^2} \tag{2.16}$$

Example 6 If $X = \{1, 2, \dots, 7\}$, $A = 0.7/1 + 0.2/2 + 0.6/4 + 0.5/5 + 1/6$ and $B = 0.2/1 + 0.6/4 + 0.8/5 + 1/7$, then:

$$l(A, B) = 0.37 \quad q(A, B) = 0.49$$

Now the basic set-theoretic and algebraic operations on fuzzy sets will be discussed. They are clearly crucial for both theory and practice.

2.2 Basic Set-Theoretic Operations on Fuzzy Sets

Similarly as in the conventional (non-fuzzy) set theory, the basic set-theoretic operations in fuzzy set theory are also the complement, intersection and union which will be defined below.

The *complement* of a fuzzy set A in X, written $\neg A$, is defined as

$$\mu_{\neg A}(x) = 1 - \mu_A(x); \forall x \in X \tag{2.17}$$

and the complement corresponds to the negation "not."

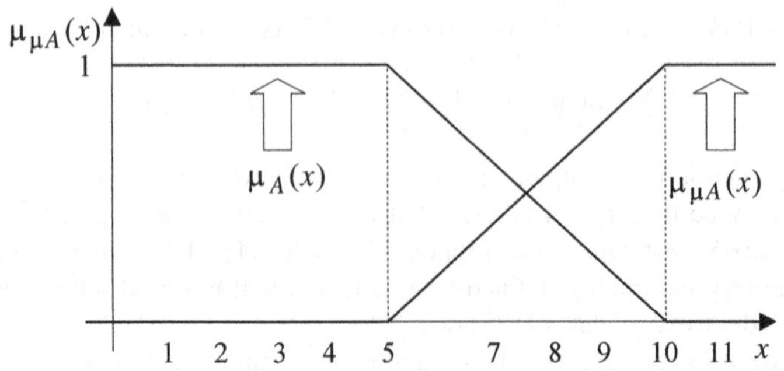

Fig. 2.3 Complement of a fuzzy set

Example 7 If $X = \{1, 2, 3\}$ and $A = 0.1/1 + 0.7/2 + 1/3$, then $\neg A = 0.9/1 + 0.3/2$.

The idea of the complement can be portrayed as in Fig. 2.3.

This definition is the simplest and most widely used, another definition, for $X = [0, 1]$, is

$$\mu_{\neg A}(x) = \mu_A(1 - x); \forall x \in [0, 1] \tag{2.18}$$

The *intersection* of two fuzzy sets A and B in X, written $A \cap B$, is defined as

$$\mu_{A \cap B}(x) = \mu_A(x) \wedge \mu_B(x); \forall x \in X \tag{2.19}$$

and the intersection corresponds to the connective "and"; "$\wedge$" is the minimum, i.e. $a \wedge b = \min(a, b)$, and can be replaced by a triangular norm (t-norm).

Example 8 If $X = \{1, 2, 3, 4\}$, and $A = 0.2/1 + 0.5/2 + 0.8/3 + 1/4$ and $B = 1/1 + 0.8/2 + 0.5/3 + 0.2/4$, then by (2.19) $A \cap B = 0.2/1 + 0.5/2 + 0.5/3 + 0.2/4$ is obtained.

The intersection can be illustrated as in Fig. 2.4 where $\mu_{A \cap B}(x)$ is shown in bold line.

The *union* of two fuzzy sets A and B in X, written $A + B$, is defined as

$$\mu_{A+B}(x) = \mu_A(x) \vee \mu_B(x); \forall x \in X \tag{2.20}$$

and the union of two fuzzy sets corresponds to the connective "or"; "$\vee$" is the maximum operation, i.e. $a \vee b = \max(a, b)$, and can be replaced by s triangular co-norm (t-conorm or s-norm).

Example 9 If $X = \{1, 2, 3, 4\}$, and $A = 0.2/1 + 0.5/2 + 0.8/3 + 1/4$ and $B = 1/1 + 0.8/2 + 0.5/3 + 0.2/4$, then $A + B = 1/1 + 0.8/2 + 0.8/3 + 1/4$.

The union can be portrayed as in Fig. 2.5 in which $\mu_{A+B}(x)$ is shown in bold line.

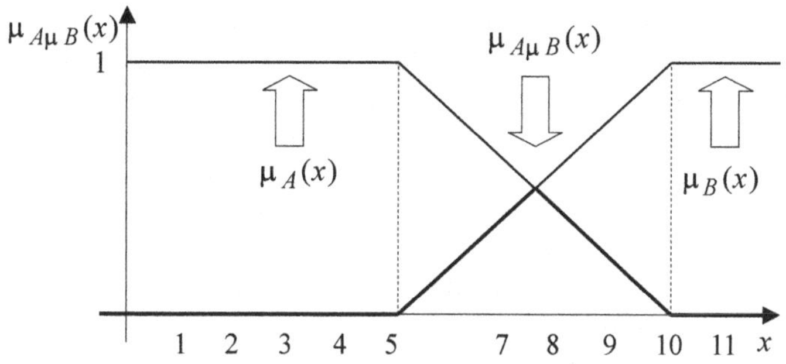

Fig. 2.4 Intersection of two fuzzy sets

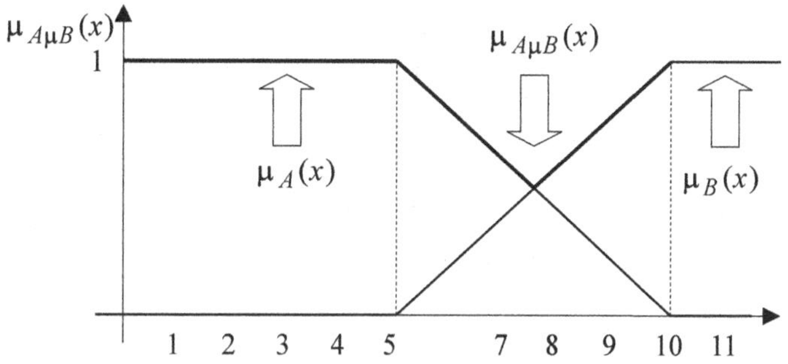

Fig. 2.5 Union of two fuzzy sets

The above definitions of the basic operations are widely used, and justified. However, other definitions are often employed too. In particular, for the intersection and union the t-norms and t-conorms (s-norms) are popular (cf. Klement, Mesiar and Pap's [9] book).

A *t-norm* is defined as

$$t : [0, 1] \times [0, 1] \longrightarrow [0, 1] \tag{2.21}$$

such that, for each $a, b, c \in [0, 1]$:

1. it has 1 as the unit element, i.e. $t(a, 1) = a$,
2. it is monotone, i.e. $a \le b \Longrightarrow t(a, c) \le t(b, c)$,
3. it is commutative, i.e. $t(a, b) = t(b, a)$, and
4. it is associative, i.e. $t[a, t(b, c)] = t[t(a, b), c]$.

Some more relevant examples of t-norms are:

- the minimum (which is the most widely used)

$$t(a, b) = a \wedge b = \min(a, b) \tag{2.22}$$

- the algebraic product

$$t(a, b) = a \cdot b \tag{2.23}$$

- the Łukasiewicz t-norm

$$t(a, b) = \max(0, a + b - 1) \tag{2.24}$$

An s-*norm* (or a t-*conorm*) is defined as

$$s : [0, 1,] \times [0, 1] \longrightarrow [0, 1] \tag{2.25}$$

such that, for each $a, b, c \in [0, 1]$:

1. it has 0 as the unit element, i.e. $s(a, 0) = a$,
2. it is monotone, i.e. $a \leq b \Longrightarrow s(a, c) \leq s(b, c)$,
3. it is commutative, i.e. $s(a, b) = s(b, a)$, and
4. it is associative, i.e. $s[a, s(b, c)] = s[s(a, b), c]$.

Some more relevant examples of s-norms are:

- the maximum (which is the most widely used)

$$s(a, b) = a \vee b = \max(a, b) \tag{2.26}$$

- the probabilistic product

$$s(a, b) = a + b - ab \tag{2.27}$$

- the Łukasiewicz s-norm

$$s(a, b) = \min(a + b, 1) \tag{2.28}$$

Notice that a t-norms is *dual* to an s-norms in that $s(a, b) = 1 - t(1 - a, 1 - b)$.

While defining the basic concepts and operations on fuzzy sets there have always used a linguistic interpretation to provide semantics. This is very relevant and has great implications for the development of fuzzy sets related tools and techniques, also for our purposes.

As to some other operations on fuzzy sets one should also mention the following ones.

The *product of a scalar* $a \in R$ *and a fuzzy set* A in X, written aA, is defined as

$$\mu_{aA}(x) = a\mu_A(x); \forall x \in X \tag{2.29}$$

where, by necessity, $0 \leq a \leq 1/\mu_A(x)$, for each $x \in X$.

The $k - th$ *power* of a fuzzy set A in X, written A^k, is defined as

$$\mu_{A^k}(x) = [\mu_A(x)]^k; \forall x \in X \tag{2.30}$$

where $k \in R$ and, evidently, $0 \leq [\mu_A(x)]^k \leq 1$.

The *adequacy* of operations on fuzzy sets, i.e. whether they do reflect the real human perception of their essence, whether they really reflect the semantics of "not," "and" and "or," etc. Diverse approaches have been used to find and justify a particular definition. These approaches may be classified as (cf. Dubois and Prade [4], Kacprzyk [6], Zimmermann [17]):

- *intuitive* as, the original Zadeh's [14] works in which it is shown by a rational argument that the operations defined are proper,
- *axiomatic* whose line of reasoning is to assume some set of plausible condition to be fulfilled, and then to show using some analytic tools that definitions assumed are the only possible ones;
- *experimental* whose essence is to device some psychological tests for a group of inquired individuals, and then use the responses to find which operation is best justified;

but we will not discuss this in this book, details can be found in any book on fuzzy sets.

2.3 Fuzzy Relations

A very important concept is that of a *fuzzy relation* which makes it possible to represent and process imprecisely specified relationships and dependences between variables.

A *fuzzy relation* R between two (nonfuzzy) sets $X = \{x\}$ and $Y = \{y\}$ is defined as a fuzzy set in the Cartesian product $X \times Y$, i.e.

$$\begin{aligned} R &= \{(\mu_R(x, y), (x, y))\} = \\ &= \{\mu_R(x, y)/(x, y)\}; \forall (x, y) \in X \times Y \end{aligned} \tag{2.31}$$

where $\mu_R(x, y) : X \times Y \longrightarrow [0, 1]$ is the membership function of the fuzzy relation R, and $\mu_R(x, y) \in [0, 1]$ is the degree to which the elements $x \in X$ and $y \in Y$ are in relation R between each other.

The above fuzzy relation is defined in the Cartesian product of (non-fuzzy!) two sets, X and Y, and is called a binary fuzzy relation. In general, a fuzzy relation may be defined in the Cartesian product of k sets, $X_1 \times \cdots \times X_k$, and is then called a k-ary fuzzy relation. In this perspective, a fuzzy set is a unary fuzzy relation.

Example 10 If $X = \{\text{horse, donkey}\}$ and $Y = \{\text{mule, cow}\}$, then the fuzzy relation R labeled "similarity" may be exemplified by

$$\begin{aligned} R &= \text{"similarity"} = \\ &= 0.8/(\text{horse, mule}) + 0.4/(\text{horse, cow}) + \\ &\quad +0.9/(\text{donkey, mule}) + 0.2/(\text{donkey, cow}) \end{aligned}$$

to be read as: the horse and the mule are similar (with respect to "our own" subjective aspects!) to degree 0.8, i.e. to a very high extent, the horse and the cow are similar to degree 0.4, i.e. to quite a low extent, etc.

Obviously, a fuzzy relation R in $X \times Y$ for X and Y of a sufficiently low dimensionality may be conveniently represented in the matrix form exemplified, for the fuzzy relation R = "similarity" in Example 10, by

R = "similarity" =

	y = mule	cow
x = horse	0.8	0.4
donkey	0.9	0.2

The fuzzy relation is clearly a fuzzy set so that all definitions, properties, operations, etc. on fuzzy sets hold. We will therefore present below some more specific ones.

The *max-min composition* of two fuzzy relations R in $X \times Y$ and S in $Y \times Z$, written $R \circ_{max-min} S$ is defined as a fuzzy relation in $X \times Z$ such that

$$\mu_{R \circ_{max-min} S}(x, y) = \\ = \max_{y \in Y}[\mu_R(x, y) \wedge \mu_S(y, z)]; \forall x \in X, z \in Z \quad (2.32)$$

and since this type of composition is the most widely used, also here, then it will be briefly denoted as $R \circ S$.

Example 11 If $X = \{1, 2\}$, $Y = \{1, 2, 3\}$ and $Z = \{1, 2, 3, 4\}$, and the fuzzy relations R and S are as below. Its resulting max-min composition, $R \circ S$, is then:

$R \circ S =$

	y = 1	2	3
x = 1	0.3	0.8	1
2	0.9	0.7	0.4

$\circ$

	z = 1	2	3	4
y = 1	0.7	0.6	0.4	0.1
2	0.4	1	0.7	0.2
3	0.5	0.9	0.6	0.8

=

	z = 1	2	3	4
x = 1	0.5	0.9	0.7	0.8
2	0.7	0.7	0.7	0.4

This max-min composition of fuzzy relations is the original Zadeh's definition (cf. Zadeh [13]), and is certainly the most widely used. However, since the "min" ($\wedge$) and "max" ($\vee$) are just specific examples of the t-norm and t-conorm (s-norm), then one can well define a much more general type of composition given below.

The *s-t–norm composition* of two fuzzy relations R in $X \times Y$ and S in $Y \times Z$, written $R \circ_{s-t} S$, is defined as a fuzzy relation in $X \times Z$ such that

$$\mu_{R \circ_{s-t} S}(x, z) = s_{y \in Y}[\mu_R(x, y)\, t\, \mu_S(y, z)], \qquad \text{for each } x \in X, z \in Z \quad (2.33)$$

Fuzzy relations, similarly as their non-fuzzy counterparts, play a crucial role in virtually all aspects of the theory and applications of fuzzy sets, notably in rule based fuzzy modeling. An important issue is related to so-called fuzzy relational equations. For a lack of space, we will not discuss them and will refer the interested reader to the literature, e.g. Klir and Yuan [10].

Finally, let us mention two concepts concerning the fuzzy sets that are related to fuzzy relations.

The *Cartesian product* of two fuzzy sets A in X and B in Y, written $A \times B$, is defined as a fuzzy set in $X \times Y$ such that

$$\mu_{A\times B}(x, y) = [\mu_A(x) \wedge \mu_B(y)]; \forall x \in X, y \in Y \tag{2.34}$$

A fuzzy relation R in $X \times Y \times \cdots \times Z$ is said to be *decomposable* if and only if it can be represented as

$$\begin{aligned} \mu_R(x, y, \ldots, z) = \mu_{R_x}(x) \wedge \mu_{R_y}(y) \wedge \ldots \\ \ldots \wedge \mu_{R_z}(z); \forall x \in X, y \in Y, \ldots, z \in Z \end{aligned} \tag{2.35}$$

where $\mu_{R_x}(x), \mu_{R_y}(y), \ldots, \mu_{R_z}(z)$ are projections of the fuzzy relation $\mu_R(x, y)$ on $X, Y, \ldots, Z$, respectively, defined as

$$\mu_{R_x}(x) = \sup_{\{y,\ldots,z\}\in Y\times Z} \mu_R(x, y, \ldots, z), ; \forall x \in X \tag{2.36}$$

The famous Zadeh's *extension principle* (cf. Zadeh [13]) is one of the most important and powerful tools in fuzzy sets theory and it addresses the following fundamental issue:

> If there is some relationship (e.g., a function) between "conventional" (nonfuzzy) entities (e.g., variables taking on nonfuzzy values, then what is its equivalent relationship between fuzzy entities (e.g., variables taking on fuzzy values)?

The extension principle makes it therefore possible, for instance, to extend some known conventional models, algorithms, etc. involving non-fuzzy variables to the case of fuzzy variables.

Let $A_1, \ldots, A_n$ be fuzzy sets in $X_1 = \{x_1\}$, ..., $X_n = \{x_n\}$, respectively, and

$$f : X_1 \times \cdots \times X_n \longrightarrow Y \tag{2.37}$$

be some (non-fuzzy) function such that $y = f(x_1, \ldots, x_n)$.

Then, according to the *extension principle*, the fuzzy set B in $Y = \{y\}$ induced by the fuzzy sets $A_1, \ldots, A_n$ via the function f (2.37) is

$$\mu_B(y) = \max_{(x_1,\ldots,x_n)\in X_1\times\ldots\times X_n : y=f(x_1,\ldots,x_n)} \bigwedge_{i=1}^{n} \mu_{A_i}(x_i) \tag{2.38}$$

Example 12 Suppose that: $X_1 = \{1, 2, 3\}$, $X_2 = \{1, 2, 3, 4\}$, f represents the addition, i.e. $y = x_1 + x_2$, $A_1 = 0.1/1 + 0.6/2 + 1/3$ and $A_2 = 0.6/1 + 1/2 + 0.5/3 + 0.1/4$, then

$$B = A_1 + A_2 = 0.1/2 + 0.6/3 + 0.6/4 + 1/5 + 0.5/6 + 0.1/7$$

and notice that "+" is used here in both the arithmetic sense (the sum of real and fuzzy numbers—cf. Sect. 2.4) and in the set-theoretic sense but this should not lead to confusion.

Equivalently, the extension principle (2.38) may also be written in terms of the α-cuts (2.10). Namely, if—for simplicity—$f : X \longrightarrow Y$, $X = \{x\}$, $Y = \{y\}$, and A_α, for each $\alpha \in (0, 1]$, are α-cuts of A, then the fuzzy set B in Y, induced by A via the extension principle is given as

$$B = f(A) = f(\sum_{\alpha \in (0,1]} \alpha \cdot A_\alpha) = \sum_\alpha \alpha f(A_\alpha) \tag{2.39}$$

which is clearly implied by the representation theorem (Theorem 2.12).

Notice that thanks to the extension principle one can extend our known non-fuzzy tools and techniques (their related algorithms and procedures) to their fuzzy counterparts.

2.4 Fuzzy Numbers and Fuzzy Arithmetic

The same fundamental role as non-fuzzy (real, integer, ...) numbers play in conventional models, the fuzzy numbers play in fuzzy models. They are of utmost importance for this book too.

A *fuzzy number* is defined as a fuzzy set in R, the real line. Usually, but not always, it is assumed to be a normal and convex fuzzy set. For example, the membership function of a fuzzy number "more or less 6" may be as shown in Fig. 2.6, i.e. as a bell-shaped function.

For our purposes operations on fuzzy numbers are the most relevant, and their definitions may readily be obtained by applying the extension principle (2.38).

Suppose therefore that A and B are two fuzzy numbers in $R = \{x\}$ characterized by their membership functions $\mu_A(x)$ and $\mu_B(x)$, respectively. Then, the extension principle yields the following definitions of the four basic *arithmetic operations* on fuzzy numbers:

- *addition*

$$\mu_{A+B}(z) = \max_{x+y=z} [\mu_A(x) \wedge \mu_B(y)]; \forall z \in R \tag{2.40}$$

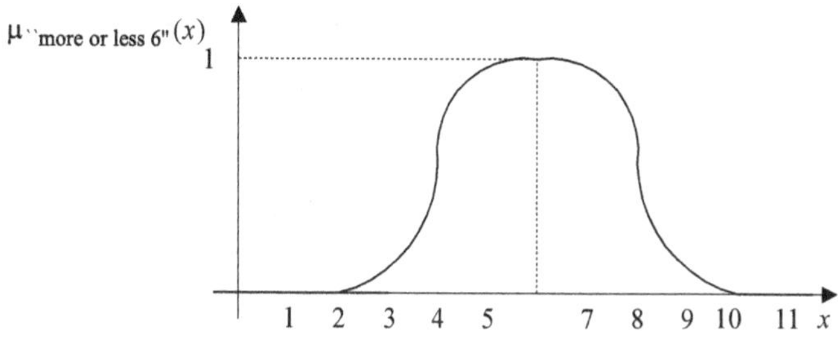

Fig. 2.6 Membership function of a fuzzy number "more or less 6"

- *subtraction*

$$\mu_{A-B}(z) = \max_{x-y=z} [\mu_A(x) \wedge \mu_B(y)]; \forall z \in R \tag{2.41}$$

- *multiplication*

$$\mu_{A \cdot B}(z) = \max_{x \cdot y=z} [\mu_A(x) \wedge \mu_B(y)]; \forall z \in R \tag{2.42}$$

- *division*

$$\mu_{A/B}(z) = \max_{x/y=z, y \neq 0} [\mu_A(x) \wedge \mu_B(y)]; \forall z \in R \tag{2.43}$$

- *minimum*

$$\mu_{\min(A,B)}(z) = \max_{\min(x,y)=z} [\mu_A(x) \wedge \mu_B(y)]; \forall z \in R \tag{2.44}$$

- *maximum*

$$\mu_{\max(A,B)}(z) = \max_{\max(x,y)=z} [\mu_A(x) \wedge \mu_B(y)]; \forall z \in R \tag{2.45}$$

The following one-argument operations on fuzzy numbers may also often be relevant:

- the *opposite* of a fuzzy number

$$\mu_{-A}(x) = \mu_A(-x); \forall x \in R \tag{2.46}$$

- the *inverse* of a fuzzy number

$$\mu_{A^{-1}}(x) = \mu_A(\frac{1}{x}); \forall x \in R \setminus \{0\} \tag{2.47}$$

In practice, however, such general definitions of fuzzy numbers and operations on them is seldom used. Normally, a further simplification is made, namely the fuzzy numbers are assumed to be *triangular* and, eventually, *trapezoid* fuzzy numbers whose membership functions are sketched in Fig. 2.7.

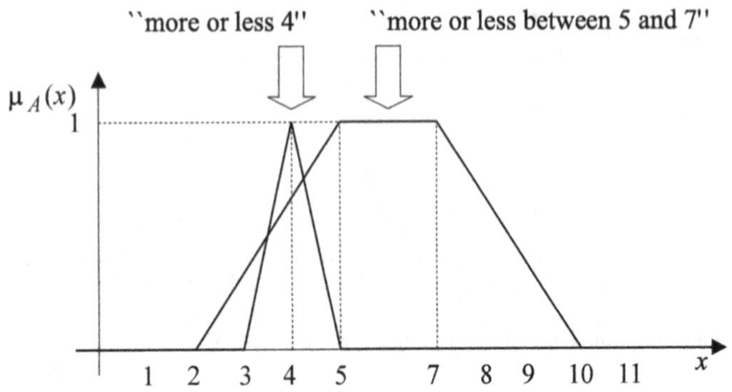

Fig. 2.7 Triangular ("*about* 4") and trapezoid ("*more or less* between 6 and 7") fuzzy numbers

For triangular and trapezoid fuzzy numbers, for the four basic operations, i.e. the addition, subtraction, multiplication and division, special formulas can be devised whose calculation is simpler that (2.40)–(2.43). The trapezoid fuzzy numbers, sometimes their special case the triangular fuzzy numbers, will be used in this work, and we will show specific forms of the arithmetic operations employed when they will be needed.

For other aspects of the fuzzy numbers, both of a general form, of the LR form, and triangular and trapezoid, we can refer the reader to a rich classic literature exemplified by Dubois and Prade [4], Klir and Yuan [10], Zimmermann [16, 17], etc.

More information on fuzzy numbers, notably the triangular and trapezoid ones, and on operations thereon, will be presented in Sect. 4.2, p. 47.

2.5 Fuzzy Events and Their Probabilities

Fuzziness and randomness are meant, in our perspective, as two aspects of *imperfect information*. Fuzziness is meant to concern entities and relations which are not crisply defined, with gradual transition between the elements belonging and not belonging to a class. Randomness concerns situations in which the event is well defined but its occurrence is uncertain.

However, in practice in many situations fuzziness and randomness can occur jointly like as when one asks about the probability of a *cold weather* tomorrow or of a *high inflation* in the next year, one has an imprecise (fuzzy) event—here cold weather and high inflation, respectively. In our context, an example may be when we would ask about the probability of a reliable delivery of a product that is highly perishable. For dealing with such often encountered problems, we need a concept of a *fuzzy event*, and of a *probability of a fuzzy event.*

The first, and still widely used approach is here by Zadeh [15]. He starts with the concept of a *fuzzy event* which is a fuzzy set A in $X = \{x\} = \{x_1, \dots, x_n\}$ and with a Borel measurable membership function. It is assumed that the probabilities of the (non-fuzzy) elementary events $x_1, \ldots, x_n \in X$ are known and equal to $p(x_1), \ldots, p(x_n) \in [0, 1]$, respectively, with $p(x_1) + \cdots + p(x_n) = 1$.

We then have some important properties:

- two fuzzy events A and B in X are *independent* if and only if

$$p(AB) = p(A)p(B) \tag{2.48}$$

- the *conditional probability* of a fuzzy event A in X with respect to a fuzzy event B in X is denoted $p(A \mid B)$ and defined as

$$p(A \mid B) = \frac{p(AB)}{p(B)}; \; p(B) > 0 \tag{2.49}$$

- and if the fuzzy events A and A are independent, then

$$p(A \mid B) = p(A) \tag{2.50}$$

and notice that they are analogous to their non-fuzzy counterparts.

The *(non-fuzzy) probability of a fuzzy event* A in $X = \{x_1, \ldots, x_n\}$ is denoted $p(A)$ and defined by Zadeh [15] as

$$p(A) = \sum_{i=1}^{n} \mu_A(x_i)p(x_i) \tag{2.51}$$

i.e. as the expected value of the membership function of A, $\mu_A(x)$.

Example 13 Suppose that $X = \{1, 2, \ldots, 5\}$, $p(x_1) = 0.1$, $p(x_2) = 0.1$, $p(x_3) = 0.1$, $p(x_4) = 0.3$, $p(x_5) = 0.4$, and $A = 0.1/2 + 0.5/3 + 0.7/4 + 0.9/5$. Then

$$p(A) = 0.1 \times 0.1 + 0.1 \times 0.5 + 0.3 \times 0.7 + 0.4 \times 0.9 = 0.73$$

Notice that the above (non-fuzzy) probability of a fuzzy event (2.51) satisfies:

1. $p(\emptyset) = 0$,
2. $p(\neg A) = 1 - p(A)$,
3. $p(A + B) = p(A) + p(B) - p(A \cap B)$,
4. and

$$\begin{aligned} p(\sum_{i=1}^{r} A_i) = \sum_{i=1}^{r} p(A_i) - \sum_{j=1}^{r} \sum_{k=1,k<j}^{r} p(A_j \cap A_k) + \\ + \sum_{j=1}^{r} \sum_{k=1,k<j}^{r} \sum_{l=1,l<k}^{r} p(A_j \cap A_k \cap A_l) + \cdots \\ \cdots + (-1)^{r+1} p(A_1 \cap A_2 \cap \ldots \cap A_r) \end{aligned}$$

so it does make sense to term the expression (2.51) a "probability."

The above Zadeh' [15] classic definition of a (non-fuzzy) probability of a fuzzy event is by far the most popular and most widely used. The reason is simple: there is a natural ordering of real numbers so that it is easy to use the results obtained as a non-fuzzy probability. However, this may be viewed counter-intuitive, and there are also some approaches to a fuzzy probability of a fuzzy event, see, e.g., Kacprzyk [6] or Klir and Yuan [10], but we will not consider them.

2.6 Defuzzification of Fuzzy Sets

In many applications one arrives at a fuzzy result. However, in it is a crisp (non-fuzzy) result that should be applied. A notable example is here fuzzy control (cf. Kacprzyk [6]).

Suppose that there is a fuzzy set A defined in $X = \{x_1, x_2, \ldots, x_n\}$, i.e. $A = \mu_A(x_1)/x_1 + \mu_A(x_2)/x_2 + \ldots + \mu_A(x_n)/x_n$. One needs to find a crisp number $a \in [x_1, x_n]$ which "best" represents A. Notice that it is assumed here that A is defined in a finite universe of discourse, but its corresponding defuzzified number a need not be in general any of the finite values of X but should be between the lowest and highest elements of X (evidently, this requires some ordering of x_i's but this is clearly satisfied as x_i's are in virtually all practical cases just real numbers).

The most commonly used defuzzification procedure is certainly the *center-of-area*, also called the *center-of-gravity*, method whose essence is

$$a = \frac{\sum_{i=1}^{n} x_i \mu_A(x_i)}{\sum_{i=1}^{n} \mu_A(x_i)} \tag{2.52}$$

The above defuzzification (2.52) is however often too complex if our analysis involves, e.g., some optimization (cf. Kacprzyk and Orlovski [7, 8]). In such a case one needs to resort to an even simpler defuzzification method which simply assumes that the defuzzified value of a fuzzy value is $x_i \in X = \{x_1, \ldots, x_n\}$ for which $\mu_A(x)$ takes on its maximum values, i.e.

$$\mu_A(a) = \max_{x_i \in X} \mu_A(x) \tag{2.53}$$

with an obvious extension that if the A determined in (2.53) is not unique, then one takes, say, the mean value of such equivalent a's or selects one of those a's randomly.

There is a whole array of other defuzzification procedures, and the reader can be referred to virtually all books on fuzzy sets already cited.

This concludes or brief introduction to the theory of fuzzy sets. We have basically limited our attention to elements which can be of use for our purposes and referring the interested reader for more details to the literature, mainly to classic books that are widely available.

References

1. K.T. Atanassov, *Intuitionistic Fuzzy Sets* (Springer Physica-Verlag, 1999)
2. B. Bede, *Mathematics of Fuzzy Sets and Fuzzy Logic* (Springer, New York and Heidelberg, 2013)
3. R.E. Bellman, L.A. Zadeh, Decision-making in a fuzzy environment. Manag. Sci. **17**, 141–164 (1970)
4. D. Dubois, H. Prade, *Fuzzy Sets and Systems: Theory and Applications* (Academic Press, New York, 1980)
5. J.A. Goguen, L-fuzzy sets. J. Math. Anal. Appl. **18**, 145–174 (1967)
6. J. Kacprzyk, *Multistage Fuzzy Control: A Model-Based Approach to Control and Decision-Making* (Wiley, Chichester, 1997)
7. J. Kacprzyk, S.A. Orlovski, Fuzzy optimization and mathematical programming: a brief introduction and survey, in *Optimization Models Using Fuzzy Sets and Possibility Theory*, ed. by J. Kacprzyk, S.A. Orlovski (Reidel, Dordrecht/Boston/Lancaster, 1987), pp. 50–72
8. J. Kacprzyk, S.A. Orlovski (eds.), *Optimization Models Using Fuzzy Sets and Possibility Theory* (Reidel, Dordrecht/Boston/Lancaster, 1987)
9. E.P. Klement, R. Mesiar, E. Pap, *Triangular Norms* (Springer, 2013)
10. G.J. Klir, B. Yuan, *Fuzzy Sets and Fuzzy Logic: Theory and Applications* (Prentice-Hall, New Jersey, 1996)
11. J.M. Mendel, General type-2 fuzzy logic systems made simple: a tutorial. IEEE Trans. Fuzzy Syst. **22**, 1162–1182 (2014)
12. M. Wygralak, *Intelligent Counting Under Information Imprecision—Applications to Intelligent Systems and Decision Support* (Springer, 2013)
13. L.A. Zadeh, Outline of a new approach to the analysis of complex systems and decision processes. IEEE Trans. Syst. Man Cybern. **3**(1), 28–44 ?(1973)
14. L.A. Zadeh, Fuzzy sets. Inf. Control **8**, 338–353 (1965)
15. L.A. Zadeh, Probability measures of fuzzy events. J. Math. Anal. Appl. **23**, 421–427 (1968)
16. H.J. Zimmermann, *Fuzzy Set Decision Making and Expert Systems* (Kluwer, Dordrecht, 1987)
17. H.J. Zimmermann, *Fuzzy Set Theory and Its Applications* (Kluwer, Dordrecht, 2001)

Chapter 3
A Brief Introduction to Fuzzy Optimization and Fuzzy Mathematical Programming

The purpose of this chapter is to briefly summarize main elements, issues and developments in fuzzy optimization. A brief account of the very essence of the fuzzy transportation problem, and its extended fuzzy transshipment problem, is then presented. We basically concentrate on some more traditional and classic approaches which best serve our very purpose to provide a point of departure for our discussion in this book.

3.1 Introductory Remarks

Optimization is part of *decision making* problems that can be viewed as that there is:

- a set of feasible options (alternatives, variants, ...),
- a representation of preferences over the set of options as, e.g., by pairwise comparisons, preference orderings, utility functions, etc.,
- a choice (rationality) criterion determining which options should be chosen (e.g. those with the highest value of a utility functions).

For practice, the set of feasible options is often described by a system on equations and/or inequalities, and the problem is then termed *mathematical programming*.

Developments of fuzzy sets theory have quickly resulted in the emergence of *fuzzy optimization* for a more adequate, effective and efficient handling of optimization problems in which usually some specifications result from human judgments and assessments that are subjective and imprecise (in natural language).

We will present now a brief account of the state of the art of fuzzy optimization and mathematical programming. We will present major concepts, ideas and developments, and refer the reader to more relevant literature, mostly the classic one which can better serve the purpose of the presentation of the essence, and details of

A. Kaur et al., *Fuzzy Transportation and Transshipment Problems*, Studies in Fuzziness and Soft Computing 385,
https://doi.org/10.1007/978-3-030-26676-9_3

some newer approaches, notably those related to the transportation and transshipment problems will be shown in more detail in next chapters. For more details, we refer the reader to more extensive surveys on fuzzy optimization and fuzzy mathematical programming as, e.g., Kacprzyk and Orlovski [9], Fedrizzi et al. [6], etc., general volumes on fuzzy optimization as, e.g., Kacprzyk and Orlovski [10], Delgado et al. [5] or Lodwick and Kacprzyk [11], as well as some more specific ones as, e.g., Fedrizzi et al. [7] or Kacprzyk [8], to just list a few more classic ones.

3.2 Main Approaches to Fuzzy Optimization

The class of fuzzy optimization problems considered here may be stated as: $X = \{x\}$ is a set of options, and the objective function is $F : X \longrightarrow L(R)$, where $L(R)$ is a family of fuzzy sets defined in R, the real line; i.e. $F(x)$ is a fuzzy number yielding an imprecise (fuzzy) evaluation of option $x \in X$. The set of feasible options is imprecisely specified by a fuzzy set C in X such that $C(x) \in [0, 1]$ stands for the degree of feasibility, from 1 for fully feasible to 0 for fully infeasible, through all intermediate values.

The optimization problem is then stated as

$$\widetilde{\max}_{x \widetilde{\in} C} F(x) \tag{3.1}$$

to be stated in words as: find a *possibly high* ($\widetilde{\max}$) value of F over the x's "belonging" ($\widetilde{\in}$) to the (fuzzy) feasible set C.

This general problem formulation may be formally stated in various ways, notably via Bellman and Zadeh's [1] general approach to decision making under fuzziness, and via the representation of the fuzzy feasible set by α-cuts.

3.3 Bellman and Zadeh's General Approach to Decision Making Under Fuzziness

We have an explicitly specified fuzzy feasible set, called a *fuzzy constraint*, and an explicitly specified fuzzy set of options which attain the goal, called a *fuzzy goal*.

The fuzzy goal $G(x)$ is usually assumed described as

$$G(x) = \begin{cases} 1 & \text{for } f(x) \geq \overline{f} \\ g(x) & \text{for } \underline{f} < f(x) < \overline{f} \\ 0 & \text{for } f(x) \leq \underline{f} \end{cases} \tag{3.2}$$

to be read as: we are fully satisfied $[G(x) = 1]$ with the x's for which the objective (performance) function $f(x)$ attains at least a satisfaction level $\overline{f}$, we are partially satisfied (to degree $0 < G(x) = g(x) < 1$) with the x's for which $f(x)$ is between the satisfaction level $\overline{f}$ and the lowest admissible level $\underline{f}$, and we are fully dissatisfied $[G(x) = 0]$ with the x's for which $f(x)$ is below the lowest admissible level $\underline{f}$; this is very intuitively appealing.

The problem is now

satisfy the fuzzy constraint **and** attain the fuzzy goal

which may be written in terms of a *fuzzy decision*, D, given as

$$D(x) = C(x) \wedge G(x) = \min[C(x), G(x)] \tag{3.3}$$

and "$\wedge$", i.e. the minimum, may be replaced by, e.g., a t-norm.

The fuzzy decision $D(x)$ (3.3) yields a fuzzy set of options and to be implemented, we should determine a crisp, non-fuzzy option (or options) which best satisfies the fuzzy constraint and attain the fuzzy goal, which is given as an optimal (maximizing) decision, $x^* \in X$, specified as

$$D(x^*) = \sup_{x \in X} D(x) = \sup_{x \in X}[C(x) \wedge G(x)] \tag{3.4}$$

The interpretation of a fuzzy constraint, fuzzy goal and fuzzy decision can be portrayed as in Fig. 3.1.

This can be extended for multiple fuzzy constraints, $C^1(x), \ldots, C^m(x)$, and multiple fuzzy goals, $G^1(x), \ldots, G^n(x)$, yielding

$$D(x) = C^1(x) \wedge \ldots \wedge C^m(x) \wedge G^1(x) \wedge \ldots \wedge G^n(x) \tag{3.5}$$

with an optimal (maximizing) decision, $x^* \in X$, given as

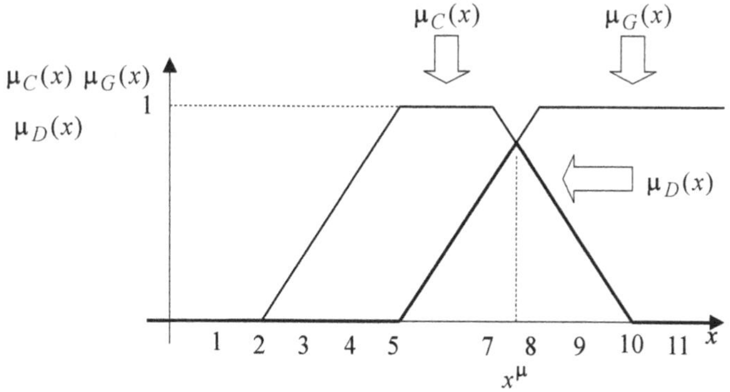

Fig. 3.1 Fuzzy goal, fuzzy constraint, fuzzy decision, and the optimal (maximizing) decision

$$D(x^*) = \\ = \sup_{x \in X} [C^1(x) \wedge \ldots \wedge C^m(x) \wedge G^1(x) \wedge \ldots \wedge G^n(x)] \tag{3.6}$$

Moreover, if the fuzzy constraint is defined in $X = \{x\}$, $C(x)$, and the fuzzy goal is defined in $Y = \{y\}$, $G(y)$, as it is usual in practice, and there is a function $h : X \longrightarrow Y$, $y = h(x)$, $\overline{G}[h(x)]$ is the fuzzy goal in X induced by $G(y)$ in Y, then then fuzzy decision (3.3) becomes

$$D(x) = C(x) \wedge \overline{G}[h(x)] \tag{3.7}$$

and we seek an optimal (maximizing) decision $x^* \in X$ such that

$$D(x^*) = \sup_{x \in X} D(x) = \sup_{x \in X} (C(x) \wedge \overline{G}[h(x)]) \tag{3.8}$$

And analogously, for multiple fuzzy constraints and multiple fuzzy goals defined in X and Y, respectively, we obtain the fuzzy decision

$$D(x) = C^1(x) \wedge \ldots \wedge C^m(x) \wedge \overline{G}^1[h(x)] \wedge \ldots \wedge \overline{G}^n[h(x)] \tag{3.9}$$

and an optimal (maximizing) decision to be found, $x^* \in X$, is given by

$$D(x^*) = \max_{x \in X} D(x) = \max_{x \in X}(C^1(x) \wedge \ldots \\ \ldots \wedge C^m(x) \wedge \overline{G}^1[h(x)] \wedge \ldots \wedge \overline{G}^n[h(x)]) \tag{3.10}$$

Notice that in the above approach the values of the objective function, $f(x)$, are non-fuzzy [cf. (3.2)], and only its maximization is imprecisely specified. In Orlovski's [14] approach the values of the objective function are fuzzy, characterized by membership functions $g : X \times R \longrightarrow [0, 1]$ such that for each value of $x \in X$ the objective function may take on different real values, with different degrees of membership from [0, 1]. This is done by using the following sets:

$$N = \{(x, r) : (x, r) \in X \times R, g(x, r) > G(x)\} \tag{3.11}$$

$$N_x = \{r : r \in R, (x, r) \in N\} \tag{3.12}$$

$$X^0 = \{x : x \in X, N_x \neq \emptyset\} \tag{3.13}$$

the fuzzy decision is defined as

$$D(x) = \begin{cases} C(x) \wedge \inf_{r \in N} G(x) & \text{for } x \in X^0 \\ C(x) & \text{otherwise} \end{cases} \tag{3.14}$$

and an optimal (maximizing) decision $x^* \in X$ sough is

$$D(x^*) = \max_{x \in X} D(x) \tag{3.15}$$

3.4 Using the α-cuts of the Fuzzy Feasible Set

Usually, while dealing with fuzzy systems, we replace fuzzy sets their equivalent α-cuts (α-level sets). If A is a fuzzy set in X, then its α-cut (α-level set) is

$$A_\alpha = \{x \in X : A(x) \geq \alpha\}, \qquad \forall \alpha \in [0, 1] \tag{3.16}$$

The α-cuts (α-level sets) may also be employed to obtain an equivalent of a fuzzy optimization problem as proposed in the approach of Orlovski [13] shown above. The problem considered is again as in (3.1), i.e.

$$\widetilde{\max}_{x \tilde{\in} C} f(x) \tag{3.17}$$

in which the maximization and inclusion should be meant in a fuzzy way. Notice that, for simplicity, we assume here a non-fuzzy objective function $f(x)$ instead of a fuzzy one $F(x)$ as in (3.1).

First, for the fuzzy feasible set C we derive its α-cuts, $C_\alpha = \{x \in X : C(x) \geq \alpha\}$, for each $\alpha \in (0, 1]$. Then, for each $\alpha \in (0, 1]$ such that $C_\alpha \neq \emptyset$, we introduce the following (non-fuzzy) set

$$N(\alpha) = \{x \in X : f(x) = \sup_{x \in C_\alpha} f(x)\} \tag{3.18}$$

A so-called *solution 1* to problem (3.17) is defined as the following fuzzy set

$$S_1(x) = \begin{cases} \sup_{x \in N(\alpha)} \alpha & \text{for } x \in \cup_{\alpha > 0} N(\alpha) \\ 0 & \text{otherwise} \end{cases} =$$
$$= \begin{cases} C(x) & \text{for } x \in \cup_{\alpha > 0} N(\alpha) \\ 0 & \text{otherwise} \end{cases} \tag{3.19}$$

and the fuzzy maximum value of $f(x)$ over the fuzzy feasible set C is defined as:

$$f(r) = \sup_{x \in f^{-1}(r)} S_1(x) = \sup_{x \in f^{-1}(r)} \sup_{x \in N(\alpha)} \alpha; \forall r \in R \tag{3.20}$$

A so-called *solution 2* to problem (3.17) is then defined. For $f(x)$ and $C(x)$ we first define the set of Pareto maximal elements, P, as a (nonfuzzy) subset $P \subseteq X$ such that $x \in P$ if and only if there exists no $y \in X$ for which

$$\begin{cases} \text{either} \\ f(y) > f(x) \text{ and } C(y) \geq C(x) \\ \text{or} \\ f(y) \geq f(x) \text{ and } C(y) > C(x) \end{cases} \tag{3.21}$$

and then *solution 2* is defined as a fuzzy set

$$S_2(x) = \begin{cases} C(x) & \text{for } x \in P \\ 0 & \text{otherwise} \end{cases} \tag{3.22}$$

It may then be shown [13] that this solution yields the same fuzzy maximal value of $f(x)$ over C as solution 1, i.e. for each $r \in R$:

$$f(r) = \sup_{x \in f^{-1}(r)} S_1(x) = \sup_{x \in f^{-1}(r)} \sup_{x \in N(\alpha)} \alpha \tag{3.23}$$

As to more interesting properties of solution 2, one may mention here $P \subset \cup_{\alpha>0} N(\alpha)$ which implies $S_2(x) \leq S_1(x)$, for each $x \in X$, i.e. solution 2 is a subset of solution 1.

Among other more relevant approaches in which $f(x)$ and $C(x)$ are dealt with separately one may mention those by Negoita and Ralescu [12] or Yager [19].

3.5 Fuzzy Mathematical Programming

In mathematical programming the feasible set is given as (a set of) equalities and/or inequalities, and a general mathematical programming problem can be written as

$$\begin{cases} & \max_{x \in R^n} f(x) \\ \text{subject to:} & \\ & g_i(x) \leq b_i;\ i = 1, \ldots, m \end{cases} \tag{3.24}$$

where $x = [x_1, \ldots, x_n]^T \in R^n$ is a vector of decision variables, $f : R^n \longrightarrow R$ is an objective function, $g_i : R^n \longrightarrow R$ are constraints, and $b_i \in R$ are the so-called right-hand sides; the maximization may be replaced by the minimization, and "$\leq$" by "$\geq$".

Of a particularly importance in practice is *linear programming* in which both the objective function and constraints are linear functions, i.e. (3.24) becomes

$$\begin{cases} & \max_{x_j \geq 0} f(x) = cx = \sum_{j=1}^{n} c_j x_j \\ \text{subject to:} & (Ax)_i = \sum_{j=1}^{n} a_{ij} x_j \leq b_i;\ i = 1, \ldots, m \end{cases} \tag{3.25}$$

and a large part of our next discussions will be devoted to fuzzy linear programming.

3.6 Fuzzy Linear Programming

In the general linear programming problem formulation (3.25) we may readily point out the following elements that may be fuzzified:

- the coefficients (costs) in the objective function $f(x)$, i.e. $c = [c_1, \ldots, c_m]$,
- the coefficients a_{ij} in the so-called technological matrix $A = [a_{ij}]$, $i = 1, \ldots, m$, $j = 1, \ldots, n$, and
- the right-hand sides $b = [b_1, \ldots, b_m]^T$

The above leads to the following basic types of fuzzy linear programming:

- problems with fuzzy constraints,
- problems with a fuzzy objective function (fuzzy goal),
- problems with fuzzy costs c_i's, and
- problems with fuzzy coefficients a_{ij}'s and b_i's.

3.7 Fuzzy Linear Programming with Fuzzy Constraints

In this case the fuzzy linear programming problem may be generally written as:

$$\begin{cases} & \max_{x \in R^n} cx \\ \text{subject to:} & \\ & Ax \tilde{\leq} b;\ x \geq 0 \end{cases} \tag{3.26}$$

where "$\tilde{\leq}$" denotes that the left-hand side should be essentially less than or equal to the right-hand side with the understanding that this should be possibly well satisfied.

Usually, "$\tilde{\leq}$" is formalized by allowing the i-th constraint in (3.26) to be violated to some extent via a degree of satisfaction of the i-th constraints given as

$$i(x) = \begin{cases} 1 & \text{if } (Ax)_i < b_i \\ h_i[(Ax)_i] & \text{if } b_i \leq (Ax)_i \leq b_i + t_i \\ 0 & \text{if } (Ax) > b_i + t_i \end{cases} \tag{3.27}$$

where $h_i(.) \in (0, 1)$ is such that the higher the violation of the i-th constraint the lower the value of h_i; in practice, $h_i(.)$ is a linear function. Moreover, t_i is the maximum vialation of the i-th constraint.

The first traditional method for solving problem (3.26) is due to Tanaka et al. [16] in which the solution of (3.26) may be replaced by: find an optimal pair $(\alpha^*, x^*) \in [0, 1] \times R^n$ such that

$$\alpha^* \wedge f(x^*) = \sup_{\alpha \in [0,1]} [\alpha \wedge \max_{x \in X_\alpha} f(x)] \tag{3.28}$$

where $f : R^n \longrightarrow [0, 1]$ is a continuous objective functions, and $X_\alpha = \{x \in R^n \mid \bigwedge_{i=1,\ldots,m} i(x) \geq \alpha\}, \forall \alpha \in (0, 1]$.

Under some mild assumptions on the continuity of f and uniqueness of α^*, an optimal solution sought—(α^*, x^*)—is iteratively obtained by:

Step 1: Assume $k = 1$ and an $\alpha_1 \in (0, 1]$.
Step 2: Compute $f_k = \max_{x \in X_{\alpha_k}} f(x)$.
Step 3: Compute $\epsilon_k = \alpha_k - f_k$. If $\mid \epsilon_k \mid > \epsilon$, then go to **Step 4**, otherwise go to **Step 5**; $\epsilon \in [0, 1]$ is a required precision.
Step 4: Compute $\alpha_{k+1} = \alpha_k - r_k \epsilon_k$, where $r_k \geq 0$ is selected so that $0 \leq \alpha_{k+1} \leq 1$. Set $k := k + 1$ and go to **Step 2**.
Step 5: Let $\alpha^* = \alpha$ and find an optimal $x^* \in R^n$ such that

$$f(x^*) = \max_{x \in X_{\alpha^*}} f(x) \tag{3.29}$$

Another basic approach to the solution of (3.26) is due to Zimmermann [20] who, first, by putting $e_j = -c_j$ into the objective function in problem (3.28), replaces the maximization by the minimization, and then replaces problem (3.26) by its following fuzzified version

$$\begin{cases} ex = \sum_{j=1}^n e_j x_j \tilde{\leq} z \\ Ax = \sum_{j=1}^n a_{ij} x_j \tilde{\leq} b_i \text{ for } i = 1, \ldots, m \\ x_j \geq 0 \qquad\qquad\qquad \text{for } j = 1, \ldots, n \end{cases} \tag{3.30}$$

to be meant as follows: ex should be "essentially smaller than or equal to" an aspiration level z, and the constraints' left-hand sides, Ax, should be "essentially smaller than or equal to" b_i's; both to be satisfied as well as possible (to the highest possible extent).

The first step is a reflection of "$\tilde{\leq}$" which stands for "essentially smaller than or equal to". First, we introduce the $(m + 1) \times n$ matrix $H = [h_{kj}]$ formed by adding to the original matrix $A = [a_{ij}]$ the row vector $[e_j]$ before the first row of A.

We donote the k-th row of Hx, the product of H and vector x, by

$$(Hx)_k = \sum_{j=1}^n h_{kj} x_j \tag{3.31}$$

and define the function

$$g_k[(Hx)_k] = \begin{cases} 1 & \text{for } (Hx)_k \leq w_k \\ 1 - \frac{(Hx)_k - w_k}{t_k} & \text{for } w_k < (Hx)_k \leq w_k + t_k \\ 0 & \text{for } (Hx)_k \leq w_k + t_k \end{cases} \tag{3.32}$$

where $w = [w_1, \ldots, w_{m+1}]^T = [z, b_1, \ldots, b_m]^T$, and the t_k's are some admissible violations of the respective constraints.

We wish to satisfy all the constraints, and hence the objective function, a fuzzy decision in the sense of (3.3), is:

$$D(x) = \bigwedge_{k=1}^{m+1} g_k[(Hx)_k] \tag{3.33}$$

and we wish to satisfy all the constraints to the highest possible extent [cf. the optimal decision (3.4)], i.e. we seek an optimal $x^* \in R^n$ such that

$$D(X^*) = \sup_{x \in R^n} D(x) \tag{3.34}$$

Each optimal solution, (λ^*, x^*), of the following linear programming problem

$$\begin{cases} & \max_{\lambda \in (0,1]} \lambda \\ \text{subject to:} & \\ & \lambda \leq w'_k - (Hx)'_k \quad k = 1, \ldots, m+1 \\ & x_j \geq 0 \qquad\qquad j = 1, \ldots, n \end{cases} \tag{3.35}$$

where $w'_k = \frac{w_k}{t_k}$ and $(Hx)'_k = \frac{(Hx)_k}{t_k}$, $k = 1, \ldots, m+1$, is also an optimal solution to problem (3.34).

The third basic approach to the solution of problem (3.26) was proposed in Verdegay [18]. It employs the so-called *representation theorem* (2.12), p. 20 which states that a fuzzy set can be uniquely represented by all its α-cuts.

First, if the membership functions of the fuzzy constraints in problem (3.26) are strictly monotone and continuous, which is not a very limiting constraint, then the α-cuts of the set of constraints are

$$C_\alpha = \{x \in R^n \mid \sum_{j=1}^{n} a_{ij} x_j \leq g^{-1}(\alpha); x_k \geq 0; i = 1, \ldots, m; j = 1, \ldots, m\} \tag{3.36}$$

where the $g_i^{-1}(\alpha)$'s are the inverse functions of the $g_i(.)$'s defined by (3.27).

Then, if C denotes the set of fuzzy constraints in problem (3.26), the representation theorem states that

$$C = \sum_{\alpha \in (0,1]} \alpha C_\alpha \tag{3.37}$$

A fuzzy solution (for all $\alpha \in (0, 1]$!) to problem (3.26) can be therefore obtained by solving the following parametric linear programming problem

$$\begin{cases} & \max_{x \in R^n} cx \\ \text{subject to:} & \\ & x \in C_\alpha; \forall \alpha \in (0, 1] \end{cases} \tag{3.38}$$

or, more explicitly:

$$\begin{cases} & \max_{x \in R^n} cx \\ \text{subject to:} & \\ & (Ax)_i \leq g_i^{-1}(\alpha); \forall \alpha \in (0, 1]; i = 1, \ldots, m \end{cases} \tag{3.39}$$

And for the practically most relevant case of linear fuzzy constraints, problem (3.39) becomes

$$\begin{cases} & \max_{x \geq 0} cx \\ \text{subject to:} & \\ & Ax \leq b + t(1 - \alpha); \forall \alpha \in (0, 1] \end{cases} \tag{3.40}$$

where $t = [t_1, \ldots, t_m]^T$ is a vector of admissible violations of the particular constraints.

Thus, if $x^*(\alpha)$ is an optimal solution to problem (3.39), then from it (for each $\alpha \in (0, 1]$) a fuzzy optimal solution to problem (3.37) can be obtained.

3.8 Fuzzy Coefficients in the Objective Function

Now the constraints are non-fuzzy and the coefficients in the objective function, c_j, are fuzzy numbers given by the membership functions $j : R^n \longrightarrow [0, 1]$, $j = 1, \ldots, n$, which can be written as

$$\begin{cases} & \max_{x \in R; x \geq 0} cx \\ \text{subject to:} & \\ & Ax \leq b \end{cases} \tag{3.41}$$

where c_j, $j = 1, \ldots, n$, are assumed to be fuzzy.

We will sketch three approaches to the formulation and solution of such problems which are due to: Delgado et al. [5], Tanaka et al. [17], and Rommelfanger et al. [15]; cf. also Chanas and Kuchta [2, 3].

In Delgado et al. [5] the coefficients in the objective function of problem (3.41), $c = [c_1, \ldots, c_j, \ldots, c_n]$, are fuzzy sets (numbers) such that, for c_j:

$$j(x) = \begin{cases} 0 & \text{if } \overline{c}_j \leq x \text{ or } x \leq \underline{c}_j \\ \underline{h}_j(x) & \text{if } \underline{c}_j < x \leq c_j \\ \overline{h}_j(x) & \text{if } c_j < x \leq \overline{c}_j \end{cases} \tag{3.42}$$

where $[\overline{c}_j, \underline{c}_j]$ is the support of the fuzzy number c_j, and $\underline{h}_j(.)$ and $\overline{h}_j(.)$ are continuous and strictly increasing and decreasing, respectively, functions such that $\underline{h}_j(c_j) = \overline{h}_j(c_j) = 1$.

Then, using a fuzzy objective function as defined in Verdegay [18] and including the $(1-\alpha)$-cuts of each cost coefficient, for each $\alpha \in (0, 1]$, we have for each $x \in R$ and for $j = 1, \ldots, n$

$$j(x) \geq 1 - \alpha \Longleftrightarrow \underline{h}_j^{-1}(1-\alpha) \leq x \leq \overline{h}_j^{-1}(1-\alpha) \tag{3.43}$$

and if we denote $= \Phi_j(.) = \underline{(h)}_j(.)$ and $\Psi(.) = \overline{h}_j(.)$, then we obtain

$$\Phi_j(1-\alpha) \leq x \leq \Psi_j(1-\alpha), \qquad \text{for each } x \in R \tag{3.44}$$

As shown in Verdegay [18], a fuzzy solution to problem (3.41) can be found from the parametric solution of the following multi-objective linear program

$$\begin{cases} & \max_{x\in R; x\geq 0}\{c^1 x, c^2 x, \ldots c^{2^n} x\} \\ \text{subject to:} & \\ & Ax \leq b \end{cases} \tag{3.45}$$

where $c^k \in E(1-\alpha), \alpha \in (0, 1], k = 1, 2, \ldots, 2^n$, and $E(1-\alpha)$ is the set of vectors in R^n each of whose components is either on the upper bound, $\Psi_j(1-\alpha)$, or on the lower bound, $\Phi_j(1-\alpha)$, of the respective $(1-\alpha)$-cuts.

On the other hand, in Tanaka et al. [17] a possibilistic approach is proposed for a more general problem formulations, with fuzzy constraints in addition to fuzzy coefficients in the objective function as in (3.41), which is based on the two relevant facts. First, if the c_j's, $j = 1, \ldots, n$, are triangular fuzzy numbers with membership function given as $\delta(\underline{c}_j, c_j, \overline{c}_j)$, then the value of the objective function, $z = c_1 x_1 + c_2 x_2 + \cdots + c_n x_n$, is also a fuzzy number with the membership function

$$G(y) = \begin{cases} \frac{1-(|2y-(\overline{c}+\underline{c})|x)}{(\overline{c}-\underline{c})x} & \text{for } x > 0, y > 0 \\ 1 & \text{for } x = 0, y > 0 \\ 0 & \text{for } x = 0, y = 0 \end{cases} \tag{3.46}$$

where $\overline{c} = [\overline{c}_1, \overline{c}_2, \ldots, \overline{c}_n]$, $\underline{c} = [\underline{c}_1, \underline{c}_2, \ldots, \underline{c}_n]$, and the $\underline{c}_j$'s and $\overline{c}_j$'s are defined as in (3.42).

The second aspect is that the maximization in (3.41) concerns now a fuzzy function, which may be written as $\widetilde{\max}_{x\in R; xge0}$, and this is meant as

$$\max_{x\in R; x\geq 0} (w_1 \overline{c} x + w_2 \underline{c} x)$$

where $w_1, w_2 \in [0, 1]$, $w_1 + w_2 = 1$, are some weights.

Then, the solution of problem (3.41) is obtained by solving the following auxiliary linear programming problem

$$\begin{cases} \max_{x\in R; x\geq 0}(w_1\overline{c}x + w_2\underline{c}x) \\ \text{subject to:} \\ Ax \leq b \end{cases} \tag{3.47}$$

The solution of problem (3.41) is also discussed by Rommelfanger et al. [15], termed a *stratified piecewise reduction* approach, and is based on the basic Zimmermann's [20] approach. Though the problem is given as (3.41), the imprecision (fuzziness) of coefficients is modeled by nested intervals. Each of such intervals, say interval k, is assigned a membership degree, or a possibility degree $\alpha_k \in [0, 1]$, $k = 1, 2, \ldots, p$, where p is the number of such intervals. Then, each fuzzy coefficient c_j, $j = 1, 2, \ldots, n$, in the objective function of (3.41) is defined as the fuzzy set

$$c_j = \frac{[\underline{c}_j, \overline{c}_j]^k]}{\alpha_k}; k = 1, 2, \ldots, p \tag{3.48}$$

such that, for all $\alpha_1, \alpha_2 \in [0, 1]$, and $j = 1, 2, \ldots, n$, there holds

$$\alpha_1 \geq \alpha_2 \Longrightarrow [\underline{c}_j, \overline{c}_j]^1 \subseteq [\underline{c}_j, \overline{c}_j]^2 \tag{3.49}$$

Clearly, fuzzy coefficients given as (3.42) may also be equivalently formulated by using α-cuts as shown by (3.48). Then, the solution of problem (3.41) is obtained by solving the following auxiliary linear programming problem

$$\begin{cases} \max_{\lambda\in(0,1]} \lambda \\ \text{subject to:} \\ f_1[\underline{c}^\alpha x] \geq \lambda \\ f_2[\overline{c}^\alpha x] \geq \lambda \\ Ax \leq b; x \geq 0 \end{cases} \tag{3.50}$$

where:

- if $\overline{z}^\alpha_{\min} \leq \underline{c}^\alpha x \leq z^{*\alpha}_{\min}$, then

$$f_1[\underline{c}^\alpha] = \frac{\underline{c}^\alpha x - \overline{z}^\alpha_{\min}}{z^{*\alpha}_{\min} - \overline{z}^\alpha_{\min}}$$

- if $\overline{z}^\alpha_{\max} \leq \overline{c}^\alpha x \leq z^{*\alpha}_{\max}$, then

$$f_2[\overline{c}^\alpha] = \frac{\overline{c}^\alpha x - \overline{z}^\alpha_{\max}}{z^{*\alpha}_{\min} - \overline{z}^\alpha_{\max}}$$

where, if we denote $\mathcal{Z} = \{x \in R^n : Ax \leq b, x \geq 0\}$, then:

$$\begin{aligned} z^{*\alpha}_{\min} &= \underline{c}^{\alpha}(x^*_{\min}) = \max\{\underline{c}^{\alpha}x \mid x \in \mathcal{Z}\} \\ z^{*\alpha}_{\max} &= \overline{c}^{\alpha}(x^*_{\max}) = \max\{\overline{c}^{\alpha}x \mid x \in \mathcal{Z}\} \\ \overline{z}^{\alpha}_{\min} &= \underline{c}^{\alpha}(x^*_{\max}) = \min\{\underline{c}^{\alpha}x \mid x \in \mathcal{Z}\} \\ \overline{z}^{\alpha}_{\min} &= \overline{c}^{\alpha}(x^*_{\min}) = \min\{\overline{c}^{\alpha}x \mid x \in \mathcal{Z}\} \end{aligned}$$

The solution of problem (3.41) may be then found as the intersection of solutions obtained by solving problem (3.50) for each $\alpha_k \in, k = 1, 2, \dots, p$.

To conclude the description of this approach, notice that problem (3.50) may be rewritten as

$$\begin{cases} & \max_{\lambda \in (0,1]} \lambda \\ \text{subject to:} & \\ & (z^{*\alpha}_{\min} - \overline{z}^{\alpha}_{\min})\lambda - \underline{c}^{\alpha}x \leq \overline{z}^{\alpha}_{\min} \\ & f_2[\overline{c}^{\alpha}x] \geq \lambda \\ & (z^{*\alpha}_{\max} - \overline{z}^{\alpha}_{\max})\lambda - \underline{c}^{\alpha}x \leq \overline{z}^{\alpha}_{\max} \\ & Ax \leq b;\, x \geq 0 \end{cases} \tag{3.51}$$

Clearly, $f_1[\underline{c}^{\alpha}x^*] = f_2[\overline{c}^{\alpha}x^8] = \lambda^*$, where (λ^*, x^*) is an optimal solution of problem (3.51).

3.9 Fuzzy Coefficients in the Technological Matrix

In this case the coefficients in problem (3.41) in the so-called technological matrix, A, and in the right-hand sides, b, are fuzzy numbers assumed, for simplicity, to be in the LR form. On the other hand, the coefficients in the objective function, c, are nonfuzzy (i.e. real numbers).

Now the fuzzy linear programming problem is

$$\begin{cases} & \max_{x \in R; x \geq 0} cx \\ \text{subject to:} & \\ & \sum_{j=1}^{n} a_{ij}x_j \leq b_i;\, i = 1, \dots, m \end{cases} \tag{3.52}$$

and is discussed in Tanaka et al. [17].

First, in problem (3.52) the fuzziness is in the coefficients, and not in the allowable violation of the constraints so that problem (3.51) is different than problem (3.26) with a fuzzy constraint set.

For solving problem (3.52), Tanaka et al. [17] solve the following auxiliary conventional linear programming problem

$$\begin{cases} & \max_{x\in R; x\geq 0} cx \\ \text{subject to:} & \\ & [(1-\frac{\beta}{2})(\underline{a}_i + a_i) + \frac{\beta}{2}(a_i - \underline{a}_i)]x \leq \\ & \leq (1-\frac{\beta}{2})(b_i + \underline{b}_i) + \frac{\beta}{2}(b_i - \underline{b}_i) \\ & [\frac{\beta}{2}(\overline{a}_i + a_i) + (1-\frac{\beta}{2})(\overline{a}_i - a_i)]x \leq \\ & \leq \frac{\beta}{2}(b_i + \overline{b}_i) + (1-\frac{\beta}{2})(\overline{b}_i - b_i) \end{cases} \tag{3.53}$$

where $\beta \in [0, 1]$ is a degree of optimism to be specified a priori.

As a prerequisite for obtaining problem (3.52), the following ordering relation between the triangular fuzzy numbers is assumed

$$a \tilde{\geq}_\beta b \Longleftrightarrow (a + \overline{a})_k \geq (b + \overline{b})_k \quad \& \quad (a - \underline{a})_k \geq (b - \underline{b})_k \tag{3.54}$$

for each $k \in [\beta, 1]$, where $(a + \overline{a})_k$ and $(a - \underline{a})_k$ are the upper and lower bounds, respectively, of the k-cut of a.

Notice that in (3.54) no fuzziness in respect to the satisfaction of constraints is involved. This is accounted for in a model of fuzzy linear programming proposed in Delgado et al. [4] the point of departure of which is the following problem

$$\begin{cases} & \max_{x\in R; x\geq 0} cx \\ \text{subject to:} & \\ & \sum_{j=1}^{n} a_{ij}x_j \tilde{\leq} b_i;\ i = 1, \ldots, m \end{cases} \tag{3.55}$$

where "$\tilde{\geq}$" means that some violation of "$\geq$" may be allowed; as before, the a_{ij}'s and b_j's are fuzzy, given as triangular fuzzy numbers.

The above violation, for the i-th constraint, is expressed by a fuzzy number t_i termed a margin of violation tolerance. Then, the set of constraints in (3.55) is replaced by

$$\sum_{j=1}^{n} a_{ij}x_j > b_i + t_i(1-\alpha), \qquad \alpha \in [0, 1], i = 1, \ldots, m \tag{3.56}$$

where "$>$" is some relation between two fuzzy numbers, preserving only the ranking under the multiplication by a positive scalar.

Therefore, the problem considered becomes

$$\begin{cases} & \max_{x\in R; x\geq 0} cx \\ \text{subject to:} & \\ & \sum_{j=1}^{n} a_{ij}x_j \leq b_i + t_i(1-\alpha) \\ & \alpha \in [0, 1], i = 1, \ldots, m \end{cases} \tag{3.57}$$

from which a fuzzy optimal solution of (3.55) can be obtained.

This concludes are short account of basic classic problems in fuzzy optimization and fuzzy mathematical programming which are of use as a point of departure for our book.

Now we will briefly outline the essence of a fuzzy transportation and transshipment problems.

References

1. R.E. Bellman, L.A. Zadeh, Decision-making in a fuzzy environment. Manag. Sci. **17**, 141–164 (1970)
2. S. Chanas, D. Kuchta, A concept of the optimal solution of the transportation problem with fuzzy cost coefficients. Fuzzy Sets Syst. **82**, 299–305 (1996)
3. S. Chanas, D. Kuchta, Fuzzy integer transportation problem. Fuzzy Sets Syst. **98**, 291–298 (1998)
4. M. Delgado, J.-L. Verdegay, M.A. Vila, A general model for fuzzy linear programming. Fuzzy Sets Syst. **29**, 21–29 (1989)
5. D. Delgado, J. Kacprzyk, J.L. Verdegay, M.A. Villa, *Fuzzy Optimization: Recent Advances* (Physica-Verlag, Heidelberg, 1994)
6. M. Fedrizzi, J. Kacprzyk, J.-L. Verdegay, A survey of fuzzy optimization and mathematical programming, in *Interactive Fuzzy Optimization*, eds. by M. Fedrizzi, J. Kacprzyk, M. Roubens (Springer, Berlin, 1991), pp. 15–28
7. M. Fedrizzi, J. Kacprzyk, M. Roubens (eds.), *Interactive Fuzzy Optimization* (Springer, Berlin/New York, 1991)
8. J. Kacprzyk, *Multistage Fuzzy Control: A Model-Based Approach to Control and Decision-Making* (Wiley, Chichester, 1997)
9. J. Kacprzyk, S.A. Orlovski, Fuzzy optimization and mathematical programming: a brief introduction and survey, in *Optimization Models Using Fuzzy Sets and Possibility Theory*, ed. by J. Kacprzyk, S.A. Orlovski (Reidel, Dordrecht/Boston/Lancaster, 1987), pp. 50–72
10. J. Kacprzyk, S.A. Orlovski (eds.), *Optimization Models Using Fuzzy Sets and Possibility Theory* (Reidel, Dordrecht/Boston/Lancaster, 1987)
11. W.A. Lodwick, J. Kacprzyk (eds.), *Fuzzy Optimization: Recent Advances and Applications* (Springer, Heidelberg/New York, 2010)
12. C.V. Negoita, D. Ralescu, On fuzzy optimization. Kybernetes **6**, 193–205 (1977)
13. S.A. Orlovski, On programming with fuzzy constraint sets. Kybernetes **6**, 197–201 (1977)
14. S.A. Orlovski, On formalization of a general fuzzy mathematical programming problem. Fuzzy Sets Syst. **3**, 311–321 (1980)
15. H.R. Rommelfanger, R. Hanuscheck, J. Wolf, Linear programming with fuzzy objectives. Fuzzy Sets Syst. **29**, 31–48 (1989)
16. H. Tanaka, T. Okuda, K. Asai, On fuzzy mathematical programming. J. Cybern. **3–4**, 37–46 (1974)
17. H. Tanaka, H. Ichihashi, K. Asai, A formulation of fuzzy linear programming problems based on comparison of fuzzy numbers. Control. Cybern. **4**, 185–194 (1984)
18. J.L. Verdegay, Fuzzy mathematical programming, in *Fuzzy Information and Decision Processes*, eds. by M.M. Gupta, E. Sanchez (Amsterdam, North-Holland, 1982), pp. 231–237
19. R.R. Yager, Mathematical programming with fuzzy constraints and a preference on the objective. Kybernetes **9**, 109–114 (1979)
20. H.J. Zimmermann, Description and optimization of fuzzy systems. Int. J. Gen. Syst. **2**, 209–215 (1976)

Chapter 4
New Methods for Solving Fully Fuzzy Transportation Problems with Trapezoidal Fuzzy Parameters

If we take into account more recent and more comprehensive literature, then, presumably, only the methods presented by Basirzadeh [1], Gani et al. [6], Pandian and Natarajan [18, 19], and Stephen Dinagar and Palanivel [21, 22] can be considered as those which are meant to find the fuzzy optimal solution of the fully fuzzy transportation problems. These are original and powerful methods but, as always is the case in science, they suffer from some shortcomings, and may not work well in general. An attempt to overcome these weaknesses will be undertaken in this chapter. Namely, we will present two new methods for solving the fully fuzzy transportation problems. We will discuss their advantages over the existing methods, i.e. Basirzadeh [1], Gani et al. [6], Pandian and Natarajan [18, 19], Stephen Dinagar and Palanivel [21, 22]. We will solve some nontrivial existing fully fuzzy transportation problem, and then show the application of proposed methods in real life problems in which the fuzzy optimal solution of an existing real life fuzzy transportation problem is considered to be the one that is well justified, and we show its derivation by using the proposed new methods.

4.1 Preliminaries

In this section we will present some basic definitions and properties related mostly to the fuzzy numbers, notably their triangular version.

4.1.1 Basic Definitions Related to Fuzzy Numbers

In this section, some basic definitions are presented which will be needed for our next considerations.

A. Kaur et al., *Fuzzy Transportation and Transshipment Problems*, Studies in Fuzziness and Soft Computing 385,
https://doi.org/10.1007/978-3-030-26676-9_4

Definition 1 Let X be a set of objects (elements), in the classical sense, that is of some *universe of discourse*. Then, the set of ordered pairs

$$\tilde{A} = \{(x, \mu_{\tilde{A}}(x)) : x \in X\}$$

where $\mu_{\tilde{A}} : X \to [0,]$, is called a *fuzzy set* in X, and $\mu_{\tilde{A}}(x)$ is called the *membership function* of $\tilde{a}$.

Definition 2 Let $\tilde{A}$ be a fuzzy set in X and $\lambda \in [0, 1]$ be a real number. Then, the (classical) set $A^{\lambda} = \{x \in X : \mu_{\tilde{A}}(x) \geq \lambda\}$ is called a λ-cut of $\tilde{A}$.

Definition 3 A fuzzy set $\tilde{A} = \{(x, \mu_{\tilde{A}}(x)) : x \in X\}$ is called a *normalized fuzzy set* if and only if

$$\sup_{x \in X} \mu_{\tilde{A}}(x) = 1.$$

Definition 4 A fuzzy set $\tilde{A}$ is called a *convex fuzzy set* if and only if

$$\mu_{\tilde{A}}(\alpha x_1 + (1 - \alpha)x_2) \geq \min\{\mu_{\tilde{A}}(x_1), \mu_{\tilde{A}}(x_2)\}, \ \forall\, x_1, x_2 \in X, \ \alpha \in [0, 1].$$

Definition 5 A convex normalized fuzzy set $\tilde{A} = \{(x, \mu_{\tilde{A}}(x)) : x \in \mathbb{R}\}$ defined on the real line $\mathbb{R}$ is called a *fuzzy number* if and only if $\mu_{\tilde{A}}(x)$ is piecewise continuous in $\mathbb{R}$.

Definition 6 A fuzzy number $\tilde{A}$ defined on the universal set of real numbers $\mathbb{R}$ is said to be a *non-negative fuzzy number* if and only if $\mu_{\tilde{A}}(x) = 0, \ \forall\, x < 0$.

Definition 7 A fuzzy number $\tilde{A}$ defined on the universal set of real numbers $\mathbb{R}$, denoted as $\tilde{A} = (a, b, c)$, is said to be a *triangular fuzzy number* if its membership function, $\mu_{\tilde{A}}(x)$, is given by

$$\mu_{\tilde{A}}(x) = \begin{cases} \frac{(x-a)}{(b-a)}, & \text{for } a \leq x < b \\ 1, & \text{for } x = b \\ \frac{(x-c)}{(b-c)}, & \text{for } b < x \leq c \\ 0, & \text{otherwise} \end{cases} \tag{4.1}$$

where $a \leq b \leq c \leq \infty$.

Definition 8 A fuzzy number $\tilde{A}$ defined on the universal set of real numbers $\mathbb{R}$, denoted as $\tilde{A} = (a, b, c, d)$, is said to be a *trapezoidal fuzzy number* if its membership function, $\mu_{\tilde{A}}(x)$, is given by

$$\mu_{\tilde{A}}(x) = \begin{cases} \frac{(x-a)}{(b-a)}, & \text{for } a \le x < b \\ 1, & \text{for } b \le x \le c \\ \frac{(x-d)}{(c-d)}, & \text{for } c < x \le d \\ 0, & \text{otherwise} \end{cases} \tag{4.2}$$

where $a \le b \le c \le d \le \infty$.

Obviously, if $b = c$, then a trapezoidal fuzzy number (a, b, c, d) collapses to the triangular fuzzy number and is denoted either as (a, b, b, d) or (a, b, d), or (a, c, d).

Definition 9 Let $\tilde{A} = (a, b, c, d)$ be a trapezoidal fuzzy number. Then, its λ-cut, A^λ, $\lambda \in [0, 1]$, is defined as follows:

$$A^\lambda = [a + (b - a)\lambda, d - (d - c)\lambda],\ 0 \le \lambda \le 1. \tag{4.3}$$

Definition 10 A trapezoidal fuzzy number $\tilde{A} = (a, b, c, d)$ is said to be a *non-negative trapezoidal fuzzy number* if and only if $a \ge 0$.

Definition 11 A trapezoidal fuzzy number $\tilde{A} = (a, b, c, d)$ is said to be the *zero trapezoidal fuzzy number* if and only if $a = b = c = d = 0$.

Definition 12 Two trapezoidal fuzzy numbers $\tilde{A}_1 = (a_1, b_1, c_1, d_1)$ and $\tilde{A}_2 = (a_2, b_2, c_2, d_2)$ are said to be *equal*, denoted $\tilde{A}_1 = \tilde{A}_2$, if and only if $a_1 = a_2, b_1 = b_2, c_1 = c_2$ and $d_1 = d_2$.

4.1.2 Arithmetic Operations on the Trapezoid Fuzzy Numbers

In this Section the sum and product, between two trapezoidal fuzzy numbers, defined on universal set of real numbers $\mathbb{R}$, are presented, which will be relevant for our discussion.

Namely, these two basic arithmetic operations on the trapezoid fuzzy numbers are defined as:

- if $\tilde{A}_1 = (a_1, b_1, c_1, d_1)$ and $\tilde{A}_2 = (a_2, b_2, c_2, d_2)$ be two trapezoidal fuzzy numbers, then their *sum*, $\tilde{A}_1 \oplus \tilde{A}_2$ is defined as

$$\tilde{A}_1 \oplus \tilde{A}_2 = (a_1 + a_2, b_1 + b_2, c_1 + c_2, d_1 + d_2) \tag{4.4}$$

- if $\tilde{A}_1 = (a_1, b_1, c_1, d_1)$ and $\tilde{A}_2 = (a_2, b_2, c_2, d_2)$ be two non-negative trapezoidal fuzzy numbers, then their product, $\tilde{A}_1 \otimes \tilde{A}_2$, is defined as

$$\tilde{A}_1 \otimes \tilde{A}_2 \simeq (a_1 \times a_2, b_1 \times b_2, c_1 \times c_2, d_1 \times d_2) \tag{4.5}$$

  and notice that there is $\simeq$ in the definition of the product (4.5).

For our purposes, in addition to the above two basic arithmetic operations, the following operation of the *product of a scalar and a trapezoidal fuzzy number* may often be important. Suppose that $\tilde{A} = (a, b, c, d)$ is a trapezoidal fuzzy number and $\gamma \in [0, 1]$ is a scalar. Then, the product of γ and $\tilde{A}$, denoted $\gamma \times \tilde{A}$ (as it is customary, $\times$ can be omitted but we will leave it for the clarity of the definition), is defined as

$$\gamma \times \tilde{A} = \begin{cases} (\gamma \times a, \gamma \times b, \gamma \times c, \gamma \times d) & \text{for } \gamma \geq 0 \\ (\gamma \times d, \gamma \times c, \gamma \times b, \gamma \times a) & \text{for } \gamma \leq 0 \end{cases} \tag{4.6}$$

4.2 A Fuzzy Linear Programming Formulation of the Balanced Fully Fuzzy Transportation Problem

The fuzzy linear programming formulation of a balanced fully fuzzy transportation problem can be written as follows, cf. Liu and Kao [12]:

$$\begin{cases} \min\limits_{\tilde{x_{ij}}} \sum\limits_{i=1}^{m} \sum\limits_{j=1}^{n} (\tilde{c}_{ij} \otimes \tilde{x}_{ij}) \\ \text{subject to:} \\ \sum\limits_{j=1}^{n} \tilde{x}_{ij} = \tilde{a}_i, \text{ for } i = 1, 2, \ldots, m \\ \sum\limits_{i=1}^{m} \tilde{x}_{ij} = \tilde{b}_j, \text{ for } j = 1, 2, \ldots, n \\ \sum\limits_{i=1}^{m} \tilde{a}_i = \sum\limits_{j=1}^{n} \tilde{b}_j \end{cases} \tag{4.7}$$

where $\tilde{x}_{ij}$ is a non-negative fuzzy number, m is the total number of sources, n is the total number of destinations, $\tilde{a}_i$ is the fuzzy availability of the product at the i-th source, i.e. (S_i), $\tilde{b}_j$ is the fuzzy demand of the product at the j-th destination, i.e. (D_j), $\tilde{c}_{ij}$ is the fuzzy cost for transporting one unit of the product quantity, from the i-th source, (S_i), to the j-th destination, (D_j), and $\tilde{x}_{ij}$ is the fuzzy quantity of the product that should be transported from the i-th source, (S_i), to the j-th destination, (D_j).

We should bear in mind that in the above problem formulation, as well as later on, the "min." is clearly not meant in the classic optimization sense, because the minimization is to proceed here over the set of fuzzy numbers, which cannot directly be solved by using the traditional mathematical programming tools and techniques. In each case, we will clarify how the minimization/maximization is to be meant.

That is, the problem (4.7) basically states that:

- the total fuzzy transportation cost

$$\sum_{i=1}^{m}\sum_{j=1}^{n}\tilde{c}_{ij}\otimes\tilde{x}_{ij}$$

should be minimized, that is

$$\min_{\tilde{x}_{ij}}\sum_{i=1}^{m}\sum_{j=1}^{n}\tilde{c}_{ij}\otimes\tilde{x}_{ij} \tag{4.8}$$

- the total fuzzy amounts of the product sent from the sources should balance the total fuzzy demand:

$$\sum_{j=1}^{n}\tilde{x}_{ij}=\tilde{a}_i \tag{4.9}$$

for $i=1,2,\ldots,m$;

- the total fuzzy demands of the products received by the destinations should balanced the total fuzzy availability:

$$\sum_{i=1}^{m}\tilde{x}_{ij}=\tilde{b}_j \tag{4.10}$$

for $j=1,2,\ldots,n$; and

- the total fuzzy availability of the product in question, $\sum_{i=1}^{m}\tilde{a}_i$ should not be exceeded, to be more specifically, to be balanced by the total fuzzy demand for this product, $\sum_{j=1}^{n}\tilde{c}_{ij}\sum_{j=1}^{n}\tilde{b}_j$), that is:

$$\sum_{i=1}^{m}\tilde{a}_i=\sum_{j=1}^{n}\tilde{b}_j \tag{4.11}$$

The following definition is relevant for our discussion in tis book:

Definition 13 If

$$\sum_{i=1}^{m}\tilde{a}_i=\sum_{j=1}^{n}\tilde{b}_j \tag{4.12}$$

then the fully fuzzy transportation problem is said to be the *balanced fully fuzzy transportation problem* otherwise it is called the *unbalanced fully fuzzy transportation problem.*

4.3 Existing Methods for Finding a Fuzzy Optimal Solution of the Fully Fuzzy Transportation Problem

As we have already mentioned in the beginning of this chapter, in the more recent literature, presumably, only the methods presented by Basirzadeh [1], Gani et al. [6], Pandian and Natarajan [18, 19], and Stephen Dinagar and Palanivel [21, 22] can be mentioned as those which make it possible to find a fuzzy optimal solution of the fully fuzzy transportation problems. In the well known method by Liu and Kao [12] all the parameters are represented by non-negative trapezoidal fuzzy numbers. However, in other well known methods mentioned above, that is, those of Basirzadeh [1], Gani et al. [6], Pandian and Natarajan [18, 19], and Stephen Dinagar and Palanivel [21, 22], in a general case, neither the cost parameters nor the obtained fuzzy optimal solution (that is, the fuzzy quantity of the product that should be transported from different sources to different destinations to minimize the total fuzzy transportation cost) need to be, by an explicit assumption, non-negative fuzzy numbers so that the obtained minimum total fuzzy transportation cost need not be a non-negative fuzzy number either. This fact, which may eventually be feasible from a formal point of view of the algorithms involved, is unfortunately not consistent with the very essence of the transportation problem as such. Clearly, in real life problems there is no physical meaning of a negative value of the cost and a negative quantity of the product transported. Therefore, for this natural and obvious reason, it makes little sense to apply the existing methods by Basirzadeh [1], Gani et al. [6], Pandian and Natarajan [18, 19], and Stephen Dinagar and Palanivel [21, 22] for solving fully fuzzy transportation problems. In this section we will therefore only discuss the method by Liu and Kao [12] for solving fully fuzzy transportation problems in which all the parameters are represented by non-negative trapezoidal fuzzy numbers.

4.4 Liu and Kao's Method

The method of Liu and Kao [12], proposed for solving the fully fuzzy transportation problems with inequality and equality constraints in which the parameters are either represented by triangular fuzzy numbers or trapezoidal fuzzy numbers, is known but—to make our discussion self contained, and help the readers more easily follow our next considerations—will be briefly presented below.

4.4.1 Fully Fuzzy Transportation Problems with the Inequality Constraints

A fuzzy optimal solution of the fully fuzzy transportation problems with inequality constraints with m sources and n destinations can be obtained by applying the Liu and Kao's [12] algorithm through the consecutive execution of the following steps:

Step 1. Find the α-cuts of $\tilde{c}_{ij}$, $\tilde{a}_i$ and $\tilde{b}_j$, respectively, that is:

$$\begin{cases} \left[(c_{ij})_\alpha^L, (c_{ij})_\alpha^U\right] \text{ for } \tilde{c}_{ij} \\ \left[(a_i)_\alpha^L, (a_i)_\alpha^U\right] \text{ for } \tilde{a}_i \\ \left[(b_j)_\alpha^L, (b_j)_\alpha^U\right] \text{ for } \tilde{b}_j \end{cases} \tag{4.13}$$

Step 2. Check if

$$\sum_{i=1}^{m} (a_i)_{\alpha=0}^L \geq \sum_{j=1}^{n} (b_j)_{\alpha=0}^U \tag{4.14}$$

and then we have the following two cases:

Case (i) If

$$\sum_{i=1}^{m} (a_i)_{\alpha=0}^L \geq \sum_{j=1}^{n} (b_j)_{\alpha=0}^U \tag{4.15}$$

then the fully fuzzy transportation problems with inequality constraints with m sources and n destinations considered, (4.7), is feasible and **go to Step 3**.

Case (ii) If

$$\sum_{i=1}^{m} (a_i)_{\alpha=0}^L < \sum_{j=1}^{n} (b_j)_{\alpha=0}^U \tag{4.16}$$

then the fully fuzzy transportation problems with inequality constraints with m sources and n destinations considered, (4.7), is infeasible and **STOP**.

Step 3 Solve the optimization problem (4.7), to find:

- the left end point $(x_{ij})_\alpha^L$ of the α-cut of the fuzzy decision variable $\tilde{x}_{ij}$ and
- the left end point (Z_α^L) of the α-cut of the minimum total fuzzy transportation cost

$$\tilde{Z} = \sum_{i=1}^{m} \sum_{j=1}^{n} (\tilde{c}_{ij} \otimes \tilde{x}_{ij}) \tag{4.17}$$

both for different values of $\alpha \in [0, 1]$.

Therefore, we obtain:

$$Z^L_\alpha = \min_{\substack{(c_{ij})^L_\alpha \le c_{ij} \le (c_{ij})^U_\alpha \\ (a_i)^L_\alpha \le a_i \le (a_i)^U_\alpha \\ (b_j)^L_\alpha \le b_j \le (b_j)^U_\alpha \\ \forall i=1,\ldots,m;\, j=1,\ldots,n}} \begin{cases} \min\limits_{\substack{i=1,2,\ldots,m \\ j=1,2,\ldots,n}} \sum\limits_{i=1}^{m} \sum\limits_{j=1}^{n} c_{ij} x_{ij} \\ \text{subject to:} \\ \sum\limits_{j=1}^{n} x_{ij} \le a_i, \text{ for } i = 1, 2, \ldots, m \\ \sum\limits_{i=1}^{m} x_{ij} \ge b_j, \text{ for } j = 1, 2, \ldots, n \\ x_{ij} \ge 0, \text{ for all } i = 1, 2, \ldots, m;\ j = 1, 2, \ldots, n \end{cases} \tag{4.18}$$

Step 4 Solve the optimization problem (4.8), to find:

- the right end point $(x_{ij})^U_\alpha$ of the α-cut of the fuzzy decision variable $\tilde{x}_{ij}$, and
- the right end point (Z^U_α) of the α-cut of the minimum total fuzzy transportation cost (4.17), i.e.

$$\tilde{Z} = \sum_{i=1}^{m} \sum_{j=1}^{n} (\tilde{c}_{ij} \otimes \tilde{x}_{ij})$$

both for different values of $\alpha \in [0, 1]$.
Therefore, we obtain:

$$Z^U_\alpha = \max_{\substack{(c_{ij})^L_\alpha \le c_{ij} \le (c_{ij})^U_\alpha \\ (a_i)^L_\alpha \le a_i \le (a_i)^U_\alpha \\ (b_j)^L_\alpha \le b_j \le (b_j)^U_\alpha \\ \forall i=1,\ldots,m;\, j=1,\ldots,n}} \begin{cases} \min\limits_{\substack{i=1,2,\ldots,m \\ j=1,2,\ldots,n}} \sum\limits_{i=1}^{m} \sum\limits_{j=1}^{n} c_{ij} x_{ij} \\ \text{subject to:} \\ \sum\limits_{j=1}^{n} x_{ij} \le a_i, \text{ for } i = 1, 2, \ldots, m \\ \sum\limits_{i=1}^{m} x_{ij} \ge b_j, \text{ for } j = 1, 2, \ldots, n \\ x_{ij} \ge 0, \text{ for all } i = 1, 2, \ldots, m;\ j = 1, 2, \ldots, n \end{cases} \tag{4.19}$$

Step 5 Use the values of

$$\begin{cases} (x_{ij})^L_\alpha \ \text{ and } \ (x_{ij})^U_\alpha \\ Z^L_\alpha \quad\ \text{ and } \ Z^U_\alpha \end{cases}$$

obtained in Step 3, i.e. by solving (4.18), and in Step 4, i.e. by solving (4.18), to find the α-cuts:

$$\begin{cases} \left[(x_{ij})^L_\alpha, (x_{ij})^U_\alpha\right] \\ \left[Z^L_\alpha, Z^U_\alpha\right] \end{cases} \tag{4.20}$$

corresponding to, respectively:

- the optimal fuzzy quantity of the product, $\tilde{x}_{ij}$, and
- the minimum total fuzzy transportation cost given by (4.17), i.e.

$$\tilde{Z} = \sum_{i=1}^{m} \sum_{j=1}^{n} (\tilde{c}_{ij} \otimes \tilde{x}_{ij})$$

4.4.2 *Fully Fuzzy Transportation Problems with Equality Constraints*

A fuzzy optimal solution of a fully fuzzy transportation problem with equality constraints, with m sources and n destinations,

Step 1 Follow Step 1 and Step 2 of the Liu and Kao's [12] method, presented in Sect. 4.4, for the case of the fully fuzzy transportation problem with the inequality constraints, to check if the fully fuzzy transportation problem under consideration (now with the equality constraints) is feasible, (in the sense shown in Sect. 4.4.1).

Step 2 If the fully fuzzy transportation problem is feasible, then in order to find:

1. the left end point $(x_{ij})^L_\alpha$ of the α-cut of the fuzzy decision variable $\tilde{x}_{ij}$ and
2. the left end point (Z^L_α) of α-cut of the minimum total fuzzy transportation cost (4.17), i.e.

$$\tilde{Z} = \sum_{i=1}^{m} \sum_{j=1}^{n} (\tilde{c}_{ij} \otimes \tilde{x}_{ij})$$

corresponding to different values of $\alpha \in [0, 1]$, solve the following problem:

$$Z^L_\alpha = \min_{\substack{(c_{ij})^L_\alpha \le c_{ij} \le (c_{ij})^U_\alpha \\ (a_i)^L_\alpha \le a_i \le (a_i)^U_\alpha \\ (b_j)^L_\alpha \le b_j \le (b_j)^U_\alpha \\ \forall i=1,\dots,m;\, j=1,\dots,n}} \left\{ \begin{array}{l} \min\limits_{\substack{i=1,2,\dots,m \\ j=1,2,\dots,n}} \sum\limits_{i=1}^{m} \sum\limits_{j=1}^{n} c_{ij} x_{ij} \\ \text{subject to:} \\ \sum\limits_{j=1}^{n} x_{ij} = a_i, \text{ for } i = 1, 2, \dots, m \\ \sum\limits_{i=1}^{m} x_{ij} = b_j, \text{ for } j = 1, 2, \dots, n \\ x_{ij} \ge 0, \text{ for all } i = 1, 2, \dots, m;\ j = 1, 2, \dots, n \end{array} \right. \tag{4.21}$$

Step 3 To find:

- the right end point $(x_{ij})^U_\alpha$ of the α-cut of the fuzzy decision variable $\tilde{x}_{ij}$, and
- the right end point (Z^U_α) of the α-cut of the minimum total fuzzy transportation cost (4.17), i.e.

$$\tilde{Z} = \sum_{i=1}^{m}\sum_{j=1}^{n}(\tilde{c}_{ij} \otimes \tilde{x}_{ij})$$

corresponding to different values of $\alpha \in [0, 1]$, solve the following problem:

$$Z_\alpha^U = \max_{\substack{(c_{ij})_\alpha^L \le c_{ij} \le (c_{ij})_\alpha^U \\ (a_i)_\alpha^L \le a_i \le (a_i)_\alpha^U \\ (b_j)_\alpha^L \le b_j \le (b_j)_\alpha^U \\ \forall i=1,\ldots,m;\, j=1,\ldots,n}} \begin{cases} \min\limits_{\substack{i=1,2,\ldots,m \\ j=1,2,\ldots,n}} \sum_{i=1}^{m}\sum_{j=1}^{n} c_{ij}x_{ij} \\ \text{subject to:} \\ \sum_{j=1}^{n} x_{ij} = a_i, \text{ for } i = 1, 2, \ldots, m \\ \sum_{i=1}^{m} x_{ij} = b_j, \text{ for } j = 1, 2, \ldots, n \\ x_{ij} \ge 0, \text{ for all } i = 1, 2, \ldots, m;\ j = 1, 2, \ldots, n \end{cases} \tag{4.22}$$

Step 4 Use the values of

$$\begin{cases} (x_{ij})_\alpha^L \text{ and } (x_{ij})_\alpha^U \\ Z_\alpha^L \quad \text{and } Z_\alpha^U \end{cases}$$

obtained from Step 2, i.e. by solving (4.21), and Step 3, i.e. by solving (4.22), respectively, to find the α-cuts:

$$\begin{cases} \left[(x_{ij})_\alpha^L, (x_{ij})_\alpha^U\right] \\ \left[Z_\alpha^L, Z_\alpha^U\right] \end{cases} \tag{4.23}$$

corresponding to, respectively:

- the optimal fuzzy quantity of the product, $\tilde{x}_{ij}$, and
- the minimum total fuzzy transportation cost given by (4.17), i.e.

$$\tilde{Z} = \sum_{i=1}^{m}\sum_{j=1}^{n}(\tilde{c}_{ij} \otimes \tilde{x}_{ij})$$

As it can be seen, the use of Liu and Kao's [12] method makes it possible to solve the fully fuzzy transportation problem with the equality constraints, too.

4.5 A Critical Analysis of the Existing Methods

Let us first briefly summarize the essence of the previous section, Sect. 4.3. Basically, we have first mentioned the main methods proposed in the literature for finding a fuzzy optimal solution of the fully fuzzy transportation problem. Among more relevant approaches known in the more recent literature, we have cited the methods proposed by Basirzadeh [1], Gani et al. [6], Pandian and Natarajan [18, 19], and Stephen Dinagar and Palanivel [21, 22]. For practical reasons, we have in particular paid attention to the presentation of the well known method by Liu and Kao [12] in which all the parameters are represented by non-negative trapezoidal fuzzy numbers. We have also pointed out that in many other well known methods, exemplified in particular by those proposed by Basirzadeh [1], Gani et al. [6], Pandian and Natarajan [18, 19], and Stephen Dinagar and Palanivel [21, 22], there is a problem that, in a general case, neither the cost parameters nor the obtained fuzzy optimal solution (that is, the fuzzy quantity of the product that should be transported from different sources to different destinations to minimize the total fuzzy transportation cost) need necessarily to be non-negative fuzzy numbers which implies that the obtained minimum total fuzzy transportation cost need not necessarily be a non-negative fuzzy number either.

Of course, from a theoretical, analytic point of view this fact does not play any role as it does not make any impact on the properties of the problem formulation and solution algorithm, it is not in line with the very essence of the transportation problem. Clearly, in real life problems there is no physical meaning of a negative value of the cost and a negative quantity of the product transported. Therefore, just due to this obvious reason, it makes little sense to apply the existing methods for solving fully fuzzy transportation problems proposed by Basirzadeh [1], Gani et al. [6], Pandian and Natarajan [18, 19], and Stephen Dinagar and Palanivel [21, 22] in real life problems.

Therefore, after the above general remarks, we have discussed in more detail only the well known method of Liu and Kao [12] which makes it possible to solve fully fuzzy transportation problems with both the inequality and equality constraints in which all the parameters are represented by non-negative trapezoidal fuzzy numbers; of course a special case of a trapezoid fuzzy number is a triangular fuzzy number so that this method can accommodate both types of fuzzy numbers.

To present the above mentioned issues, we will now provide a critical analysis of the existing methods for solving the fully fuzzy transportation problem, namely the ones mentioned above, that is, due to Basirzadeh [1], Gani et al. [6], Pandian and Natarajan [18, 19], and Stephen Dinagar and Palanivel [21, 22].

This can be presented in a concise form as follows:

First, we consider the extended methods by Julien [7], and by Parra et al. [20]. Taking these methods as the object of interest, Liu and Kao [12] have presented and solved first the fuzzy transportation problem, presented in Example 14, given below, by using those Julien's [7], and Parra et al. [20] methods.

Example 14 Suppose that we have a simple fuzzy transportation problem with inequality constraints with $m = 2$ sources and $n = 2$ destinations. The availabilities and demands for a product, and the costs are assumed to be given as triangular fuzzy numbers; notice that some of them are real (or integer, to be more specific) numbers which are clearly special cases of the triangular fuzzy numbers. They are as follows:

- Availabilities:

$$\begin{cases} \tilde{a}_1 = (2, 3, 5) \\ a_2 = 5 \end{cases}$$

- Demands:

$$\begin{cases} b_1 = 4 \\ \tilde{b}_2 = (1, 3, 6) \end{cases}$$

- Costs:

$$\begin{cases} c_{11} = 1 \; c_{12} = 3 \\ c_{21} = 7 \; c_{22} = 2 \end{cases}$$

where the triangular fuzzy numbers are given as triples (a, b, c), i.e. by the left most, central and right most values; notice that the costs are all real numbers, as are some availabilities and demands so that we have a fuzzy but not fully fuzzy transportation problem.

As shown by Liu and Kao [12], the optimal solution of fuzzy transportation problem, presented in Example 14, is:

$$\begin{cases} Z_0^L = 18 \\ Z_0^U = 17 \end{cases}$$

which constitute, respectively, the values of left end point and right end point of the α-cut

$$[Z_\alpha^L, Z_\alpha^U]$$

of the minimum total fuzzy transportation cost $\tilde{Z}$ at $\alpha = 0$.

Furthermore, Liu and Kao [12] pointed out that

$$Z_0^L > Z_0^U$$

which contradicts the obvious condition

$$Z_\alpha^U \geq Z_\alpha^L; \forall \alpha \in [0, 1]$$

which constitute, respectively, the values of left end point and right end point of the α-cut

$$[Z_\alpha^L, Z_\alpha^U]$$

of the minimum total fuzzy transportation cost $\tilde{Z}$ at $\alpha = 0$. additional fuzzy transportation problem.

Then, Liu and Kao [12] went a step further by assuming that we have the fully fuzzy transportation problem, and proposed a new method for solving that problem, both with the inequality and equality constraints, as we have presented in Sects. 4.4.1 and 4.4.2, respectively.

The Liu and Kao's [12] method can, unfortunately, yield some results that contradict some natural conditions, too. To best show these possibly counter intuitive assumptions, properties and results of their method, it is good to show another example, Example 15 given below, of a fuzzy transportation problem, solved in the aforementioned source Liu and Kao's [12] work; notice this is now a fully fuzzy transportation problem.

Example 15 Consider a simple fully fuzzy transportation problem with equality constraints with $m = 2$ sources and $n = 3$ destinations. Now we have the following fuzzy availabilities and demands for a product, given as the triangular fuzzy numbers, and the real and fuzzy costs, which are assumed to be as follows:

- Fuzzy availabilities:

$$\begin{cases} \tilde{a}_1 = (70, 90, 100) \\ \tilde{a}_2 = (40, 60, 70, 80) \end{cases}$$

- Fuzzy demands:

$$\begin{cases} \tilde{b}_1 = (30, 40, 50, 70) \\ \tilde{b}_2 = (20, 30, 40, 50) \\ \tilde{b}_3 = (40, 50, 80) \end{cases}$$

- Costs:

$$\begin{cases} c_{11} = 10 & c_{12} = 50 \; c_{13} = 80 \\ \tilde{c}_{21} = (60, 70, 80, 90) & c_{22} = 60 \; c_{23} = 20 \end{cases}$$

In their work, Liu and Kao [12] claimed that by solving the fully fuzzy transportation problem, as presented in this Example 15, at $\alpha = 0$, the optimal solution obtained is:

$$\begin{cases} x_{11}^L = 50 \\ x_{11}^U = 30 \end{cases}$$

Since, unfortunately,

$$x_{11}^L > x_{11}^U$$

then the same criticism, pointed out by Liu and Kao [12] with respect to the methods of Julien [7], and Parra et al. [20], as shown in Example 14, also occurs in the case of Liu and Kao's [12] method itself.

Moreover, by continuing in the same spirit, Liu and Kao [12] claimed that while solving the fully fuzzy transportation problem, presented in Example 15, the values of Z_α^L and Z_α^U for $\alpha = 0.9$ are equal to 3680, and for $\alpha = 1$ the solution is infeasible. They also claimed that, since $Z_\alpha^L = Z_\alpha^U = 3680$ for $\alpha = 0.9$, for higher values of α the obtained solution was infeasible. Therefore, the obtained fuzzy optimal solution would be a triangular fuzzy number with the maximum membership degree equal 0.9.

However, there may also exist a fully fuzzy transportation problem in which

$$Z_\alpha^L \neq Z_\alpha^U$$

for $\alpha = 0.9$ but the solution is infeasible for $\alpha = 1$. In such a situation, there will exist an infinite number of values of α between 0.9 and 1. Unfortunately, in such a case it is neither possible to find the maximum degree of membership nor to find the transportation cost corresponding to which this maximum degree of membership will exist. That is, in such a case the obtained solution can be represented by an interval but not a triangular fuzzy number.

In turn, Stephen Dinagar and Palanivel [22] proposed a method for solving the fully fuzzy transportation problem and solved, to illustrate their method, the fully fuzzy transportation problem presented in Example 16.

Example 16 Consider a fully fuzzy transportation problem having three sources and four destinations: Stephen Dinagar and Palanivel [22]. The fuzzy availabilities, fuzzy demands and fuzzy costs are as follows:

- The fuzzy availabilities:

$$\begin{cases} \tilde{a}_1 = (0, 2, 4, 6) \\ \tilde{a}_2 = (2, 4, 9, 13) \\ \tilde{a}_3 = (2, 4, 6, 8) \end{cases}$$

- The fuzzy demands:

$$\begin{cases} \tilde{b}_1 = (1, 3, 5, 7) \\ \tilde{b}_2 = (0, 2, 4, 6) \\ \tilde{b}_3 = (1, 3, 5, 7) \\ \tilde{b}_4 = (1, 3, 5, 7) \end{cases}$$

- The fuzzy costs:

$$\begin{cases} \tilde{c}_{11} = (-2, 0, 2, 8) & \tilde{c}_{12} = (-2, 0, 2, 8) & \tilde{c}_{13} = (-2, 0, 2, 8) & \tilde{c}_{14} = (-1, 0, 1, 4) \\ \tilde{c}_{21} = (4, 8, 12, 16) & \tilde{c}_{22} = (4, 7, 9, 12) & \tilde{c}_{23} = (2, 4, 6, 8) & \tilde{c}_{24} = (1, 3, 5, 7) \\ \tilde{c}_{31} = (2, 4, 9, 13) & \tilde{c}_{32} = (0, 6, 8, 10) & \tilde{c}_{33} = (0, 6, 8, 10) & \tilde{c}_{34} = (4, 7, 9, 12) \end{cases}$$

It is now easy to notice an obvious fact that there exist a negative part in the trapezoidal fuzzy numbers

$$\tilde{c}_{11}\ \tilde{c}_{12}\ \tilde{c}_{13}\ \tilde{c}_{14}$$

representing the fuzzy costs for transporting one unit of the quantity of the product from the first source to all destinations, that implies that the transportation cost can be negative.

Moreover, Stephen Dinagar and Palanivel [21] claimed that while solving the fully fuzzy transportation problem, presented in Example 16, the following results are obtained:

(a) The fuzzy optimal quantities of the product that should be transported from the second source to the third destination, the third source to the first destination and the third source to the third destination are, respectively:

$$(-5, -1, 6, 12)\ (-5, -1, 3, 7)\ (-11, -3, 6, 12)$$

(b) The minimum total fuzzy transportation cost is

$$(-122, -2, 139, 257)$$

so that it is clear that there exist a negative part in all the obtained trapezoidal fuzzy numbers. Similarly, there also exists a negative part in the results obtained by using the existing methods proposed by Basirzadeh [1], Gani et al. [6], Pandian and Natarajan [18, 19] or Stephen Dinagar and Palanivel [22].
However, for obvious reasons, in real life there is no physical meaning of the negative cost and the negative quantity of a product. Therefore, it can be concluded that it makes little sense to apply these existing methods, i.e. those due to by Basirzadeh [1], Gani et al. [6], Pandian and Natarajan [18, 19] or Stephen Dinagar and Palanivel [22], for the solution of fully fuzzy transportation problems.
And going further, the fuzzy optimal solution, obtained by using the existing methods mentioned above, i.e. due to by Basirzadeh [1], Gani et al. [6], Pandian and Natarajan [18, 19] or Stephen Dinagar and Palanivel [22], in general do not exactly satisfy the constraints in the sense that by putting the values of fuzzy decision variables, obtained by using these methods, in the left hand side of the constraints, the correct (feasible) values at right hand side may not be obtained. For instance, as shown by Stephen Dinagar and Palanivel [21, 22], by solving the fully fuzzy transportation problem presented in Example 16, the obtained fuzzy optimal quantity of the product that should be transported from the second source to the third and fourth destinations are

$$(-5, -1, 6, 12)\ (1, 3, 5, 7)$$

respectively.
It is obvious that the sum of these two fuzzy quantities is not exactly equal to the fuzzy availability

$$\tilde{a}_2 = (2, 4, 9, 13)$$

of the product at second source, that is, the constraint

$$\sum_{j=1}^{4} \tilde{x}_{2j} = \tilde{a}_2$$

is not satisfied.

4.6 On Some New Methods for Solving the Fully Fuzzy Transportation Problem

As we have indicated in the previous section, the well known and often employed existing methods proposed by Basirzadeh [1], Gani et al. [6], Pandian and Natarajan [18, 19] or Stephen Dinagar and Palanivel [21, 22], exhibit some shortcoming which basically boil down to the fact that they can yield results that are not consistent with the very preconditions for the transportation problems, to be more specific, related to the obvious non-negativity of the costs, quantity of products, etc.

In this section, we will propose a novel approach to overcome these counter intuitive properties of the traditional methods be presenting two new methods or solving such fully fuzzy transportation problems. They are based on a fuzzy linear programming formulation with a specific tabular representation employed for convenience and clarity. We assume that all the parameters are represented by trapezoidal fuzzy numbers.

We will also discuss some more general issues related to conceptual, algorithmic and implementation advantages of the two new proposed methods over some well known methods from the literature, notably those due to the already cited ones by Basirzadeh [1], Gani et al. [6], Pandian and Natarajan [18, 19] or Stephen Dinagar and Palanivel [22], and also a more general one by Stephen Dinagar and Palanivel [21].

4.6.1 A New Method Based on a Fuzzy Linear Programming Formulation

In this section we will present a new method which is based on a formulation of the fully fuzzy transportation problem as a fuzzy linear programming problem.

The steps of the proposed new method are as follow:

Step 1. Find:

- the total fuzzy availability $\sum_{i=1}^{m} \tilde{a}_i$, and
- the total fuzzy demand $\sum_{j=1}^{n} \tilde{b}_j$.

Let

$$\begin{cases} \sum_{i=1}^{m} \tilde{a}_i = (a, b, c, d) \\ \sum_{j=1}^{n} \tilde{b}_j = (a', b', c', d') \end{cases}$$

Use Definition 12 to check if the fully fuzzy transportation problem is balanced or not, i.e., if

$$\sum_{i=1}^{m} \tilde{a}_i = \sum_{j=1}^{n} \tilde{b}_j$$

or

$$\sum_{i=1}^{m} \tilde{a}_i \neq \sum_{j=1}^{n} \tilde{b}_j$$

Now, we have:

Case (i) If the problem is balanced, i.e., if

$$\sum_{i=1}^{m} \tilde{a}_i = \sum_{j=1}^{n} \tilde{b}_j$$

then **go to Step 2**.

Case (ii) If the problem is not balanced, i.e. if

$$\sum_{i=1}^{m} \tilde{a}_i \neq \sum_{j=1}^{n} \tilde{b}_j$$

then convert the unbalanced problem into the balanced problem.

The comparison of fuzzy numbers is still an open problem though there are maybe hundreds of papers on this topic. In this work we follow the path employed by many authors, exemplified by Campos and Verdegay [2], Chiang [3], Ebrahimnejad et al. [4], Ganesan and Veeramani [5], Gani et al. [6], Kheirfam and Verdegay [8], Kumar et al. [9], Liu and Kao [12], Mahadavi-Amiri and Nasseri [13, 14], Maleki et al. [15], Mukherjee and Basu [16], Pandian and Natarajan [17–19], Stephen Dinagar and Palanivel [21, 22], to just name a few.

Namely, these authors use quite a natural, yet powerful property that:

If

$$\begin{aligned} &\min_{1\le w\le h}\left\{\Re\left(\sum_{i=1}^{m}\sum_{j=1}^{n} a'_{ij}a^{w}_{ij}, \sum_{i=1}^{m}\sum_{j=1}^{n} b'_{ij}b^{w}_{ij}, \sum_{i=1}^{m}\sum_{j=1}^{n} c'_{ij}c^{w}_{ij}, \sum_{i=1}^{m}\sum_{j=1}^{n} d'_{ij}d^{w}_{ij}\right)\right\} \\ &= \\ &\Re\left(\sum_{i=1}^{m}\sum_{j=1}^{n} a'_{ij}a^{\eta}_{ij}, \sum_{i=1}^{m}\sum_{j=1}^{n} b'_{ij}b^{\eta}_{ij}, \sum_{i=1}^{m}\sum_{j=1}^{n} c'_{ij}c^{\eta}_{ij}, \sum_{i=1}^{m}\sum_{j=1}^{n} d'_{ij}d^{\eta}_{ij}\right) \end{aligned} \tag{4.24}$$

then there also holds

$$\min_{1\leq w\leq h}\left\{\left(\sum_{i=1}^{m}\sum_{j=1}^{n}a'_{ij}a^{w}_{ij},\sum_{i=1}^{m}\sum_{j=1}^{n}b'_{ij}b^{w}_{ij},\sum_{i=1}^{m}\sum_{j=1}^{n}c'_{ij}c^{w}_{ij},\sum_{i=1}^{m}\sum_{j=1}^{n}d'_{ij}d^{w}_{ij}\right)\right\} = \left(\sum_{i=1}^{m}\sum_{j=1}^{n}a'_{ij}a^{\eta}_{ij},\ \sum_{i=1}^{m}\sum_{j=1}^{n}b'_{ij}b^{\eta}_{ij},\ \sum_{i=1}^{m}\sum_{j=1}^{n}c'_{ij}c^{\eta}_{ij},\ \sum_{i=1}^{m}\sum_{j=1}^{n}d'_{ij}d^{\eta}_{ij}\right) \tag{4.25}$$

where

$$\Re\left(\sum_{i=1}^{m}\sum_{j=1}^{n}a'_{ij}a^{\eta}_{ij},\sum_{i=1}^{m}\sum_{j=1}^{n}b'_{ij}b^{\eta}_{ij},\sum_{i=1}^{m}\sum_{j=1}^{n}c'_{ij}c^{\eta}_{ij},\sum_{i=1}^{m}\sum_{j=1}^{n}d'_{ij}d^{\eta}_{ij}\right) = \frac{\sum_{i=1}^{m}\sum_{j=1}^{n}a'_{ij}a^{\eta}_{ij}+\sum_{i=1}^{m}\sum_{j=1}^{n}b'_{ij}b^{\eta}_{ij}+\sum_{i=1}^{m}\sum_{j=1}^{n}c'_{ij}c^{\eta}_{ij}+\sum_{i=1}^{m}\sum_{j=1}^{n}d'_{ij}d^{\eta}_{ij}}{4} \tag{4.26}$$

is Liou and Wang's [11] ranking index of a trapezoidal fuzzy number

$$\left(\sum_{i=1}^{m}\sum_{j=1}^{n}a'_{ij}a^{\eta}_{ij},\sum_{i=1}^{m}\sum_{j=1}^{n}b'_{ij}b^{\eta}_{ij},\ \sum_{i=1}^{m}\sum_{j=1}^{n}c'_{ij}c^{\eta}_{ij},\ \sum_{i=1}^{m}\sum_{j=1}^{n}d'_{ij}d^{\eta}_{ij}\right)$$

Then, using the above properties, the fuzzy optimal solution of the fuzzy linear programming problem (4.7) can be obtained by solving the following crisp linear programming problem:

$$\begin{cases}\min\limits_{a_{ij},b_{ij},c_{ij},d_{ij}}\left(\dfrac{\sum_{i=1}^{m}\sum_{j=1}^{n}a'_{ij}a_{ij}+\sum_{i=1}^{m}\sum_{j=1}^{n}b'_{ij}b_{ij}+\sum_{i=1}^{m}\sum_{j=1}^{n}c'_{ij}c_{ij}+\sum_{i=1}^{m}\sum_{j=1}^{n}d'_{ij}d_{ij}}{4}\right)\\ \text{subject to:}\\ \sum_{j=1}^{n}a_{ij}=a_i;\text{ for } i=1,2,\ldots,m\\ \sum_{j=1}^{n}b_{ij}=b_i;\text{ for } i=1,2,\ldots,m\\ \sum_{j=1}^{n}c_{ij}=c_i;\text{ for } i=1,2,\ldots,m\\ \sum_{j=1}^{n}d_{ij}=d_i;\text{ for } i=1,2,\ldots,m\\ \sum_{i=1}^{m}a_{ij}=a'_j;\text{ for } j=1,2,\ldots,n\\ \sum_{i=1}^{m}b_{ij}=b'_j;\text{ for } j=1,2,\ldots,n\\ \sum_{i=1}^{m}c_{ij}=c'_j;\text{ for } j=1,2,\ldots,n\\ \sum_{i=1}^{m}d_{ij}=d'_j;\text{ for } j=1,2,\ldots,n\end{cases} \tag{4.27}$$

where $a_{ij}, b_{ij} - a_{ij}, c_{ij} - b_{ij}, d_{ij} - c_{ij} \geq 0; \forall i = 1, 2, \ldots, m; \ j = 1, 2, \ldots, n,$

Step 7. Solve the crisp linear programming problem, obtained in Step 6, i.e. (4.27), to find the optimal solution:

$$\{a_{ij}, b_{ij}, c_{ij}, d_{ij}\}$$

Step 8. Find the fuzzy optimal solution, $\{\tilde{x}_{ij}\}$, by putting the values of $a_{ij}, b_{ij}, c_{ij}, d_{ij}$ obtained by solving the problem in Step 7 into $\tilde{x}_{ij}$, that is $\tilde{x}_{ij} = (a_{ij}, b_{ij}, c_{ij}, d_{ij})$.

Step 9. Find the minimum total fuzzy transportation cost by putting the values of $\tilde{x}_{ij}$ obtained above in Step 8 into

$$\sum_{i=1}^{m}\sum_{j=1}^{n}(\tilde{c}_{ij} \otimes \tilde{x}_{ij})$$

Remark 1 If

$$\sum_{i=1}^{m} \tilde{a}_i \neq \sum_{j=1}^{n} \tilde{b}_j$$

that is,

$$(a, b, c, d) \neq (a', b', c', d')$$

and neither:

Case (a): $a \leq a', b - a \leq b' - a', c - b \leq c' - b', d - c \leq d' - c'$

nor

Case (b): $a \geq a', b - a \geq b' - a', c - b \geq c' - b', d - c \geq d' - c'$

is satisfied, then there may exist infinitely many non-negative trapezoidal fuzzy numbers,

$$\begin{cases} (a_1, b_1, c_1, d_1) \\ (a_1', b_1', c_1', d_1') \end{cases}$$

such that

$$(a, b, c, d) \oplus (a_1, b_1, c_1, d_1) = (a', b', c', d') \oplus (a_1', b_1', c_1', d_1')$$

but the aim is to be find the trapezoidal fuzzy numbers

$$\begin{cases} (a_1, b_1, c_1, d_1) \\ (a_1', b_1', c_1', d_1') \end{cases}$$

which satisfy:

1. (a_1, b_1, c_1, d_1) and (a_1', b_1', c_1', d_1') are non-negative trapezoidal fuzzy numbers,
2. it holds

$$(a, b, c, d) \oplus (a_1, b_1, c_1, d_1) = (a', b', c', d') \oplus (a_1', b_1', c_1', d_1'),$$

3. if there exist two non-negative trapezoidal fuzzy numbers, (x, y, z, w) and (x', y', z', w') such that

$$(a, b, c, d) \oplus (x, y, z, w) = (a', b', c', d') \oplus (x', y', z', w')$$

then

$$\Re(x, y, z, w) \geq \Re(a_1, b_1, c_1, d_1)$$

and

$$\Re(x', y', z', w') \geq \Re(a_1', b_1', c_1', d_1').$$

4.6.2 Method Based on the Tabular Representation

In this section we present a new method for the determination of an optimal solution to the fully fuzzy transportation problem which is based on the tabular representation of the fully fuzzy transportation problem considered.

The consecutive steps of the proposed method are as follows:

Step 1. Perform Step 1 of the method for the solution of the fully fuzzy transportation based on a fuzzy linear programming formulation which is presented in the previous section, i.e. Sect. 4.6.1 to obtain a balanced fully fuzzy transportation problem.

Step 2. Represent the balanced fully fuzzy transportation problem obtained in Step 1 in the tabular form as shown in Table 4.1.

Step 3. Split Table 4.1 into four tables representing the four specific crisp transportation problems, i.e., into Tables 4.2, 4.3, 4.4 and 4.5, respectively, that is: where:

$$\lambda_{ij} = \frac{(a_{ij}' + b_{ij}' + c_{ij}' + d_{ij}')}{4}; i = 1, 2, \ldots, m; \ j = 1, 2, \ldots, n.$$

where:

$$\rho_{ij} = \frac{(b_{ij}' + c_{ij}' + d_{ij}')}{4}; i = 1, 2, \ldots, m; \ j = 1, 2, \ldots, n.$$

Table 4.1 Tabular representation of a balanced fully fuzzy transportation problem

Destinations → Sources↓	D_1	D_2	...	D_j	...	D_n	Availability ($\tilde{a}_i$)
S_1	$\tilde{c}_{11}$	$\tilde{c}_{12}$	...	$\tilde{c}_{1j}$	...	$\tilde{c}_{1n}$	$\tilde{a}_1$
$\vdots$	$\vdots$	$\vdots$	$\vdots$	$\vdots$	$\vdots$	$\vdots$	$\vdots$
S_i	$\tilde{c}_{i1}$	$\tilde{c}_{i2}$	...	$\tilde{c}_{ij}$	...	$\tilde{c}_{in}$	$\tilde{a}_i$
$\vdots$	$\vdots$	$\vdots$	$\vdots$	$\vdots$	$\vdots$	$\vdots$	$\vdots$
S_m	$\tilde{c}_{m1}$	$\tilde{c}_{m2}$	...	$\tilde{c}_{mj}$	...	$\tilde{c}_{mn}$	$\tilde{a}_m$
Demand ($\tilde{b}_j$)	$\tilde{b}_1$	$\tilde{b}_2$	...	$\tilde{b}_j$	...	$\tilde{b}_n$	$\sum_{i=1}^{m} \tilde{a}_i = \sum_{j=1}^{n} \tilde{b}_j$

where:

$$\delta_{ij} = \frac{(c'_{ij} + d'_{ij})}{4}; i = 1, 2, \ldots, m; j = 1, 2, \ldots, n.$$

where:

$$\xi_{ij} = \frac{d'_{ij}}{4}; i = 1, 2, \ldots, m; j = 1, 2, \ldots, n.$$

Step 4. Solve the crisp transportation problems represented by the above shown tables, i.e. Tables 4.2, 4.3, 4.4 and 4.5, to find the optimal solution, respectively:

$$\{ \{a_{ij}\} \ \{b_{ij} - a_{ij}\} \ \{c_{ij} - b_{ij}\} \ \{d_{ij} - c_{ij}\}$$

Step 5. Solve the problems mentioned in Step 4, to find the values of a_{ij}, b_{ij}, c_{ij} and d_{ij}.

Step 6. Find the fuzzy optimal solution $\{\tilde{x}_{ij}\}$ by putting the values of

$$a_{ij}, b_{ij}, c_{ij}, d_{ij}$$

into

$$\tilde{x}_{ij} = (a_{ij}, b_{ij}, c_{ij}, d_{ij})$$

Step 7. Find the minimum total fuzzy transportation cost by putting the values of $\tilde{x}_{ij}$ into

$$\sum_{i=1}^{m} \sum_{j=1}^{n} (\tilde{c}_{ij} \otimes \tilde{x}_{ij})$$

Remark 2 Since, in the transportation problems negative parameters have no physical meaning, then in the new methods proposed all the parameters be assumed to be non-negative trapezoidal fuzzy numbers.

Table 4.2 Tabular representation of the first crisp transportation problem from (4.27)

Destinations → Sources ↓	D_1	D_2	$\cdots$	D_j	$\cdots$	D_n	a_i
S_1	λ_{11}	λ_{12}	$\cdots$	λ_{1j}	$\cdots$	λ_{1n}	a_1
$\vdots$	$\vdots$	$\vdots$	$\vdots$	$\vdots$	$\vdots$	$\vdots$	$\vdots$
S_i	λ_{i1}	λ_{i2}	$\cdots$	λ_{ij}	$\cdots$	λ_{in}	a_i
$\vdots$	$\vdots$	$\vdots$	$\vdots$	$\vdots$	$\vdots$	$\vdots$	$\vdots$
S_m	λ_{m1}	λ_{m2}	$\cdots$	λ_{mj}	$\cdots$	λ_{mn}	a_m
a'_j	a'_1	a'_2	$\cdots$	a'_j	$\cdots$	a'_n	$\sum_{i=1}^{m} a_i = \sum_{j=1}^{n} b_j$

Table 4.3 Tabular representation of the second crisp transportation problem from (4.27)

Destinations → Sources ↓	D_1	D_2	$\cdots$	D_j	$\cdots$	D_n	$b_i - a_i$
S_1	ρ_{11}	ρ_{12}	$\cdots$	ρ_{1j}	$\cdots$	ρ_{1n}	$b_1 - a_1$
$\vdots$	$\vdots$	$\vdots$	$\vdots$	$\vdots$	$\vdots$	$\vdots$	$\vdots$
S_i	ρ_{i1}	ρ_{i2}	$\cdots$	ρ_{ij}	$\cdots$	ρ_{in}	$b_i - a_i$
$\vdots$	$\vdots$	$\vdots$	$\vdots$	$\vdots$	$\vdots$	$\vdots$	$\vdots$
S_m	ρ_{m1}	ρ_{m2}	$\cdots$	ρ_{mj}	$\cdots$	ρ_{mn}	$b_m - a_m$
$b'_j - a'_j$	$b'_1 - a'_1$	$b'_2 - a'_2$	$\cdots$	$b'_j - a'_j$	$\cdots$	$b'_n - a'_n$	$\sum_{i=1}^{m}(b_i - a_i) = \sum_{j=1}^{n}(b'_j - a'_j)$

Table 4.4 Tabular representation of the third crisp transportation problem from (4.27)

Destinations → Sources ↓	D_1	D_2	$\cdots$	D_j	$\cdots$	D_n	$c_i - b_i$
S_1	δ_{11}	δ_{12}	$\cdots$	δ_{1j}	$\cdots$	δ_{1n}	$c_1 - b_1$
$\vdots$	$\vdots$	$\vdots$	$\vdots$	$\vdots$	$\vdots$	$\vdots$	$\vdots$
S_i	δ_{i1}	δ_{i2}	$\cdots$	δ_{ij}	$\cdots$	δ_{in}	$c_i - b_i$
$\vdots$	$\vdots$	$\vdots$	$\vdots$	$\vdots$	$\vdots$	$\vdots$	$\vdots$
S_m	δ_{m1}	δ_{m2}	$\cdots$	δ_{mj}	$\cdots$	δ_{mn}	$c_m - b_m$
$c'_j - b'_j$	$c'_1 - b'_1$	$c'_2 - b'_2$	$\cdots$	$c'_j - b'_j$	$\cdots$	$c'_n - b'_n$	$\sum_{i=1}^{m}(c_i - b_i) = \sum_{j=1}^{n}(c'_j - b'_j)$

Table 4.5 Tabular representation of the fourth crisp transportation problem from (4.27)

Destinations → Sources ↓	D_1	D_2	$\cdots$	D_j	$\cdots$	D_n	$d_i - c_i$
S_1	ξ_{11}	ξ_{12}	$\cdots$	ξ_{1j}	$\cdots$	ξ_{1n}	$d_1 - c_1$
$\vdots$	$\vdots$	$\vdots$	$\vdots$	$\vdots$	$\vdots$	$\vdots$	$\vdots$
S_i	ξ_{i1}	ξ_{i2}	$\cdots$	ξ_{ij}	$\cdots$	ξ_{in}	$d_i - c_i$
$\vdots$	$\vdots$	$\vdots$	$\vdots$	$\vdots$	$\vdots$	$\vdots$	$\vdots$
S_m	ξ_{m1}	ξ_{m2}	$\cdots$	ξ_{mj}	$\cdots$	ξ_{mn}	$d_m - c_m$
$d'_j - c'_j$	$d'_1 - c'_1$	$d'_2 - c'_2$	$\cdots$	$d'_j - c'_j$	$\cdots$	$d'_n - c'_n$	$\sum_{i=1}^{m}(d_i - c_i)= \sum_{j=1}^{n}(d'_j - c'_j)$

4.6.3 *Advantages of the Proposed Methods over the Existing Methods*

In this section we present a new method for the determination of an optimal solution to the fully fuzzy transportation problem which is based on the tabular representation of the fully fuzzy transportation problem considered.

To be more specific, we will concentrate on showing the advantages of the proposed method over the well known existing methods existing methods proposed by Basirzadeh [1], Gani et al. [6], Liu and Kao [12], Pandian and Natarajan [18, 19], and Stephen Dinagar and Palanivel [21, 22], that is, we will discuss:

1. The advantages of the new methods proposed over the existing method proposed by Liu and Kao [12] can be summarized as follows:

   (a) First, since in the new methods proposed the constraints $b - a \geq 0$, $c - b \geq 0$ and $d - c \geq 0$ are used, then as a result the constraint, $d \geq a$ (or $x^U \geq x^L$) will always be satisfied. Therefore, by using the new methods proposed, the problem with the existing method by Liu and Kao [12] is resolved.
   (b) Second, since while on solving the fully fuzzy transportation problems by using the new methods proposed the obtained minimum total fuzzy transportation cost will always be a fuzzy number, then the problems with the existing method by Liu and Kao [12] will also be resolved.

2. In Sect. 4.3 it is pointed out that while solving the fully fuzzy transportation problems by using the existing methods mentioned, i.e. those due to Basirzadeh [1], Gani et al. [6], Liu and Kao [12], Pandian and Natarajan [18, 19], and Stephen Dinagar and Palanivel [21, 22], we obtain the negative minimum total fuzzy transportation cost and the negative quantity of the product. This is infeasible in real life, and also not consistent with the very assumption underlying all kinds of transportation problems. Moreover, it has also been pointed out that fuzzy optimal solution, obtained by using the existing methods mentioned above, does not

satisfy the constraints exactly. Since in the new methods proposed all the cost parameters are represented by non-negative trapezoidal fuzzy numbers and also the constraint should hold, then while solving the fully fuzzy transportation problems by using these new methods the obtained minimum total fuzzy transportation cost and fuzzy quantity of the product will be either zero or non-negative. Also, the fuzzy optimal solution, obtained by using the new methods mentioned above, will always exactly satisfy all the constraints.

Therefore, while applying the new methods proposed, all problems with the existing methods developed by Basirzadeh [1], Gani et al. [6], Liu and Kao [12], Pandian and Natarajan [18, 19], and Stephen Dinagar and Palanivel [21, 22], pointed out in Sect. 4.4, will be resolved.

4.7 An Illustrative Example

In this section, to illustrate the new methods proposed, and the existing fully fuzzy transportation problem due to Liu and Kao [12], presented in Example 15, is solved by the proposed methods.

4.7.1 Fuzzy Optimal Solution Using the Method Based on Fuzzy Linear Programming Formulation

Using the new method proposed, based on a fuzzy linear programming formulation, the fuzzy optimal solution of the fully fuzzy transportation problem, presented in Example 15, can be obtained as follows:

Step 1. The total fuzzy availability = (110, 150, 160, 180) and total fuzzy demand = (90, 120, 140, 200). Since, the total fuzzy availability $\neq$ total fuzzy demand, then we have an unbalanced fully fuzzy transportation problem.

Now, as described in the new method proposed (using Case (c) of Step 1 presented in Sect. 4.6.1, the unbalanced fully fuzzy transportation problem can be converted into a balanced fully fuzzy transportation problem by introducing a dummy source S_3 with the fuzzy availability (0, 0, 10, 50) and a dummy destination D_4 with the fuzzy demand (20, 30, 30, 30).

Step 2. By assuming the fuzzy cost transporting for one unit quantity of the product from the dummy source S_3 to all destinations and from all sources to the dummy destination D_4 to be the zero trapezoidal fuzzy number, that is:

$$\tilde{c}_{14} = \tilde{c}_{24} = \tilde{c}_{31} = \tilde{c}_{32} = \tilde{c}_{33} = \tilde{c}_{34} = (0, 0, 0, 0)$$

the balanced fully fuzzy transportation problem, obtained in Step 1, can be formulated as the following fuzzy linear programming problem:

$$\begin{cases} \min\{(10,10,10,10)\otimes\tilde{x}_{11}\oplus(50,50,50,50)\otimes\tilde{x}_{12}\oplus(80,80,80,80)\otimes \\ \tilde{x}_{13}\oplus(0,0,0,0)\otimes\tilde{x}_{14}\oplus(60,70,80,90)\otimes\tilde{x}_{21}\oplus(60,60,60,60)\otimes\tilde{x}_{22} \\ \oplus(20,20,20,20)\otimes\tilde{x}_{23}\oplus(0,0,0,0)\otimes\tilde{x}_{24}\oplus(0,0,0,0)\otimes\tilde{x}_{31}\oplus(0,0,0,0) \\ \otimes\tilde{x}_{32}\oplus(0,0,0,0)\otimes\tilde{x}_{33}\oplus(0,0,0,0)\otimes\tilde{x}_{34}\} \\ \text{subject to:} \\ \tilde{x}_{11}\oplus\tilde{x}_{12}\oplus\tilde{x}_{13}\oplus\tilde{x}_{14}=(70,90,90,100) \\ \tilde{x}_{21}\oplus\tilde{x}_{22}\oplus\tilde{x}_{23}\oplus\tilde{x}_{24}=(40,60,70,80) \\ \tilde{x}_{31}\oplus\tilde{x}_{32}\oplus\tilde{x}_{33}\oplus\tilde{x}_{34}=(0,0,10,50) \\ \tilde{x}_{11}\oplus\tilde{x}_{21}\oplus\tilde{x}_{31}=(30,40,50,70) \\ \tilde{x}_{12}\oplus\tilde{x}_{22}\oplus\tilde{x}_{32}=(20,30,40,50) \\ \tilde{x}_{13}\oplus\tilde{x}_{23}\oplus\tilde{x}_{33}=(40,50,50,80) \\ \tilde{x}_{14}\oplus\tilde{x}_{24}\oplus\tilde{x}_{34}=(20,30,30,30 \\ \\ \text{where: } \tilde{x}_{11},\tilde{x}_{12},\tilde{x}_{13},\tilde{x}_{14},\tilde{x}_{21},\tilde{x}_{22},\tilde{x}_{23},\tilde{x}_{24},\tilde{x}_{31},\tilde{x}_{32},\tilde{x}_{33},\tilde{x}_{34} \\ \text{are non-negative trapezoidal fuzzy numbers} \end{cases} \tag{4.28}$$

Step 3. By assuming $\tilde{x}_{ij}=(a_{ij},b_{ij},c_{ij},d_{ij})$ and using the arithmetic operations, defined in Sect. 4.1.2, the fuzzy linear programming problem, obtained in Step 2, can be converted into the following fuzzy linear programming problem:

$$\begin{cases} \min\{10a_{11}+50a_{12}+80a_{13}+60a_{21}+60a_{22}+20a_{23}, \\ 10b_{11}+50b_{12}+80b_{13}+70b_{21}+60b_{22}+20b_{23}, \\ 10c_{11}+50c_{12}+80c_{13}+80c_{21}+60c_{22}+20c_{23}, \\ 10d_{11}+50d_{12}+80d_{13}+90d_{21}+60d_{22}+20d_{23}\} \\ \text{subject to:} \\ \left(\sum_{j=1}^{4}a_{1j},\sum_{j=1}^{4}b_{1j},\sum_{j=1}^{4}c_{1j},\sum_{j=1}^{4}d_{1j}\right)=(70,90,90,100) \\ \left(\sum_{j=1}^{4}a_{2j},\sum_{j=1}^{4}b_{2j},\sum_{j=1}^{4}c_{2j},\sum_{j=1}^{4}d_{2j}\right)=(40,60,70,80) \\ \left(\sum_{j=1}^{4}a_{3j},\sum_{j=1}^{4}b_{3j},\sum_{j=1}^{4}c_{3j},\sum_{j=1}^{4}d_{3j}\right)=(0,0,10,50) \\ \left(\sum_{i=1}^{3}a_{i1},\sum_{i=1}^{3}b_{i1},\sum_{i=1}^{3}c_{i1},\sum_{i=1}^{3}d_{i1}\right)=(30,40,50,70) \\ \left(\sum_{i=1}^{3}a_{i2},\sum_{i=1}^{3}b_{i2},\sum_{i=1}^{3}c_{i2},\sum_{i=1}^{3}d_{i2}\right)=(20,30,40,50) \\ \left(\sum_{i=1}^{3}a_{i3},\sum_{i=1}^{3}b_{i3},\sum_{i=1}^{3}c_{i3},\sum_{i=1}^{3}d_{i3}\right)=(40,50,50,80) \\ \left(\sum_{i=1}^{3}a_{i4},\sum_{i=1}^{3}b_{i4},\sum_{i=1}^{3}c_{i4},\sum_{i=1}^{3}d_{i4}\right)=(20,30,30,30) \\ \text{where: } (a_{ij},b_{ij},c_{ij},d_{ij}) \text{ is a non-negative trapezoidal fuzzy number.} \end{cases} \tag{4.29}$$

Step 4. By using Definitions 10 and 12, the fuzzy linear programming problem, obtained in Step 3, can be converted into the following fuzzy linear programming problem (4.30):

$$\left\{\begin{array}{l}\min\{10a_{11}+50a_{12}+80a_{13}+60a_{21}+60a_{22}+20a_{23},\\ 10b_{11}+50b_{12}+80b_{13}+70b_{21}+60b_{22}+20b_{23},\\ 10c_{11}+50c_{12}+80c_{13}+80c_{21}+60c_{22}+20c_{23},\\ 10d_{11}+50d_{12}+80d_{13}+90d_{21}+60d_{22}+20d_{23}\}\\ \text{subject to:}\\ \sum_{j=1}^{4}a_{1j}=70,\quad \sum_{j=1}^{4}b_{1j}=90,\quad \sum_{j=1}^{4}c_{1j}=90,\quad \sum_{j=1}^{4}d_{1j}=100\\ \sum_{j=1}^{4}a_{2j}=40,\quad \sum_{j=1}^{4}b_{2j}=60,\quad \sum_{j=1}^{4}c_{2j}=70,\quad \sum_{j=1}^{4}d_{2j}=80\\ \sum_{j=1}^{4}a_{3j}=0,\quad \sum_{j=1}^{4}b_{3j}=0,\quad \sum_{j=1}^{4}c_{3j}=10,\quad \sum_{j=1}^{4}d_{3j}=50\\ \sum_{i=1}^{3}a_{i1}=30,\quad \sum_{i=1}^{3}b_{i1}=40,\quad \sum_{i=1}^{3}c_{i1}=50,\quad \sum_{i=1}^{3}d_{i1}=70\\ \sum_{i=1}^{3}a_{i2}=20,\quad \sum_{i=1}^{3}b_{i2}=30,\quad \sum_{i=1}^{3}c_{i2}=40,\quad \sum_{i=1}^{3}d_{i2}=50\\ \sum_{i=1}^{3}a_{i3}=40,\quad \sum_{i=1}^{3}b_{i3}=50,\quad \sum_{i=1}^{3}c_{i3}=50,\quad \sum_{i=1}^{3}d_{i3}=80\\ \sum_{i=1}^{3}a_{i4}=20,\quad \sum_{i=1}^{3}b_{i4}=30,\quad \sum_{i=1}^{3}c_{i4}=30,\quad \sum_{i=1}^{3}d_{i4}=30\\ \text{where: } a_{ij}, b_{ij}-a_{ij}, c_{ij}-b_{ij}, d_{ij}-c_{ij}\geq 0;\ i=1,2,3;\ j=1,2,3,4\end{array}\right. \tag{4.30}$$

Step 5. Using Step 6 of the method, proposed in Sect. 4.6.1, the fuzzy optimal solution of the fuzzy linear programming problem (4.30), can be obtained by solving the following crisp linear programming problem:

$$\left\{\begin{array}{l}\min \frac{1}{4}(10a_{11}+10b_{11}+10c_{11}+10d_{11}+\ +50a_{12}+50b_{12}+50c_{12}+\\ +50d_{12}+80a_{13}+80b_{13}+80c_{13}+80d_{13}+\ +60a_{21}+70b_{21}+\\ +80c_{21}+90d_{21}+\ +60a_{22}+60b_{22}+60c_{22}+60d_{22}+\ +20a_{23}+\\ +20b_{23}+20c_{23}+20d_{23})\\ \text{subject to:}\\ \sum_{j=1}^{4}a_{1j}=70;\ \sum_{j=1}^{4}b_{1j}=90;\ \sum_{j=1}^{4}c_{1j}=90;\ \sum_{j=1}^{4}d_{1j}=100;\\ \sum_{j=1}^{4}a_{2j}=40;\ \sum_{j=1}^{4}b_{2j}=60;\ \sum_{j=1}^{4}c_{2j}=70;\ \sum_{j=1}^{4}d_{2j}=80;\\ \sum_{j=1}^{4}a_{3j}=0;\ \sum_{j=1}^{4}b_{3j}=0;\ \sum_{j=1}^{4}c_{3j}=10;\ \sum_{j=1}^{4}d_{3j}=50;\\ \sum_{i=1}^{3}a_{i1}=30;\ \sum_{i=1}^{3}b_{i1}=40;\ \sum_{i=1}^{3}c_{i1}=50;\ \sum_{i=1}^{3}d_{i1}=70;\\ \sum_{i=1}^{3}a_{i2}=20;\ \sum_{i=1}^{3}b_{i2}=30;\ \sum_{i=1}^{3}c_{i2}=40;\ \sum_{i=1}^{3}d_{i2}=50;\\ \sum_{i=1}^{3}a_{i3}=40;\ \sum_{i=1}^{3}b_{i3}=50;\ \sum_{i=1}^{3}c_{i3}=50;\ \sum_{i=1}^{3}d_{i3}=80;\\ \sum_{i=1}^{3}a_{i4}=20;\ \sum_{i=1}^{3}b_{i4}=30;\ \sum_{i=1}^{3}c_{i4}=30;\ \sum_{i=1}^{3}d_{i4}=30;\\ \text{where: } a_{ij}, b_{ij}-a_{ij}, c_{ij}-b_{ij}, d_{ij}-c_{ij}\geq 0;\ i=1,2,3;\ j=1,2,3,4\end{array}\right. \tag{4.31}$$

Step 6. The optimal solution of the crisp linear programming problem, obtained in Step 5, (4.31), is:

$$\begin{cases} a_{11} = 30, b_{11} = 40, c_{11} = 40, d_{11} = 40, \\ a_{12} = 20, b_{12} = 30, c_{12} = 30, d_{12} = 40, \\ a_{13} = 0, b_{13} = 0, c_{13} = 0, d_{13} = 0, \\ a_{14} = 20, b_{14} = 20, c_{14} = 20, d_{14} = 20, \\ a_{21} = 0, b_{21} = 0, c_{21} = 0, d_{21} = 0, \\ a_{22} = 0, b_{22} = 0, c_{22} = 10, d_{22} = 10, \\ a_{23} = 40, b_{23} = 50, c_{23} = 50, d_{23} = 60, \\ a_{24} = 0, b_{24} = 10, c_{24} = 10, d_{24} = 10, \\ a_{31} = 0, b_{31} = 0, c_{31} = 10, d_{31} = 30, \\ a_{32} = 0, b_{32} = 0, c_{32} = 0, d_{32} = 0, \\ a_{33} = 0, b_{33} = 0, c_{33} = 0, d_{33} = 20, \\ a_{34} = 0, b_{34} = 0, c_{34} = 0, d_{34} = 0 \end{cases} \tag{4.32}$$

Step 7. Putting the values of $a_{ij}, b_{ij}, c_{ij}, d_{ij}, i = 1, 2, 3, j = 1, 2, 3, 4$, obtained from Step 6, into $\tilde{x}_{ij} = (a_{ij}, b_{ij}, c_{ij}, d_{ij}), i = 1, 2, 3, j = 1, 2, 3, 4$, the fuzzy optimal solution is:

$$\begin{cases} \tilde{x}_{11} = (30, 40, 40, 40) \\ \tilde{x}_{12} = (20, 30, 30, 40) \\ \tilde{x}_{13} = (0, 0, 0, 0) \\ \tilde{x}_{14} = (20, 20, 20, 20) \\ \tilde{x}_{21} = (0, 0, 0, 0) \\ \tilde{x}_{22} = (0, 0, 10, 10) \\ \tilde{x}_{23} = (40, 50, 50, 60) \\ \tilde{x}_{24} = (0, 10, 10, 10) \\ \tilde{x}_{31} = (0, 0, 10, 30) \\ \tilde{x}_{32} = (0, 0, 0, 0) \\ \tilde{x}_{33} = (0, 0, 0, 20) \\ \tilde{x}_{34} = (0, 0, 0, 0) \end{cases} \tag{4.33}$$

Step 8. Putting the values of

$$\begin{cases} \tilde{x}_{11} \ \tilde{x}_{12} \ \tilde{x}_{13} \ \tilde{x}_{14} \\ \tilde{x}_{21} \ \tilde{x}_{22} \ \tilde{x}_{23} \ \tilde{x}_{24} \\ \tilde{x}_{31} \ \tilde{x}_{32} \ \tilde{x}_{33} \ \tilde{x}_{34} \end{cases}$$

into

$$\begin{aligned} &(10, 10, 10, 10) \otimes \tilde{x}_{11} \oplus (50, 50, 50, 50) \otimes \tilde{x}_{12} \oplus (80, 80, 80, 80) \otimes \tilde{x}_{13} \oplus \\ &\oplus (0, 0, 0, 0) \otimes \tilde{x}_{14} \oplus (60, 70, 80, 90) \otimes \tilde{x}_{21} \oplus (60, 60, 60, 60) \otimes \tilde{x}_{22} \oplus \\ &\oplus (20, 20, 20, 20) \otimes \tilde{x}_{23} \oplus (0, 0, 0, 0) \otimes \tilde{x}_{24} \oplus (0, 0, 0, 0) \otimes \tilde{x}_{31} \oplus \\ &\oplus (0, 0, 0, 0) \otimes \tilde{x}_{32} \oplus (0, 0, 0, 0) \otimes \tilde{x}_{33} \oplus (0, 0, 0, 0) \otimes \tilde{x}_{34} \end{aligned}$$

we obtain the following minimum total fuzzy transportation cost:

$$(2100, 2900, 3500, 4200)$$

4.7.2 *Fuzzy Optimal Solution Using the Method Based on Tabular Representation*

By using the new method proposed, based on the tabular representation, the fuzzy optimal solution of the fully fuzzy transportation problem, considered in Example 15, can be now obtained as through the following steps:

Step 1. The balanced fully fuzzy transportation problem, obtained in Step 1 of the algorithm presented in Sect. 4.6.2, can be represented by Table 4.6.

Step 2. Using Step 3 of the method proposed in Sect. 4.6.2, Table 4.6 can be split into four crisp transportation tables, that is, Tables 4.7, 4.8, 4.9 and 4.10.

Table 4.6 Tabular representation of balanced fully fuzzy transportation problem

	D_1	D_2	D_3	D_4	$\tilde{a}_i$
S_1	(10,10,10,10)	(50,50,50,50)	(80,80,80,80)	(0,0,0,0)	(70,90,90,100)
S_2	(60,70,80,90)	(60,60,60,60)	(20,20,20,20)	(0,0,0,0)	(40,60,70,80)
S_3	(0,0,0,0)	(0,0,0,0)	(0,0,0,0)	(0,0,0,0)	(0,0,10,50)
$\tilde{b}_j$	(30,40,50,70)	(20,30,40,50)	(40,50,50,80)	(20,30,30,30)	$\sum_{i=1}^{3} \tilde{a}_i = \sum_{j=1}^{4} \tilde{b}_j$

Table 4.7 Tabular representation of the first crisp transportation problem

	D_1	D_2	D_3	D_4	a_i
S_1	10	50	80	0	70
S_2	75	60	20	0	40
S_3	0	0	0	0	0
a'_j	30	20	40	20	

Table 4.8 Tabular representation of the second crisp transportation problem

	D_1	D_2	D_3	D_4	$b_i - a_i$
S_1	7.5	37.5	60	0	20
S_2	60	45	15	0	20
S_3	0	0	0	0	0
$b'_j - a'_j$	10	10	10	10	

Table 4.9 Tabular representation of the third crisp transportation problem

	D_1	D_2	D_3	D_4	$c_i - b_i$
S_1	5	25	40	0	0
S_2	42.5	30	10	0	10
S_3	0	0	0	0	10
$c'_j - b'_j$	10	10	0	0	

Table 4.10 Tabular representation of the fourth crisp transportation problem

S_1	2.5	22.5	30	0	10
S_2	22.5	15	5	0	10
S_3	0	0	0	0	40
$d'_j - c'_j$	20	10	30	0	

Step 3. The optimal solution of crisp transportation problems, shown in Tables 4.7, 4.8, 4.9 and 4.10 are, respectively:

$$\begin{cases} a_{11} = 30, a_{12} = 20, a_{13} = 0, a_{14} = 20, \\ a_{21} = 0, a_{22} = 0, a_{23} = 40, a_{24} = 0, \\ a_{31} = 0, a_{32} = 0, a_{33} = 0, a_{34} = 0, \\ b_{11} - a_{11} = 10, b_{12} - a_{12} = 10, b_{13} - a_{13} = 0, b_{14} - a_{14} = 0, \\ b_{21} - a_{21} = 0, b_{22} - a_{22} = 0, b_{23} - a_{23} = 10, b_{24} - a_{24} = 10, \\ b_{31} - a_{31} = 0, b_{32} - a_{32} = 0, b_{33} - a_{33} = 0, b_{34} - a_{34} = 0, \\ c_{11} - b_{11} = 0, c_{12} - b_{12} = 0, c_{13} - b_{13} = 0, c_{14} - b_{14} = 0, \\ c_{21} - b_{21} = 0, c_{22} - b_{22} = 10, c_{23} - b_{23} = 0, c_{24} - b_{24} = 0, \\ c_{31} - b_{31} = 0, c_{32} - b_{32} = 10, c_{33} - b_{33} = 0, c_{34} - b_{34} = 0, \\ d_{11} - c_{11} = 10, d_{12} - c_{12} = 10, d_{13} - c_{13} = 0, d_{14} - c_{14} = 0, \\ d_{21} - c_{21} = 0, d_{22} - c_{22} = 0, d_{23} - c_{23} = 10, d_{24} - c_{24} = 0, \\ d_{31} - c_{31} = 20, d_{32} - c_{32} = 0, d_{33} - c_{33} = 20, d_{34} - c_{34} = 0. \end{cases}$$

Step 4. Due to Step 3, the values of a_{ij}, b_{ij}, c_{ij} and d_{ij} are:

$$\begin{cases} a_{11} = 30, b_{11} = 40, c_{11} = 40, d_{11} = 40, \\ a_{12} = 20, b_{12} = 30, c_{12} = 30, d_{12} = 40, \\ a_{13} = 0, b_{13} = 0, c_{13} = 0, d_{13} = 0, \\ a_{14} = 20, b_{14} = 20, c_{14} = 20, d_{14} = 20, \\ a_{21} = 0, b_{21} = 0, c_{21} = 0, d_{21} = 0, \\ a_{22} = 0, b_{22} = 0, c_{22} = 10, d_{22} = 10, \\ a_{23} = 40, b_{23} = 50, c_{23} = 50, d_{23} = 60, \\ a_{24} = 0, b_{24} = 10, c_{24} = 10, d_{24} = 10, \\ a_{31} = 0, b_{31} = 0, c_{31} = 10, d_{31} = 30, \\ a_{32} = 0, b_{32} = 0, c_{32} = 0, d_{32} = 0, \\ a_{33} = 0, b_{33} = 0, c_{33} = 0, d_{33} = 20, \\ a_{34} = 0, b_{34} = 0, c_{34} = 0, d_{34} = 0. \end{cases}$$

Step 5. By putting the values of a_{ij}, b_{ij}, c_{ij} and d_{ij} obtained in Step 4 into $\tilde{x}_{ij} = (a_{ij}, b_{ij}, c_{ij}, d_{ij})$, the fuzzy optimal solution of the he balanced fully fuzzy transportation problem represented by Table 4.6 is:

$$\begin{cases} \tilde{x}_{11} = (30, 40, 40, 40) & \tilde{x}_{12} = (20, 30, 30, 40) & \tilde{x}_{13} = (0, 0, 0, 0) & \tilde{x}_{14} = (20, 20, 20, 20) \\ \tilde{x}_{21} = (0, 0, 0, 0) & \tilde{x}_{22} = (0, 0, 10, 10) & \tilde{x}_{23} = (40, 50, 50, 60) & \tilde{x}_{24} = (0, 10, 10, 10) \\ \tilde{x}_{31} = (0, 0, 10, 30) & \tilde{x}_{32} = (0, 0, 0, 0) & \tilde{x}_{33} = (0, 0, 0, 20) & \tilde{x}_{34} = (0, 0, 0, 0) \end{cases}$$

Step 6. Putting the values of

$$\begin{cases} \tilde{x}_{11}, \tilde{x}_{12}, \tilde{x}_{13}, \tilde{x}_{14}, \\ \tilde{x}_{21}, \tilde{x}_{22}, \tilde{x}_{23}, \tilde{x}_{24}, \\ \tilde{x}_{31}, \tilde{x}_{32}, \tilde{x}_{33}, \tilde{x}_{34} \end{cases}$$

into

$$\begin{cases} (10, 10, 10, 10) \otimes \tilde{x}_{11} \oplus (50, 50, 50, 50) \otimes \tilde{x}_{12} \\ \oplus (80, 80, 80, 80) \otimes \tilde{x}_{13} \oplus (0, 0, 0, 0) \otimes \tilde{x}_{14} \\ \oplus (60, 70, 80, 90) \otimes \tilde{x}_{21} \oplus (60, 60, 60, 60) \otimes \tilde{x}_{22} \\ \oplus (20, 20, 20, 20) \otimes \tilde{x}_{23} \oplus (0, 0, 0, 0) \otimes \tilde{x}_{24} \\ \oplus (0, 0, 0, 0) \otimes \tilde{x}_{31} \oplus (0, 0, 0, 0) \otimes \tilde{x}_{32} \\ \oplus (0, 0, 0, 0) \otimes \tilde{x}_{33} \oplus (0, 0, 0, 0) \otimes \tilde{x}_{34} \end{cases}$$

the minimum total fuzzy transportation cost is (2100, 2900, 3500, 4200).

4.7.3 *Interpretation of Results*

We will now interpret the minimum total fuzzy transportation cost obtained by using the new methods proposed, as well as the fuzzy optimal solution obtained.

First, let us remind that the minimum total fuzzy transportation cost that has been obtained in Sect. 4.7.2 is (2100, 2900, 3500, 4200). This can clearly be interpreted as follows:

1. The least amount of minimum total transportation cost is 2100,
2. The most possible amount of minimum total transportation cost lies between 2900 and 3500,
3. The greatest amount of minimum total transportation cost is 4200, i.e., the minimum total transportation cost will always be greater than 2100 and lower than 4200, and the highest chance is to have the minimum total transportation cost lie between 2900 and 3500.

This interpretation of the minimum total fuzzy transportation cost, represented in the form of a piecewise linear membership function which is assumed in this book and is also the most widely used in practice, can be illustratively presented as in Fig. 4.1.

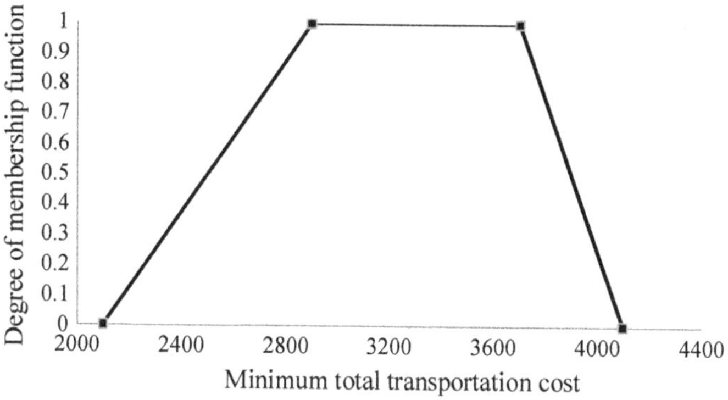

Fig. 4.1 The piecewise linear membership function of the minimum fuzzy total fuzzy transportation cost

4.8 Case Study

Liang et al. [10] have proposed a method to find the crisp optimal solution of such fuzzy transportation problems in which all the parameters are represented by triangular fuzzy numbers. Then, they have used their new method proposed to find the crisp optimal solution of a real life fuzzy transportation problem to be described in Sect. 4.8.1.

However, it is often better to find a fuzzy optimal solution than a crisp optimal solution. In this section we will therefore show how in the problem considered by Liang et al. [10] we can obtain a fuzzy optimal solution of the same real life problem using the new method proposed.

4.8.1 Description of the Problem

The Dali Company is a leading producer of soft drinks and refrigerated foods in Taiwan. Currently, Dali plans to develop the South-East Asian market and broaden the visibility of Dali products in the Chinese market. Notably, following the entry of Taiwan to the World Trade Organization, Dali plans to seek a strategic alliance with prominent international companies, to go more global.

In the domestic soft drinks market, Dali produces tea beverages to meet demand from four distribution centers in Taichung, Chiayi, Kaohsiung, and Taipei, with the production at three plants in Changhua, Toulliu and Hsinchu.

A potential availability of the products at these three plants, a forecast demand from the four distribution centers, and the unit transportation cost for each route used by Dali for the upcoming season is presented in Table 4.11, clearly taking into account some additional information, related to the environment, for instance weather forecasts, because this is what has a crucial importance for the market of soft drinks.

Table 4.11 Summarized data in the Dali Company case (in U.S. dollar)

Source (i)	Destination (j)				Supply (000 dozen bottles)
	Taichung (1)	Chiayi (2)	Kaohsiung (3)	Taipei (4)	
Changhua (1)	($8, $10, $10.8)	($20.4, $22, $24)	($8, $10, $10.6)	($18.8, $20, $22)	(7.2, 8, 8.8)
Touliu (2)	($14, $15, $16)	($18.2, $20, $22)	($10, $12, $13)	($6, $8, $8.8)	(12, 14, 16)
Hsinchu (3)	($18.4, $20, $21)	($9.6, $12, $13)	($7.8, $10, $10.8)	($14, $15, $16)	(10.2, 12, 13.8)
Demand (000 dozen bottles)	(6.2, 7, 7.8)	(8.6, 10, 11.4)	(6.5, 8, 9.5)	(7.8, 9, 10.2)	

The environmental coefficients and related parameters are usually imprecisely known, due to incomplete and highly uncertain information available, and that is why their representation by imprecise numbers with triangular possibility distributions makes much sense. For example, the available supply of the Changhua plant is (7.2, 8, 8.8) thousand dozen bottles, the forecast demand of the Taichung distribution center is (6.2, 7, 7.8) thousand dozen bottles and the transportation cost per dozen bottles from Changhua to Taichung is ($8, $10, $10.8). These values are to be meant as the traingular fuzzy numbers in which the left hand side term stands for the lowest possible, the middle term—for the most possible, and the right hand side term—for the highest possible value; $ means the U.S. dollar.

Since the transportation costs is the major expense, the management of the Dali Company has commissioned a study to reduce these costs as much as possible.

4.8.2 Results Obtained

By solving the real life problem presented in Sect. 4.8.1 and in a comprehensive form in Table 4.11 by using the proposed methods presented in Sect. 4.6, the obtained fuzzy optimal solution $\{\tilde{x}_{ij}\}$, representing the fuzzy quantity of soft drinks that should be transported from the ith source to the jth destination to minimize the total fuzzy transportation cost, is:

$$\begin{cases} \tilde{x}_{11} = (6.2, 7, 7.8) & \tilde{x}_{13} = (1, 1, 1) \\ \tilde{x}_{23} = (3.9, 4.7, 5.5) & \tilde{x}_{24} = (7.8, 9, 10.2) \; \tilde{x}_{25} = (.3, .3, .3) \\ \tilde{x}_{32} = (8.6, 9.7, 10.8) & \tilde{x}_{33} = (1.6, 2.3, 3) \quad \tilde{x}_{42} = (0, .3, .6) \end{cases}$$

and the minimum total fuzzy transportation cost is

$$(\$238.44, \$347.8, \$428.9)$$

Remark 3 Since the real life problem considered in Sect. 4.8.1 is an unbalanced problem, then to find its solution a dummy source (4) and a dummy destination (5) are introduced. Then, in the results obtained, presented in Sect. 4.8.2, $\tilde{x}_{25}$ and $\tilde{x}_{42}$

represent the fuzzy quantity of the product that should be transported from source (2) to the dummy destination (5) and from the dummy source (4) to destination (2), respectively.

4.8.3 *Interpretation of Results*

We will now interpret the minimum total fuzzy transportation cost obtained in Sect. 4.8.2 by using the proposed methods presented in Sect. 4.6. Similarly, the obtained fuzzy optimal solution will also be interpreted.

By using the methods proposed the minimum total fuzzy transportation cost is ($238.44, $347.8, $428.9), which can be physically interpreted as follows:

1. The least amount of the minimum total transportation cost is $238.44,
2. The most possible amount of minimum total transportation cost is $347.8,
3. The greatest amount of the minimum total transportation cost is $428.9 i.e., the minimum total transportation cost will always be greater than $238.44 and less than $428.9, and the highest chances are that the minimum total transportation cost will be $347.8.

4.9 Concluding Remarks

From both theoretical and algorithmic considerations, and examples solved in this chapter, it can be noticed that some shortcomings of the methods for solving the fuzzy transportation problems known from the literature can be resolved by using the new methods proposed in Sect. 4.6.

References

1. H. Basirzadeh, An approach for solving fuzzy transportation problem. Appl. Math. Sci. **5**, 1549–1566 (2011)
2. L. Campos, J.L. Verdegay, Linear programming problems and ranking of fuzzy numbers. Fuzzy Sets Syst. **32**, 1–11 (1989)
3. J. Chiang, The optimal solution of the transportation problem with fuzzy demand and fuzzy product. J. Inf. Sci. Eng. **21**(2), 439–451 (2005)
4. A. Ebrahimnejad, S.H. Nasseri, F.H. Lotfi, M. Soltanifar, A primal-dual method for linear programming problems with fuzzy variables. Eur. J. Ind. Eng. **4**, 189–209 (2010)
5. K. Ganesan, P. Veeramani, Fuzzy linear programs with trapezoidal fuzzy numbers. Ann. Oper. Res. **143**, 305–315 (2006)
6. A.N. Gani, A.E. Samuel, D. Anuradha, Simplex type algorithm for solving fuzzy transportation problem. Tamsui Oxf. J. Inf. Math. Sci. **27**, 89–98 (2011)
7. B. Julien, An extension to possibilistic linear programming. Fuzzy Sets Syst. **64**, 195–206 (1994)

8. B. Kheirfam, J.L. Verdegay, Strict sensitivity analysis in fuzzy quadratic programming. Fuzzy Sets Syst. (2011). https://doi.org/10.1016/j.fss.2011.10.019
9. A. Kumar, J. Kaur, P. Singh, A new method for solving fully fuzzy linear programming problems. Appl. Math. Model. **35**, 817–823 (2011)
10. T.F. Liang, C.S. Chiu, H.W. Cheng, Using possibilistic linear programming for fuzzy transportation planning decisions. Hsiuping J. **11**, 93–112 (2005)
11. T.S. Liou, M.J. Wang, Ranking fuzzy number with integral values. Fuzzy Sets Syst. **50**, 247–255 (1992)
12. S.T. Liu, C. Kao, Solving fuzzy transportation problems based on extension principle. Eur. J. Oper. Res. **153**, 661–674 (2004)
13. N. Mahadavi-Amiri, S.H. Nasseri, Duality in fuzzy number linear programming by the use of a certain linear ranking function. Appl. Math. Comput. **180**, 206–216 (2006)
14. N. Mahadavi-Amiri, S.H. Nasseri, Duality results and a dual simplex method for linear programming problems with trapezoidal fuzzy variables. Fuzzy Sets Syst. **158**, 1961–1978 (2007)
15. H.R. Maleki, M. Tata, M. Mashinchi, Linear programming with fuzzy variables. Fuzzy Sets Syst. **109**, 21–33 (2000)
16. S. Mukherjee, K. Basu, Application of fuzzy ranking method for solving assignment problems with fuzzy costs. Int. J. Comput. Appl. Math. **5**, 359–368 (2010)
17. P. Pandian, G. Natarajan, A new method for finding an optimal solution of fully interval integer transportation problems. Appl. Math. Sci. **4**, 1819–1830 (2010)
18. P. Pandian, G. Natarajan, A new algorithm for finding a fuzzy optimal solution for fuzzy transportation problems. Appl. Math. Sci. **4**, 79–90 (2010)
19. P. Pandian, G. Natarajan, An optimal more-for-less solution to fuzzy transportation problems with mixed constraints. Appl. Math. Sci. **4**, 1405–1415 (2010)
20. M.A. Parra, T.A. Bilbao, M.V.R. Uria, Solving the multiobjective possibilistic linear programming problem. Eur. J. Oper. Res. **117**, 79–90 (1999)
21. D. Stephen Dinagar, K. Palanivel, On trapezoidal membership functions in solving transportation problem under fuzzy environment. Int. J. Comput. Phys. Sci. **1**, 1–12 (2009)
22. D. Stephen Dinagar, K. Palanivel, The transportation problem in fuzzy environment. Int. J. Algorithms Comput. Math. **2**, 65–71 (2009)

Chapter 5
New Methods for Solving the Fully Fuzzy Transportation Problems with the LR Flat Fuzzy Numbers

In this chapter we will first of all indicate some limitations of the methods proposed in the previous chapter, Chap. 4. This will provide us with a point of departure for the proposal of novel. Modified methods which will overcome the above mentioned limitation while solving the fully fuzzy transportation problems. We will present these modifications, analyze their advantage over the method proposed in Chap. 4, and illustrate our new method on some relevant and illustrative examples of a fully fuzzy transportation problem is solved.

5.1 Preliminaries

In the literature—as indicated in many classic, more general books exemplified by those by Dubois and Prade [4] or Zimmermann [10], or more specialized books by Kacprzyk [8] or Bozhenyuk, Gerasimenko, Kacprzyk and Rozenberg [2]—it has been pointed out that the computational efforts required to solve a fuzzy linear programming problem can be reduced if the values of parameters, coefficients, etc. could be expressed by using the LR flat fuzzy numbers (cf. Chap. 2). Needless to say that crisp (nonfuzzy, real, integer, ...) numbers, number intervals, triangular and trapezoidal fuzzy numbers, etc. are all examples of the LR flat fuzzy numbers.

In this section, we will first provide a brief account of some basic definitions and arithmetic operations of the LR flat fuzzy numbers, for convenience of the reader and to make our discussion self contained.

A. Kaur et al., *Fuzzy Transportation and Transshipment Problems*, Studies in Fuzziness and Soft Computing 385,
https://doi.org/10.1007/978-3-030-26676-9_5

5.2 Basic Definitions

In this section, first some basic definitions, well known from the literature are recalled to provide a point of departure.

Definition 14 A function $L : [0, \infty) \to [0, 1]$ (or $R : [0, \infty) \to [0, 1]$) is said to be reference function of fuzzy number if and only if (cf. Dubois and Prade [4]):

1. $L(x) = L(-x)$ (or $R(x) = R(-x)$),
2. $L(0) = 1$ (or $R(0) = 1$).
3. L (or R) is non-increasing on $[0, \infty)$.

Definition 15 A fuzzy number $\widetilde{A}$ defined on the set of real numbers $\mathbb{R}$, denoted as $(m, n, \alpha, \beta)_{LR}$, is said to be an LR flat fuzzy number if its membership function $\mu_{\widetilde{A}}(x)$ is given by (cf. Dubois and Prade [4]):

$$\mu_{\widetilde{A}}(x) = \begin{cases} L(\frac{m-x}{\alpha}), & x \le m,\ \alpha > 0 \\ R(\frac{x-n}{\beta}), & x \ge n,\ \beta > 0 \\ 1, & m \le x \le n \end{cases}$$

Definition 16 Let $\widetilde{A} = (m, n, \alpha, \beta)_{LR}$ be an LR flat fuzzy number and λ be a real number from the interval $[0, 1]$. Then, the classical set $A_{\underset{\sim}{\lambda}} = \{x \in X : \mu_{\widetilde{A}}(x) \ge \lambda\} = [m - \alpha L^{-1}(\lambda), n + \beta R^{-1}(\lambda)]$, is said to be a λ-cut of $\widetilde{A}$ (cf. Dubois and Prade [4]).

Definition 17 An LR flat fuzzy number $\tilde{A} = (m, n, \alpha, \beta)_{LR}$ is said to be the zero LR flat fuzzy number if and only if $m = 0$, $n = 0$, $\alpha = 0$ and $\beta = 0$ (cf. Dubois and Prade [4]).

Definition 18 Two LR flat fuzzy numbers

$$\begin{cases} \tilde{A}_1 = (m_1, n_1, \alpha_1, \beta_1)_{LR} \\ \tilde{A}_2 = (m_2, n_2, \alpha_2, \beta_2)_{LR} \end{cases}$$

are said to be equal, i.e., $\tilde{A}_1 = \tilde{A}_2$ if and only if $m_1 = m_2, n_1 = n_2, \alpha_1 = \alpha_2$ and $\beta_1 = \beta_2$ (cf. Dubois and Prade [4]).

Definition 19 An LR flat fuzzy number $\tilde{A} = (m, n, \alpha, \beta)_{LR}$ is said to be non-negative LR flat fuzzy number if and only if $m - \alpha \ge 0$ (cf. Dehghan et al. [3]).

Definition 20 An LR flat fuzzy number $\tilde{A} = (m, n, \alpha, \beta)_{LR}$ is said to be negative (or positive, respectively) LR flat fuzzy number if and only if $n + \beta < 0$ (or $m - \alpha > 0$, respectively) (cf. Dehghan et al. [3]).

Moreover, it is worthwhile to mention the following important resulting remarks.

Remark 4 If $m = n$, then an LR flat fuzzy number $(m, n, \alpha, \beta)_{LR}$ is said to be an LR fuzzy number and is denoted as $(m, m, \alpha, \beta)_{LR}$ or $(n, n, \alpha, \beta)_{LR}$ or $(m, \alpha, \beta)_{LR}$ or $(n, \alpha, \beta)_{LR}$.

Remark 5 If $m = n$ and $L(x) = R(x) = \max\{0, 1 - x\}$, then an LR flat fuzzy number $(m, n, \alpha, \beta)_{LR}$ is said to be a triangular fuzzy number and is denoted as (a, b, c) where $a = m - \alpha$, $b = m$ (or $= n$), $c = m + \beta$ (or $n + \beta$).

Remark 6 If $m \neq n$ but $L(x) = R(x) = \max\{0, 1 - x\}$ then an LR flat fuzzy number $(m, n, \alpha, \beta)_{LR}$ is said to be a trapezoidal fuzzy number and is denoted as (a, b, c, d) where, $a = m - \alpha, b = m, c = n, d = n + \beta$.

5.3 Arithmetic Operations on the *LR* Flat Fuzzy Numbers

Now, again, to introduce a specific notation used in our book, and to make it self contained, we briefly present the basic arithmetic operations on the LR flat fuzzy numbers referring mainly to the works by Bede [1]. Dubois and Prade [4], Kacprzyk [8], etc.

For our purposes the operation of addition and multiplication of the LR flat fuzzy numbers are the most important and are defined as follows:

- If $\tilde{A}_1 = (m_1, n_1, \alpha_1, \beta_1)_{LR}$ and $\tilde{A}_2 = (m_2, n_2, \alpha_2, \beta_2)_{LR}$ are two LR flat fuzzy numbers, then their addition is defined as

$$\tilde{A}_1 \oplus \tilde{A}_2 = (m_1 + m_2, n_1 + n_2, \alpha_1 + \alpha_2, \beta_1 + \beta_2)_{LR}$$

- If $\tilde{A}_1 = (m_1, n_1, \alpha_1, \beta_1)_{LR}$ and $\tilde{A}_2 = (m_2, n_2, \alpha_2, \beta_2)_{LR}$ are two non-negative LR flat fuzzy numbers, then their multiplication is defined as

$$\tilde{A}_1 \otimes \tilde{A}_2 \simeq (m_1 m_2, n_1 n_2, m_1 m_2 - (m_1 - \alpha_1)(m_2 - \alpha_2), (n_1 + \beta_1)(n_2 + \beta_2) - n_1 n_2)_{LR} \tag{5.1}$$

We are now ready to proceed to the discussion of how to formulate and solve the fully fuzzy transportation problems in which the parameters are represented by LR fuzzy numbers or LR flat fuzzy numbers.

5.4 Solution of the Fully Fuzzy Transportation Problems with Parameters Represented by the *LR* Fuzzy Numbers or *LR* Flat Fuzzy Numbers

In this section we will thoroughly analyze the method proposed in Chap. 4 for solving fully fuzzy transportation problems with the parameters assumed to be either represented by triangular fuzzy numbers or trapezoidal fuzzy numbers. This will be done from the point of view of whether one could use the methods already proposed in Chap. 4 to solve such fully fuzzy transportation problems in which the parameters are represented by LR fuzzy numbers or LR flat fuzzy numbers, for instance the fully fuzzy transportation problem as shown in Example 17 given below for clarification, and then to show that this is not possible.

Example 17 Suppose that a company has two sources S_1 and S_2 and three destinations D_1, D_2 and D_3 for their product, the fuzzy cost for transporting one unit quantity of the product from the ith source to the jth destination is $\tilde{c}_{ij}$ where

$$[\tilde{c}_{ij}]_{2\times 3} = \begin{pmatrix} (20, 30, 10, 10)_{LR} & (60, 70, 10, 20)_{LR} & (90, 110, 10, 10)_{LR} \\ (70, 80, 10, 10)_{LR} & (80, 100, 10, 20)_{LR} & (30, 50, 10, 10)_{LR} \end{pmatrix}$$

The fuzzy availabilities of the product at the first and second source, S_1 and S_2, are $(90, 90, 20, 10)_{LR}$ and $(60, 70, 20, 10)_{LR}$, respectively, and the fuzzy demands of the product at the first, second and third destination, D_1, D_2 and D_3, are $(40, 50, 10, 20)_{LR}$, $(30, 40, 10, 10)_{LR}$ and $(50, 50, 10, 30)_{LR}$, respectively, where $L(x) = R(x) = \max\{0, 1 - x^4\}$.

The company wants to determine the fuzzy quantity of the product that should be transported from each source to each destination so that to minimize the total fuzzy transportation cost.

The definitions and properties mentioned, and the above, very illustrative example presented, will greatly clarifies the rationale for, and the essence of the new method to be proposed.

5.5 New Methods

In this section, to overcome the limitations of the methods proposed in Chap. 4, two new methods will be proposed for solving the fully fuzzy transportation problems in which all the parameters are represented by the LR flat fuzzy numbers. Moreover, the advantages of the proposed methods over the methods proposed in Chap. 4 will be discussed.

5.5.1 *Method Based on Fuzzy Linear Programming*

We propose here a new method for finding a fuzzy optimal solution of the fully fuzzy transportation problems in which the parameters are represented by the LR flat fuzzy numbers defined in Definition 15. This new method will be based on the fuzzy linear programming formulation of the fully fuzzy transportation problems the basic philosophy of which has, by the way, a long tradition that dates back probably to th first works by Verdegay and his collaborators (cf. for instance, [5–7], to just mention a few.

The consecutive steps of our new method proposed are:

Step 1 Find:

- the total fuzzy availability $\sum_{i=1}^{m} \tilde{a}_i$ and

- the total fuzzy demand $\sum_{j=1}^{n} \tilde{b}_j$.

Let:

$$\begin{cases} \sum_{i=1}^{m} \tilde{a}_i = (m, n, \alpha, \beta)_{LR} \\ \sum_{j=1}^{n} \tilde{b}_j = (m', n', \alpha', \beta')_{LR} \end{cases}$$

Use Definition 18 to examine if the problem is balanced or not, that is, if:

$$\sum_{i=1}^{m} \tilde{a}_i = \sum_{j=1}^{n} \tilde{b}_j$$

or

$$\sum_{i=1}^{m} \tilde{a}_i \neq \sum_{j=1}^{n} \tilde{b}_j$$

and then proceed according to:

Case (i) If the problem is balanced, i.e.

$$\sum_{i=1}^{m} \tilde{a}_i = \sum_{j=1}^{n} \tilde{b}_j,$$

then Go to Step 2.

Case (ii) If the problem is unbalanced, i.e.

$$\sum_{i=1}^{m} \tilde{a}_i \neq \sum_{j=1}^{n} \tilde{b}_j$$

then convert the unbalanced problem into the balanced problem as follows:

Case (a) If

$$\begin{cases} m - \alpha \leq m' - \alpha' \\ \alpha \leq \alpha' \\ n - m \leq n' - m' \\ \text{and} \\ \beta \leq \beta' \end{cases}$$

then introduce a dummy source with the fuzzy availability equal to

$$(m' - m, n' - n, \alpha' - \alpha, \beta' - \beta)_{LR}$$

and assume the fuzzy cost for transporting one unit quantity of the product from this dummy source to all destinations as equal to the zero LR flat fuzzy number. Go to Step 2.
Case (b) If

$$\begin{cases} m-\alpha \geq m'-\alpha' \\ \alpha \geq \alpha' \\ n-m \geq n'-m' \\ \text{and} \\ \beta \geq \beta' \end{cases}$$

then introduce a dummy destination with the fuzzy demand

$$(m-m', n-n', \alpha-\alpha', \beta-\beta')_{LR}$$

and assume the fuzzy cost for transporting one unit quantity of the product from all sources to this dummy destination as the zero LR flat fuzzy number. Go to Step 2.
Case (c) If neither Case (a) nor Case (b), as shown above, is satisfied, then introduce a dummy source with the fuzzy availability

$$\begin{cases} (\max\{0,(m'-\alpha')-(m-\alpha)\}+\max\{0,(\alpha'-\alpha)\}, \\ \max\{0,(m'-\alpha')-(m-\alpha)\}+\max\{0,(\alpha'-\alpha)\}+ \\ \max\{0,(n'-m')-(n-m)\}, \max\{0,(\alpha'-\alpha)\}, \\ \max\{0,(\beta'-\beta)\})_{LR}) \end{cases}$$

and a dummy destination with the fuzzy demand

$$\begin{cases} (\max\{0,(m-\alpha)-(m'-\alpha')\}+\max\{0,(\alpha-\alpha')\}, \\ \max\{0,(m-\alpha)-(m'-\alpha')\}+\max\{0,(\alpha-\alpha')\}+ \\ \max\{0,(n-m)-(n'-m')\}, \max\{0,(\alpha-\alpha')\}, \\ \max\{0,(\beta-\beta')\})_{LR}) \end{cases}$$

Assume the fuzzy cost for transporting one unit quantity of the product from the above dummy source to all destinations and from all sources to the above dummy destination as the zero LR flat fuzzy number. Go to Step 2.

Step 2 Reformulate the balanced fully fuzzy transportation problem obtained in Step 1 into the fuzzy linear programming problem given as (4.27).
Step 3 By assuming

$$\begin{cases} \tilde{c}_{ij} = (m'_{ij}, n'_{ij}, \alpha'_{ij}, \beta'_{ij})_{LR} \\ \tilde{x}_{ij} = (m_{ij}, n_{ij}, \alpha_{ij}, \beta_{ij})_{LR} \\ \tilde{a}_i = (m_i, n_i, \\ \alpha_i, \beta_i)_{LR} \\ \tilde{b}_j = (m'_j, n'_j, \alpha'_j, \beta'_j)_{LR} \end{cases}$$

the fuzzy linear programming problem, obtained in Step 2, can be written as:

$$\begin{cases} \min \sum_{i=1}^{m}\sum_{j=1}^{n}\left((m'_{ij}, n'_{ij}, \alpha'_{ij}, \beta'_{ij})_{LR} \otimes (m_{ij}, n_{ij}, \alpha_{ij}, \beta_{ij})_{LR}\right) \\ \text{subject to: } \sum_{j=1}^{n}((m_{ij}, n_{ij}, \alpha_{ij}, \beta_{ij})_{LR}) = (m_i, n_i, \alpha_i, \beta_i)_{LR};\ i = 1, 2, \ldots, m \\ \sum_{i=1}^{m}((m_{ij}, n_{ij}, \alpha_{ij}, \beta_{ij})_{LR}) = (m'_j, n'_j, \alpha'_j, \beta'_j)_{LR};\ j = 1, 2, \ldots, n \\ (m_{ij}, n_{ij}, \alpha_{ij}, \beta_{ij})_{LR} \text{ is a non-negative LR flat fuzzy number.} \end{cases} \tag{5.2}$$

Step 4 Using the arithmetic operations, defined in Sect. 4.1.2, and assuming $\sum_{i=1}^{m}\sum_{j=1}^{n}\left((m'_{ij}, n'_{ij}, \alpha'_{ij}, \beta'_{ij})_{LR} \otimes (m_{ij}, n_{ij}, \alpha_{ij}, \beta_{ij})_{LR}\right) = (m_0, n_0, \alpha_0, \beta_0)_{LR}$ the fuzzy linear programming problem, obtained in Step 3, can be written as:

$$\begin{cases} \min(m_0, n_0, \alpha_0, \beta_0)_{LR} \\ \text{subject to:} \\ \left(\sum_{j=1}^{n} m_{ij}, \sum_{j=1}^{n} n_{ij}, \sum_{j=1}^{n} \alpha_{ij}, \sum_{j=1}^{n} \beta_{ij}\right)_{LR} = (m_i, n_i, \alpha_i, \beta_i)_{LR};\ i = 1, 2, \ldots, m \\ \left(\sum_{i=1}^{m} m_{ij}, \sum_{i=1}^{m} n_{ij}, \sum_{i=1}^{m} \alpha_{ij}, \sum_{i=1}^{m} \beta_{ij}\right)_{LR} = (m'_j, n'_j, \alpha'_j, \beta'_j)_{LR};\ j = 1, 2, \ldots, n \\ \text{where: } (m_{ij}, n_{ij}, \alpha_{ij}, \beta_{ij})_{LR} \text{ is a non-negative LR flat fuzzy number} \end{cases} \tag{5.3}$$

Step 5 Using Definitions 18 and 19, the fuzzy linear programming problem, obtained in Step 4, can be converted into the following fuzzy linear programming problem:

$$\begin{cases} \min(m_0, n_0, \alpha_0, \beta_0)_{LR} \\ \text{subject to:} \\ \sum_{j=1}^{n} m_{ij} = m_i;\ i = 1, 2, \ldots, m \\ \sum_{j=1}^{n} n_{ij} = n_i;\ i = 1, 2, \ldots, m \\ \sum_{j=1}^{n} \alpha_{ij} = \alpha_i;\ i = 1, 2, \ldots, m \\ \sum_{j=1}^{n} \beta_{ij} = \beta_i;\ i = 1, 2, \ldots, m \\ \sum_{i=1}^{m} m_{ij} = m'_j;\ j = 1, 2, \ldots, n \\ \sum_{i=1}^{m} n_{ij} = n'_j;\ j = 1, 2, \ldots, n \\ \sum_{i=1}^{m} \alpha_{ij} = \alpha'_j;\ j = 1, 2, \ldots, n \\ \sum_{i=1}^{m} \beta_{ij} = \beta'_j;\ j = 1, 2, \ldots, n \\ m_{ij} - \alpha_{ij}, n_{ij} - m_{ij}, \alpha_{ij}, \beta_{ij} \geq 0;\ \forall i = 1, 2, \ldots, m;\ j = 1, 2, \ldots, n \end{cases} \tag{5.4}$$

Step 6 As discussed in Step 6 of the method, proposed in Chap. 4, Sect. 4.6.1, the fuzzy optimal solution of the fuzzy linear programming problem (5.4) can be obtained by solving the following crisp linear programming problem:

$$\left\{\begin{array}{l}\min \Re(m_0, n_0, \alpha_0, \beta_0)_{LR} \\ \text{subject to:} \\ \sum_{j=1}^{n} m_{ij} = m_i;\ i = 1, 2, \ldots, m \\ \sum_{j=1}^{n} n_{ij} = n_i;\ i = 1, 2, \ldots, m \\ \sum_{j=1}^{n} \alpha_{ij} = \alpha_i;\ i = 1, 2, \ldots, m \\ \sum_{j=1}^{n} \beta_{ij} = \beta_i;\ i = 1, 2, \ldots, m \\ \sum_{i=1}^{m} m_{ij} = m'_j;\ j = 1, 2, \ldots, n \\ \sum_{i=1}^{m} n_{ij} = n'_j;\ j = 1, 2, \ldots, n \\ \sum_{i=1}^{m} \alpha_{ij} = \alpha'_j;\ j = 1, 2, \ldots, n \\ \sum_{i=1}^{m} \beta_{ij} = \beta'_j;\ j = 1, 2, \ldots, n \\ m_{ij} - \alpha_{ij}, n_{ij} - m_{ij}, \alpha_{ij}, \beta_{ij} \geq 0;\ \forall i = 1, 2, \ldots, m;\ j = 1, 2, \ldots, n\end{array}\right. \tag{5.5}$$

Step 7 Using the result proposed by Liu and Kao [9], i.e.

$$\Re(m_0, n_0, \alpha_0, \beta_0)_{LR} = \frac{1}{2}\left(\int_0^1 (m_0 - \alpha_0 L^{-1}(\lambda))d\lambda + \int_0^1 (n_0 + \beta_0 R^{-1}(\lambda))d\lambda\right)$$

the crisp linear programming problem, obtained in Step 6, can be converted into the following crisp linear programming problem:

$$\left\{\begin{array}{l}\min \frac{1}{2}\left(\int_0^1 (m_0 - \alpha_0 L^{-1}(\lambda))d\lambda + \int_0^1 (n_0 + \beta_0 R^{-1}(\lambda))d\lambda\right) \\ \text{subject to:} \\ \sum_{j=1}^{n} m_{ij} = m_i;\ i = 1, 2, \ldots, m \\ \sum_{j=1}^{n} n_{ij} = n_i;\ i = 1, 2, \ldots, m \\ \sum_{j=1}^{n} \alpha_{ij} = \alpha_i;\ i = 1, 2, \ldots, m \\ \sum_{j=1}^{n} \beta_{ij} = \beta_i;\ i = 1, 2, \ldots, m \\ \sum_{i=1}^{m} m_{ij} = m'_j;\ j = 1, 2, \ldots, n \\ \sum_{i=1}^{m} n_{ij} = n'_j;\ j = 1, 2, \ldots, n \\ \sum_{i=1}^{m} \alpha_{ij} = \alpha'_j;\ j = 1, 2, \ldots, n \\ \sum_{i=1}^{m} \beta_{ij} = \beta'_j;\ j = 1, 2, \ldots, n \\ m_{ij} - \alpha_{ij}, n_{ij} - m_{ij}, \alpha_{ij}, \beta_{ij} \geq 0;\ \forall i = 1, 2, \ldots, m;\ j = 1, 2, \ldots, n\end{array}\right. \tag{5.6}$$

Step 8 Solve the crisp linear programming problem, obtained in Step 7, to find the optimal solution

$$\{m_{ij}, n_{ij}, \alpha_{ij}, \beta_{ij}\}$$

Step 9 Find the fuzzy optimal solution $\{\tilde{x}_{ij}\}$ by putting the values of

$$m_{ij}, n_{ij}, \alpha_{ij}, \beta_{ij}$$

into

$$\tilde{x}_{ij} = (m_{ij}, n_{ij}, \alpha_{ij}, \beta_{ij})_{LR}$$

Step 10 Find the minimum total fuzzy transportation cost by putting the values of $\tilde{x}_{ij}$ into

$$\sum_{i=1}^{m}\sum_{j=1}^{n}(\tilde{c}_{ij} \otimes \tilde{x}_{ij})$$

We have therefore obtained an improved algorithm for finding a fuzzy optimal solution of the fully fuzzy transportation problems in which the parameters are represented by the LR flat fuzzy numbers.

The following remark can here be interesting and useful.

Remark 7 Let $\tilde{A} = (m_{ij}, n_{ij}, \alpha_{ij}, \beta_{ij})_{LR}$ be an LR flat fuzzy number with $L(x) = R(x) = \max\{0, 1 - x^4\}$. Then

$$\begin{aligned}\Re(\tilde{A}) =& \frac{1}{2}((m_{ij} + n_{ij}) - \alpha_{ij}\int_0^1 L^{-1}(\lambda)d\lambda + \beta_{ij}\int_0^1 R^{-1}(\lambda)d\lambda) \\ &= \frac{1}{10}(5m_{ij} + 5n_{ij} + 4\beta_{ij} - 4\alpha_{ij}) \qquad (5.7)\end{aligned}$$

5.5.2 *Method Based on the Tabular Representation*

In this section we will present a new method for finding a fuzzy optimal solution of the fully fuzzy transportation problems represented by using the tabular representation as proposed in Sect. 4.7.2 but now with the parameters assumed in the form of the LR flat fuzzy numbers. The method is based on the tabular representation of this problem (cf. Sect. 4.6.2), and the fuzzy optimal solution can be now obtained via the following steps:

Step 1. Use Step 1 of the method proposed in Sect. 5.5.1, to obtain the balanced fully fuzzy transportation problem.
Step 2. Represent the balanced fully fuzzy transportation problem, obtained in Step 1, in the tabular form given in Table 5.1

Step 3 Split Table 5.1 from Step 2 into four crisp transportation tables, i.e., Tables 5.2, 5.3, 5.4 and 5.5.
where:

$$\eta_{ij} = \frac{1}{2}((m'_{ij} + n'_{ij}) - \alpha'_{ij} \int_0^1 L^{-1}(\lambda)d\lambda + \beta'_{ij} \int_0^1 R^{-1}(\lambda)d\lambda);$$
$$i = 1, 2, \ldots, m;\ j = 1, 2, \ldots, n$$

where:

$$\rho_{ij} = \frac{1}{2}((m'_{ij} + n'_{ij}) - m'_{ij} \int_0^1 L^{-1}(\lambda)d\lambda + \beta'_{ij} \int_0^1 R^{-1}(\lambda)d\lambda);$$
$$i = 1, 2, \ldots, m;\ j = 1, 2, \ldots, n$$

where:

$$\delta_{ij} = \frac{1}{2}(n'_{ij} + \beta'_{ij} \int_0^1 R^{-1}(\lambda)d\lambda);$$
$$i = 1, 2, \ldots, m;\ j = 1, 2, \ldots, n$$

where:

$$\xi_{ij} = \frac{1}{2}((n'_{ij} + \beta'_{ij}) \int_0^1 R^{-1}(\lambda)d\lambda);$$
$$i = 1, 2, \ldots, m;\ j = 1, 2, \ldots, n$$

Step 4. Solve the crisp transportation problems represented by, respectively, Tables 5.2, 5.3, 5.4 and 5.5, to find the optimal solutions, respectively:

$$\{m_{ij} - \alpha_{ij}\}$$
$$\{\alpha_{ij}\}$$
$$\{n_{ij} - m_{ij}\}$$
$$\{\beta_{ij}\}$$

Step 5 From the optimal solutions obtained in Step 4, find the values of m_{ij}, n_{ij}, α_{ij} and β_{ij}.
Step 6 Find the fuzzy optimal solution $\{\tilde{x}_{ij}\}$ by putting the values of m_{ij}, n_{ij}, α_{ij}, and β_{ij} into

$$\tilde{x}_{ij} = (m_{ij}, n_{ij}, \alpha_{ij}, \beta_{ij})_{LR}.$$

Step 7 Find the minimum total fuzzy transportation cost by putting the values of $\tilde{x}_{ij}$ into

Table 5.1 Tabular representation of a balanced fully fuzzy transportation problem

Destinations → Sources ↓	D_1	D_2	$\cdots$	D_j	$\cdots$	D_n	Availability $(\tilde{a}_i)$
S_1	$\tilde{c}_{11}$	$\tilde{c}_{12}$	$\cdots$	$\tilde{c}_{1j}$	$\cdots$	$\tilde{c}_{1n}$	$\tilde{a}_1$
$\vdots$	$\vdots$	$\vdots$	$\vdots$	$\vdots$	$\vdots$	$\vdots$	$\vdots$
S_i	$\tilde{c}_{i1}$	$\tilde{c}_{i2}$	$\cdots$	$\tilde{c}_{ij}$	$\cdots$	$\tilde{c}_{in}$	$\tilde{a}_i$
$\vdots$	$\vdots$	$\vdots$	$\vdots$	$\vdots$	$\vdots$	$\vdots$	$\vdots$
S_m	$\tilde{c}_{m1}$	$\tilde{c}_{m2}$	$\cdots$	$\tilde{c}_{mj}$	$\cdots$	$\tilde{c}_{mn}$	$\tilde{a}_m$
Demand$(\tilde{b}_j)$	$\tilde{b}_1$	$\tilde{b}_2$	$\cdots$	$\tilde{b}_j$	$\cdots$	$\tilde{b}_n$	$\sum_{i=1}^{m} \tilde{a}_i = \sum_{j=1}^{n} \tilde{b}_j$

Table 5.2 Tabular representation of the first crisp transportation problem

Destinations → Sources ↓	D_1	D_2	$\cdots$	D_j	$\cdots$	D_n	$m_i - \alpha_i$
S_1	η_{11}	η_{12}	$\cdots$	η_{1j}	$\cdots$	η_{1n}	$m_1 - \alpha_1$
$\vdots$	$\vdots$	$\vdots$	$\vdots$	$\vdots$	$\vdots$	$\vdots$	$\vdots$
S_i	η_{i1}	η_{i2}	$\cdots$	η_{ij}	$\cdots$	η_{in}	$m_i - \alpha_i$
$\vdots$	$\vdots$	$\vdots$	$\vdots$	$\vdots$	$\vdots$	$\vdots$	$\vdots$
S_m	η_{m1}	η_{m2}	$\cdots$	η_{mj}	$\cdots$	η_{mn}	$m_m - \alpha_m$
$m'_j - \alpha'_j$	$m'_1 - \alpha'_1$	$m'_2 - \alpha'_2$	$\cdots$	$m'_j - \alpha'_j$	$\cdots$	$m'_n - \alpha'_n$	$\sum_{i=1}^{m} (m_i - \alpha_i) = \sum_{j=1}^{n} (m'_j - \alpha'_j)$

$$\sum_{i=1}^{m} \sum_{j=1}^{n} (\tilde{c}_{ij} \otimes \tilde{x}_{ij})$$

The algorithm shown above yields the fuzzy optimal solution of the fully fuzzy transportation problems represented by using the tabular representation sought.

The following remarks are interesting:

Remark 8 If in the proposed methods all the parameters are represented by such LR flat fuzzy numbers in which $L(x) = R(x) = \max\{0, 1 - x\}$, then the methods, proposed in this chapter, will be same as the methods proposed in Chap. 4.

Remark 9 Let $\tilde{A} = (m_{ij}, n_{ij}, \alpha_{ij}, \beta_{ij})_{LR}$ be an LR flat fuzzy number with $L(x) = R(x) = \max\{0, 1 - x^4\}$. Then

Table 5.3 Tabular representation of the second crisp transportation problem

Destinations → Sources↓	D_1	D_2	$\cdots$	D_j	$\cdots$	D_n	α_i
S_1	ρ_{11}	ρ_{12}	$\cdots$	ρ_{1j}	$\cdots$	ρ_{1n}	α_1
$\vdots$	$\vdots$	$\vdots$	$\vdots$	$\vdots$	$\vdots$	$\vdots$	$\vdots$
S_i	ρ_{i1}	ρ_{i2}	$\cdots$	ρ_{ij}	$\cdots$	ρ_{in}	α_i
$\vdots$	$\vdots$	$\vdots$	$\vdots$	$\vdots$	$\vdots$	$\vdots$	$\vdots$
S_m	ρ_{m1}	ρ_{m2}	$\cdots$	ρ_{mj}	$\cdots$	ρ_{mn}	α_m
α'_i	α'_1	α'_2	$\cdots$	α'_j	$\cdots$	α'_n	$\sum_{i=1}^{m} \alpha_i = \sum_{j=1}^{n} \alpha'_j$

Table 5.4 Tabular representation of the third crisp transportation problem

Destinations → Sources ↓	D_1	D_2	$\cdots$	D_j	$\cdots$	D_n	$n'_i - m'_i$
S_1	δ_{11}	δ_{12}	$\cdots$	δ_{1j}	$\cdots$	δ_{1n}	$n_1 - m_1$
$\vdots$	$\vdots$	$\vdots$	$\vdots$	$\vdots$	$\vdots$	$\vdots$	$\vdots$
S_i	δ_{i1}	δ_{i2}	$\cdots$	δ_{ij}	$\cdots$	δ_{in}	$n_i - m_i$
	$\vdots$	$\vdots$	$\vdots$	$\vdots$	$\vdots$	$\vdots$	$\vdots$
S_m	δ_{m1}	δ_{m2}	$\cdots$	δ_{mj}	$\cdots$	δ_{mn}	$n_m - m_m$
$n'_j - m'_j$	$n'_1 - m'_1$	$n'_2 - m'_2$	$\cdots$	$n'_j - m'_j$	$\cdots$	$n'_n - m'_n$	$\sum_{i=1}^{m} (n_i - m_i) = \sum_{j=1}^{n} (n'_j - m'_j)$

Table 5.5 Tabular representation of the fourth crisp transportation problem

Destinations → Sources ↓	D_1	D_2	$\cdots$	D_j	$\cdots$	D_n	β'_i
S_1	ξ_{11}	ξ_{12}	$\cdots$	ξ_{1j}	$\cdots$	ξ_{1n}	β_1
$\vdots$	$\vdots$	$\vdots$	$\vdots$	$\vdots$	$\vdots$	$\vdots$	$\vdots$
S_i	ξ_{i1}	ξ_{i2}	$\cdots$	ξ_{ij}	$\cdots$	ξ_{in}	β_i
$\vdots$	$\vdots$	$\vdots$	$\vdots$	$\vdots$	$\vdots$	$\vdots$	$\vdots$
S_m	ξ_{m1}	ξ_{m2}	$\cdots$	ξ_{mj}	$\cdots$	ξ_{mn}	β_m
β'_j	β'_1	β'_2	$\cdots$	β'_j	$\cdots$	β'_n	$\sum_{i=1}^{m} \beta_i = \sum_{j=1}^{n} \beta'_j$

$$
\begin{aligned}
&\eta_{ij} = \\
&= \frac{1}{2}((m_{ij} + n_{ij}) - \alpha_{ij} \int_0^1 L^{-1}(\lambda)d\lambda + \beta_{ij} \int_0^1 R^{-1}(\lambda)d\lambda) = \\
&= \frac{1}{10}(5m_{ij} + 5n_{ij} + 4\beta_{ij} - 4\alpha_{ij}),
\end{aligned}
$$

$$
\begin{aligned}
&\rho_{ij} = \\
&= \frac{1}{2}((m_{ij} + n_{ij}) - m_{ij} \int_0^1 L^{-1}(\lambda)d\lambda + \beta_{ij} \int_0^1 R^{-1}(\lambda)d\lambda) = \\
&= \frac{1}{10}(m_{ij} + 5n_{ij} + 4\beta_{ij}),
\end{aligned}
$$

$$
\delta_{ij} = \frac{1}{2}(n_{ij} + \beta_{ij} \int_0^1 R^{-1}(\lambda)d\lambda) = \frac{1}{10}(5n_{ij} + 4\beta_{ij})
$$

$$
\xi_{ij} = \frac{1}{2}((n_{ij} + \beta_{ij}) \int_0^1 R^{-1}(\lambda)d\lambda) = \frac{4}{10}(n_{ij} + \beta_{ij})
$$

5.5.3 *Main Advantages of the Proposed Methods*

First, let us notice that the methods proposed in Chap. 4 can only be used to find a fuzzy optimal solution of such fully fuzzy transportation problems in which the parameters are either represented by the triangular fuzzy numbers or trapezoidal fuzzy numbers but cannot be used for solving such fully fuzzy transportation problems in which parameters are either represented by the LR fuzzy numbers or the LR flat fuzzy numbers.

However, since the triangular fuzzy numbers, trapezoidal fuzzy numbers and LR fuzzy numbers are special cases of the LR flat fuzzy numbers, then the methods proposed in this chapter can also be used for the solution of such fully fuzzy transportation problems in which the parameters are represented by the triangular fuzzy numbers, trapezoidal fuzzy numbers or LR fuzzy numbers, that is, for quite general cases.

For clarity we will now present some illustrative example.

5.6 Illustrative Example

We will now present the use of the methods proposed in the previous sections to solve the fully fuzzy transportation problem already shown in Example 17.

5.6.1 Determination of the Fuzzy Optimal Solution Using the Method Based on the Fuzzy Linear Programming

Using the proposed method based on the fuzzy linear programming (cf. Sect. 5.5.1), the fuzzy optimal solution of the fully fuzzy transportation problem shown in Example 17, can be obtained as follows:

Step 1 Suppose that we have:

- Total fuzzy availability = $(150, 160, 40, 20)_{LR}$,
- Total fuzzy demand = $(120, 140, 30, 60)_{LR}$.

Since, the total fuzzy availability is not equal to the total fuzzy demand, then our problem is an unbalanced fully fuzzy transportation problem.
Therefore, as described in the proposed method using Case (c) of Step 1 of the proposed method, discussed in Sect. 5.1, the unbalanced fully fuzzy transportation problem in question can be converted into a balanced fully fuzzy transportation problem by introducing:

- a dummy source S_3 with the fuzzy availability $(0, 10, 0, 40)_{LR}$,
- a dummy destination D_4 with fuzzy demand $(30, 30, 10, 0)_{LR}$.

Step 2 By assuming the fuzzy transportation cost for one unit of quantity of the product from the dummy source S_3 to all destinations, and from all sources to the dummy destination D_4, as the zero LR flat fuzzy number, i.e.,

$$\tilde{c}_{14} = \tilde{c}_{24} = \tilde{c}_{31} = \tilde{c}_{32} = \tilde{c}_{33} = \tilde{c}_{34} = (0, 0, 0, 0)_{LR}$$

the balanced fully fuzzy transportation problem from Step 1 can be formulated as the following fuzzy linear programming problem:

$$\begin{cases} min(20, 30, 10, 10)_{LR} \otimes \tilde{x}_{11} \oplus (60, 70, 10, 20)_{LR} \otimes \\ \otimes \tilde{x}_{12} \oplus (90, 110, 10, 10)_{LR} \otimes \tilde{x}_{13} \oplus (0, 0, 0, 0)_{LR} \otimes \\ \times \tilde{x}_{14} \oplus (70, 80, 10, 10)_{LR} \otimes tildex_{21} \oplus (80, 100, 10, 20)_{LR} \otimes \\ \otimes \tilde{x}_{22} \oplus (30, 50, 10, 10)_{LR} \otimes\ tildex_{23} \oplus (0, 0, 0, 0)_{LR} \otimes \\ \otimes \tilde{x}_{24} \oplus (0, 0, 0, 0)_{LR} \otimes \tilde{x}_{31} \oplus 0, 0, 0, 0)_{LR} \otimes \tilde{x}_{32} \oplus (0, 0, 0, 0)_{LR} \otimes \\ \otimes \tilde{x}_{33} \oplus (0, 0, 0, 0)_{LR} \otimes \tilde{x}_{34} \\ \text{subject to:} \\ \tilde{x}_{11} \oplus \tilde{x}_{12} \oplus \tilde{x}_{13} \oplus \tilde{x}_{14} = (90, 90, 20, 10)_{LR} \\ \tilde{x}_{21} \oplus \tilde{x}_{22} \oplus \tilde{x}_{23} \oplus \tilde{x}_{24} = (60, 70, 20, 10)_{LR} \\ \tilde{x}_{31} \oplus \tilde{x}_{32} \oplus \tilde{x}_{33} \oplus \tilde{x}_{34} = (0, 10, 0, 40)_{LR} \\ \tilde{x}_{11} \oplus \tilde{x}_{21} \oplus \tilde{x}_{31} = (40, 50, 10, 20)_{LR} \\ \tilde{x}_{12} \oplus \tilde{x}_{22} \oplus \tilde{x}_{32} = (30, 40, 10, 10)_{LR} \\ \tilde{x}_{13} \oplus \tilde{x}_{23} \oplus \tilde{x}_{33} = (50, 50, 10, 30)_{LR} \\ \tilde{x}_{14} \oplus \tilde{x}_{24} \oplus \tilde{x}_{34} = (30, 30, 10, 0)_{LR} \end{cases} \tag{5.8}$$

where:

$$\begin{cases} \tilde{x}_{11}\ \tilde{x}_{12}\ \tilde{x}_{13}\ \tilde{x}_{14} \\ \tilde{x}_{21}\ \tilde{x}_{22}\ \tilde{x}_{23}\ \tilde{x}_{24} \\ \tilde{x}_{31}\ \tilde{x}_{32}\ \tilde{x}_{33}\ \tilde{x}_{34} \end{cases}$$

are non-negative LR flat fuzzy numbers.

Step 3 Using Steps 4–7 of the method, proposed in Sect. 5.5.1, and Remark 7, the fuzzy linear programming problem obtained in Step 2 can be converted into the following crisp linear programming problem:

$$\begin{cases} min\frac{1}{10}(60m_{11}+190n_{11}-40\alpha_{11}+160\beta_{11}+ \\ +260m_{12}+430n_{12}-200\alpha_{12}+360\beta_{12}+410m_{13}+ \\ +590n_{13}-320\alpha_{13}+480\beta_{13}+310m_{21}+440n_{21}-240\alpha_{21}+ \\ +360\beta_{21}+360m_{22}+580n_{22}-280\alpha_{22}+480\beta_{22}+ \\ +110m_{23}+290n_{23}-80\alpha_{23}+240\beta_{23}) \\ \text{subject to:} \\ \sum_{j=1}^{4} m_{1j}=90;\ \sum_{j=1}^{4} n_{1j}=90;\ \sum_{j=1}^{4} \alpha_{1j}=20;\ \sum_{j=1}^{4} \beta_{1j}=10 \\ \sum_{j=1}^{4} m_{2j}=60;\ \sum_{j=1}^{4} n_{2j}=70;\ \sum_{j=1}^{4} \alpha_{2j}=20;\ \sum_{j=1}^{4} \beta_{2j}=10 \\ \sum_{j=1}^{4} m_{3j}=0;\ \sum_{j=1}^{4} n_{3j}=10;\ \sum_{j=1}^{4} \alpha_{3j}=0;\ \sum_{j=1}^{4} \beta_{3j}=40 \\ \sum_{i=1}^{3} m_{i1}=40;\ \sum_{i=1}^{3} n_{i1}=50;\ \sum_{i=1}^{3} \alpha_{i1}=10;\ \sum_{i=1}^{3} \beta_{i1}=20 \\ \sum_{i=1}^{3} m_{i2}=30;\ \sum_{i=1}^{3} n_{i2}=40;\ \sum_{i=1}^{3} \alpha_{i2}=10;\ \sum_{i=1}^{3} \beta_{i2}=10 \\ \sum_{i=1}^{3} m_{i3}=50;\ \sum_{i=1}^{3} n_{i3}=50;\ \sum_{i=1}^{3} \alpha_{i3}=10;\ \sum_{i=1}^{3} \beta_{i3}=30 \\ \sum_{i=1}^{3} m_{i4}=30;\ \sum_{i=1}^{3} n_{i4}=30;\ \sum_{i=1}^{3} \alpha_{i4}=10;\ \sum_{i=1}^{3} \beta_{i4}=0 \end{cases} \tag{5.9}$$

where: $m_{ij}-\alpha_{ij}, n_{ij}-m_{ij}, \alpha_{ij}, \beta_{ij} \geq 0$, for all $i=1,2,3;\ j=1,2,3,4$.

Step 4 The optimal solution of the crisp linear programming problem (5.9), obtained in Step 3, is

$$\begin{cases} m_{11}=40, n_{11}=40, \alpha_{11}=10, \beta_{11}=10, \\ m_{12}=20, n_{12}=20, \alpha_{12}=0, \beta_{12}=0, \\ m_{13}=25, n_{13}=25, \alpha_{13}=10, \beta_{13}=0, \\ m_{14}=5, n_{14}=5, \alpha_{14}=0, \beta_{14}=0, \\ m_{21}=0, n_{21}=10, \alpha_{21}=0, \beta_{21}=10, \\ m_{22}=10, n_{22}=10, \alpha_{22}=10, \beta_{22}=0, \\ m_{23}=25, n_{23}=25, \alpha_{23}=0, \beta_{23}=0, \\ m_{24}=25, n_{24}=25, \alpha_{24}=10, \beta_{24}=0, \\ m_{31}=0, n_{31}=0, \alpha_{31}=0, \beta_{31}=0, \\ m_{32}=0, n_{32}=10, \alpha_{32}=0, \beta_{32}=10, \\ m_{33}=0, n_{33}=0, \alpha_{33}=0, \beta_{33}=30, \\ m_{34}=0, n_{34}=0, \alpha_{34}=0, \beta_{34}=0. \end{cases}$$

Step 5 Putting the values of $m_{ij}, n_{ij}, \alpha_{ij}$ and β_{ij} into

$$\tilde{x}_{ij} = (m_{ij}, n_{ij}, \alpha_{ij}, \beta_{ij})_{LR}$$

the fuzzy optimal solution is therefore

$$\begin{cases} \tilde{x}_{11} = (40, 40, 10, 10)_{LR} \\ \tilde{x}_{12} = (20, 20, 0, 0)_{LR} \\ \tilde{x}_{13} = (25, 25, 10, 0)_{LR} \\ \tilde{x}_{14} = (5, 5, 0, 0)_{LR} \\ \tilde{x}_{21} = (0, 10, 0, 10)_{LR} \\ \tilde{x}_{22} = (10, 10, 10, 0)_{LR} \\ \tilde{x}_{23} = (25, 25, 0, 0)_{LR} \\ \tilde{x}_{24} = (25, 25, 10, 0)_{LR} \\ \tilde{x}_{31} = (0, 0, 0, 0)_{LR} \\ \tilde{x}_{32} = (0, 10, 0, 10)_{LR} \\ \tilde{x}_{33} = (0, 0, 0, 30)_{LR} \\ \tilde{x}_{34} = (0, 0, 0, 0)_{LR} \end{cases}$$

Step 6 Putting the values of

$$\begin{cases} \tilde{x}_{11}, \tilde{x}_{12}, \tilde{x}_{13}, \tilde{x}_{14}, \\ \tilde{x}_{21}, \tilde{x}_{22}, \tilde{x}_{23}, \tilde{x}_{24}, \\ \tilde{x}_{31}, \tilde{x}_{32}, \tilde{x}_{33}, \tilde{x}_{34} \end{cases}$$

into

$$\begin{cases} ((20, 30, 10, 10)_{LR} \otimes \tilde{x}_{11} \oplus (60, 70, 10, 20)_{LR} \otimes \tilde{x}_{12} \oplus (90, 110, 10, 10)_{LR} \otimes \\ \otimes \tilde{x}_{13} \oplus (0, 0, 0, 0)_{LR} \otimes \tilde{x}_{14} \oplus \oplus (70, 80, 10, 10)_{LR} \otimes \tilde{x}_{21} \oplus \\ \oplus (80, 100, 10, 20)_{LR} \otimes \tilde{x}_{22} \oplus (30, 50, 10, 10)_{LR} \otimes \tilde{x}_{23} \oplus (0, 0, 0, 0)_{LR} \otimes \\ \otimes \tilde{x}_{24} \oplus (0, 0, 0, 0)_{LR} \otimes \tilde{x}_{31} \oplus (0, 0, 0, 0)_{LR} \otimes \tilde{x}_{32} \oplus (0, 0, 0, 0)_{LR} \otimes \\ \otimes \tilde{x}_{33} \oplus (0, 0, 0, 0)_{LR} \otimes \tilde{x}_{34}) \end{cases}$$

the minimum total fuzzy transportation cost is $(5800, 8400, 2800, 2900)_{LR}$.

5.6.2 *Determination of the Fuzzy Optimal Solution Using the Method Based on the Tabular Representation*

Using the proposed method, based on the tabular representation, the fuzzy optimal solution of the fully fuzzy transportation problem considered in Example 17, can be obtained as follows:

Step 1 The balanced fully fuzzy transportation problem, shown in Step 1 in Sect. 5.6.1, can be represented by Table 5.6 shown:

Step 2 Using Step 3 of the method, proposed in Sect. 5.5.2, and Remark 9, Table 5.6 can be split into four crisp transportation tables shown in Tables 5.7, 5.8, 5.9, 5.10 shown.

Step 3 The optimal solution of the crisp transportation problems, shown in Table 5.7, Table 5.8, Table 5.9 and Table 5.10, respectively, are:

$$\begin{cases} m_{11}-\alpha_{11}=30, m_{12}-\alpha_{12}=20, m_{13}-\alpha_{13}=15, m_{14}-\alpha_{14}=5, \\ m_{21}-\alpha_{21}=0, m_{22}-\alpha_{22}=0, m_{23}-\alpha_{23}=25, m_{24}-\alpha_{24}=15, \\ m_{31}-\alpha_{31}=0, m_{32}-\alpha_{32}=0, m_{33}-\alpha_{33}=0, m_{34}-\alpha_{34}=0; \\ \alpha_{11}=10, \alpha_{12}=0, \alpha_{13}=10, \alpha_{14}=0, \\ \alpha_{21}=0, \alpha_{22}=10, \alpha_{23}=0, \alpha_{24}=10, \\ \alpha_{31}=0, \alpha_{32}=0, \alpha_{33}=0, \alpha_{34}=0; \\ n_{11}-m_{11}=0, n_{12}-m_{12}=0, n_{13}-m_{13}=0, n_{14}-m_{14}=0, \\ n_{21}-m_{21}=10, n_{22}-m_{22}=0, n_{23}-m_{23}=0, n_{24}-m_{24}=0, \\ n_{31}-m_{31}=0, n_{32}-m_{32}=10, n_{33}-m_{33}=0, n_{34}-m_{34}=0 \\ \beta_{11}=10, \beta_{12}=0, \beta_{13}=0, \beta_{14}=0, \\ \beta_{21}=10, \beta_{22}=0, \beta_{23}=0, \beta_{24}=0, \\ \beta_{31}=0, \beta_{32}=10, \beta_{33}=30, \beta_{34}=0 \end{cases}$$

Step 4 By solving the equations from Step 3 of the method proposed in Sect. 5.5.2, the values of $m_{ij}, n_{ij}, \alpha_{ij}$ and β_{ij} are:

$$\begin{cases} m_{11}=40, n_{11}=40, \alpha_{11}=10, \beta_{11}=10, \\ m_{12}=20, n_{12}=20, \alpha_{12}=0, \beta_{12}=0, \\ m_{13}=25, n_{13}=25, \alpha_{13}=10, \beta_{13}=0, \\ m_{14}=5, n_{14}=5, \alpha_{14}=0, \beta_{14}=0, \\ m_{21}=0, n_{21}=10, \alpha_{21}=0, \beta_{21}=10, \\ m_{22}=10, n_{22}=10, \alpha_{22}=10, \beta_{22}=0, \\ m_{23}=25, n_{23}=25, \alpha_{23}=0, \beta_{23}=0, \\ m_{24}=25, n_{24}=25, \alpha_{24}=10, \beta_{24}=0, \\ m_{31}=0, n_{31}=0, \alpha_{31}=0, \beta_{31}=0, \\ m_{32}=0, n_{32}=10, \alpha_{32}=0, \beta_{32}=10, \\ m_{33}=0, n_{33}=0, \alpha_{33}=0, \beta_{33}=30, \\ m_{34}=0, n_{34}=0, \alpha_{34}=0, \beta_{34}=0 \end{cases}$$

Step 5 Putting the values of $m_{ij}, n_{ij}, \alpha_{ij}$ and β_{ij} obtained in Step 4 into $\tilde{x}_{ij} = (m_{ij}, n_{ij}, \alpha_{ij}, \beta_{ij})_{LR}$, the fuzzy optimal solution becomes

$$\begin{cases} \tilde{x}_{11}=(40,40,10,10)_{LR}, \tilde{x}_{12}=(20,20,0,0)_{LR}, \tilde{x}_{13}=(25,25,10,0)_{LR}, \\ \tilde{x}_{14}=(5,5,0,0)_{LR}, \tilde{x}_{21}=(0,10,0,10)_{LR}, \tilde{x}_{22}=(10,10,10,0)_{LR}, \\ \tilde{x}_{23}=(25,25,0,0)_{LR}, \tilde{x}_{24}=(25,25,10,0)_{LR}, \tilde{x}_{31}=(0,0,0,0)_{LR}, \\ \tilde{x}_{32}=(0,10,0,10)_{LR}, \tilde{x}_{33}=(0,0,0,30)_{LR}, \tilde{x}_{34}=(0,0,0,0)_{LR}. \end{cases}$$

Table 5.6 Tabular representation of the balanced fully fuzzy transportation problem from Example 17

	D_1	D_2	D_3	D_4	$\tilde{a}_i$
S_1	$(20, 30, 10, 10)_{LR}$	$(60, 70, 10, 20)_{LR}$	$(90, 110, 10, 10)_{LR}$	$(0, 0, 0, 0)_{LR}$	$(90, 90, 20, 10)_{LR}$
S_2	$(70, 80, 10, 10)_{LR}$	$(80, 100, 10, 20)_{LR}$	$(30, 50, 10, 10)_{LR}$	$(0, 0, 0, 0)_{LR}$	$(60, 70, 20, 10)_{LR}$
S_3	$(0, 0, 0, 0)_{LR}$	$(0, 0, 0, 0)_{LR}$	$(0, 0, 0, 0)_{LR}$	$(0, 0, 0, 0)_{LR}$	$(0, 10, 0, 40)_{LR}$
$\tilde{b}_j$	$(40, 50, 10, 20)_{LR}$	$(30, 40, 10, 10)_{LR}$	$(50, 50, 10, 30)_{LR}$	$(30, 30, 10, 0)_{LR}$	$\sum_{i=1}^{3} \tilde{a}_i = \sum_{j=1}^{4} \tilde{b}_j$

Table 5.7 Tabular representation of the first crisp transportation table from Table 5.6

	D_1	D_2	D_3	D_4	$m_i - \alpha_i$
S_1	25	69	100	0	70
S_2	75	94	40	0	40
S_3	0	0	0	0	0
$m'_j - \alpha'_j$	30	20	40	20	

Table 5.8 Tabular representation of the second crisp transportation table from Table 5.6

	D_1	D_2	D_3	D_4	α_i
S_1	21	49	68	0	20
S_2	51	66	32	0	20
S_3	0	0	0	0	0
α'_j	10	10	10	10	

Step 6 Putting the values of $\tilde{x}_{11}, \tilde{x}_{12}, \tilde{x}_{13}, \tilde{x}_{14}, \tilde{x}_{21}, \tilde{x}_{22}, \tilde{x}_{23}, \tilde{x}_{24}, \tilde{x}_{31}, \tilde{x}_{32}, \tilde{x}_{33}, \tilde{x}_{34}$, obtained in Step 5, into

$$(20, 30, 10, 10)_{LR} \otimes \tilde{x}_{11} \oplus (60, 70, 10, 20)_{LR} \otimes \tilde{x}_{12} \oplus (90, 110, 10, 10)_{LR} \otimes \\ \otimes \tilde{x}_{13} \oplus (0, 0, 0, 0)_{LR} \otimes \tilde{x}_{14} \oplus (70, 80, 10, 10)_{LR} \otimes \tilde{x}_{21} \oplus (80, 100, 10, 20)_{LR} \otimes \\ \otimes \tilde{x}_{22} \oplus (30, 50, 10, 10)_{LR} \otimes \tilde{x}_{23} \oplus (0, 0, 0, 0)_{LR} \otimes \tilde{x}_{24} \oplus (0, 0, 0, 0)_{LR} \otimes \\ \otimes \tilde{x}_{31} \oplus (0, 0, 0, 0)_{LR} \otimes \tilde{x}_{32} \oplus (0, 0, 0, 0)_{LR} \otimes \tilde{x}_{33} \oplus (0, 0, 0, 0)_{LR} \otimes \tilde{x}_{34}$$

the minimum total fuzzy transportation cost is $(5800, 8400, 2800, 2900)_{LR}$.

Table 5.9 Tabular representation the third crisp transportation table from Table 5.6

	D_1	D_2	D_3	D_4	$n_i - m_i$
S_1	19	43	59	0	0
S_2	44	58	29	0	10
S_3	0	0	0	0	10
$n'_j - m'_j$	10	10	0	0	

Table 5.10 Tabular representation the fourth crisp transportation table from Table 5.6

	D_1	D_2	D_3	D_4	β_i
S_1	16	36	48	0	10
S_2	36	48	24	0	10
S_3	0	0	0	0	40
β'_j	20	10	30	0	

5.6.3 *Interpretation of Results*

The main purpose of this section is to provide an easily comprehensible interpretation of the results obtained in the previous sections, Sects. 5.5.3 and 5.5.3, that is, the minimum total fuzzy transportation cost, obtained by using the proposed methods, presented in the above mentioned section, that is, the one based on the fuzzy linear programming and the one based on the tabular representation.

By using the proposed methods the minimum total fuzzy transportation cost has been obtained as $(5800, 8400, 2800, 2900)_{LR}$, which can be interpreted as follows:

1. The least amount of the minimum total transportation cost is 3000;
2. The most possible amount of the minimum total transportation cost is between 5800 and 8400;
3. The greatest amount of the minimum total transportation cost is 11300;

that is, the minimum total transportation cost will always be greater than 3000 and less than 11300, while there is the highest possibility that the minimum total transportation cost will be between 5800 and 8400.

Table 5.11 Results obtained by using the methods proposed in this chapter and some methods proposed in Chap. 4

Example	Minimum total fuzzy transportation cost	
	Methods proposed in Chap. 4	Methods proposed in this chapter
2.2	(2100, 2900, 3500, 4200)	$(2900, 3500, 800, 700)_{LR}$
3.1	Not applicable	$(5800, 8400, 2800, 2900)_{LR}$

5.7 A Comparative Study

The results obtained in this chapter by using the two methods proposed in Sects. 5.6.1 and 5.6.2, i.e. the one based on the fuzzy linear programming and the one based on the tabular representation, are compared by those obtained by using some methods proposed in Chap. 4 which is shown in Table 5.11.

The results depicted in Table 5.11 can be briefly explained as follows:

1. The methods, proposed in Chap. 4, can be used only for solving such fully fuzzy transportation problems in which the parameters are either represented by triangular fuzzy numbers or trapezoidal fuzzy numbers Therefore, since in the fully fuzzy transportation problem discussed in Example 15, the parameters are represented by trapezoidal fuzzy numbers, this problem can be solved by the methods proposed in Chap. 4. However, in the fully fuzzy transportation problem discussed in Example 17 the parameters are represented by such LR flat fuzzy numbers which are neither triangular nor trapezoidal, then we cannot solve by using the methods presented in Chap. 4 the fully fuzzy transportation problem from Example 17.
2. The methods proposed in this chapter can be used for solving such fully fuzzy transportation problems in which the parameters are represented by LR flat fuzzy numbers and, as mentioned in Remark 6, the trapezoidal fuzzy numbers are obviously a special case type of the LR flat fuzzy numbers so that both the fully fuzzy transportation problems considered in Examples 15 and 17, can be solved by using the methods proposed in this chapter.

5.8 Concluding Remarks

Looking at the results of a comparison of results obtained presented in Sect. 5.7, we can see that the problems which can be solved by using the methods proposed in Chap. 4 can also be solved by the methods proposed in this chapter. One should however bear in mind that there are some problems which can only be solved by the methods proposed in this chapter. This clearly suggests that it is better to use the methods proposed in this chapter.

References

1. B. Bede, *Mathematics of Fuzzy Sets and Fuzzy Logic* (Springer, New York and Heindelberg, 2013)
2. A.V. Bozhenyuk, E.M. Gerasimenko, J. Kacprzyk, I.N. Rozenberg, *Flows in Networks Under Fuzzy Conditions* (Springer, 2017)
3. M. Dehghan, B. Hashemi, M. Ghatee, Computational methods for solving fully fuzzy linear systems. Appl. Math. Comput. **179**, 328–343 (2006)
4. D. Dubois, H. Prade, *Fuzzy Sets and Systems: Theory and Applications* (Academic Press, New York, 1980)
5. F. Jimenez, J.L. Verdegay, Interval multiobjective solid transportation problem via genetic algorithms, in *Proceedings of IPMU '96—Sixth International Conference on Information Processing and Management of Uncertainty in Knowledge-Based Systems*, vol. 2 (Granada, Spain, 1996), pp. 787–792
6. F. Jimenez, J.L. Verdegay, Obtaining fuzzy solutions to the fuzzy solid transportation problem with genetic algorithms, in Proceedings of FUZZ-IEEE '97—Sixth IEEE International Conference on Fuzzy Systems, vol. III (IEEE Press, 1997), pp. 1657–1663
7. F. Jimenez, J.-L. Verdegay, Uncertain solid transportation problems. Fuzzy Sets Syst. **100**(1–2), 45–57 (1998)
8. J. Kacprzyk, *Multistage Fuzzy Control: A Model-Based Approach to Control and Decision-Making* (Wiley, Chichester, 1997)
9. S.T. Liu, C. Kao, Solving fuzzy transportation problems based on extension principle. Eur. J. Oper. Res. **153**, 661–674 (2004)
10. H.J. Zimmermann, *Fuzzy Set Theory and Its Applications* (Kluwer, Dordrecht, 2001)

Chapter 6
New Improved Methods for Solving the Fully Fuzzy Transshipment Problems with Parameters Given as the LR Flat Fuzzy Numbers

The purpose of this chapter is to discuss in more detail Ghatee and Hashemi's [1] method for solving the problem of finding an optimal solution to the fully fuzzy transshipment problems, which is considered in this paper in terms of a fully fuzzy minimal cost flow problem, This is presumably the best known, if not the only comprehensive and constructive method for solving such a type of problems. Though this method is good, indeed, it has some limitations which will be briefly pointed out in this chapter. Then, two new method for solving the fully fuzzy transshipment problems will be proposed that are free form those limitations mentioned. This will be followed by a comparison of results obtained by the source Ghatee and Hashemi's [1] method and the two methods proposed in the previous chapters, Finally, for illustration, the two proposed methods will be applied for solving a real world transshipment problem.

6.1 Fuzzy Linear Programming Formulation of the Balanced Fully Fuzzy Transshipment Problems

Basically, Ghatee and Hashemi [1] employ a fuzzy linear programming formulation for finding the fuzzy optimal solution to the balanced fully fuzzy transshipment problems. To be more specific, the following types of nodes in the network are employed:

- **Purely source node**: A node S is said to be a *purely source node* if there exist at least one node S' such that the product may be supplied from S to S' but there does not exist any node S'' such that the product may be supplied from S'' to S; the set of all such purely source nodes is denoted by N_{PS}.

A. Kaur et al., *Fuzzy Transportation and Transshipment Problems*, Studies in Fuzziness and Soft Computing 385,
https://doi.org/10.1007/978-3-030-26676-9_6

- **Purely destination node**: A node D is said to be a *purely destination node* if there does not exist any node D' such that the product may be supplied from D to D' but there exist at least one node D'' such that product may be supplied from D'' to D; the set of all such nodes is denoted by N_{PD}.
- **Intermediate node**: The following nodes in the network are said to be the *intermediate nodes*:

1. a node S at which some quantity of the product is available to be transshipped to other nodes and also there exist some nodes such that some quantity of the product is being supplied from that nodes to node S; all such nodes are said to be *source nodes* and the set of all such nodes is denoted by N_S;
2. a node D at which some quantity of the product is required and also there exist some nodes such that the product is being supplied from node D to that nodes; all such nodes are said to be *destination nodes* and the set of all such nodes is denoted by N_D;
3. a node T at which neither any quantity of the product is available to be transshipped to other nodes nor any quantity of the product is required but there exist some nodes such that some quantity of the product is being supplied from that nodes to node T, and the same quantity of the product is being supplied from T to some other nodes; all such nodes are said to be *transition nodes* and the set of all such nodes is denoted by N_T.

Now, consider a balanced fully fuzzy transshipment problem with:

- m purely source nodes N_{PS_i}, $i = 1, \ldots, m$,
- t purely destination nodes N_{PD_j}, $j = 1, \ldots, t$, and
- $(l + q + r)$ intermediate nodes.

Moreover, suppose that among $(l + q + r)$ intermediate nodes:

- l intermediate nodes N_{S_i}, $i = 1, \ldots, l$ are the source nodes,
- q intermediate nodes N_{D_j}, $j = 1, \ldots, q$ are the destination nodes, and
- the remaining r intermediate nodes N_{T_i}, $i = 1, \ldots, r$ are the transition nodes.

Further, suppose that the *availability* of the product:

- at the ith purely source node N_{PS_i} be $\tilde{a}_i$, and
- at the ith source node N_{S_i} be $\tilde{a}_{m+i}$,

and the *demand* of the product:

- at the jth purely destination node N_{PD_j} be $\tilde{b}_j$, and
- at the jth destination node N_{D_j} be $\tilde{b}_{t+j}$.

Let $\tilde{c}_{ij}$ be the fuzzy cost for transporting one unit quantity of the product from the ith node the to jth node and $\tilde{x}_{ij}$ be the fuzzy quantity of the product that should be transported from the ith node to the jth node to minimize the total fuzzy transportation cost.

Then, fuzzy linear programming formulation of balanced fully fuzzy transshipment problem can be written, according to Ghatee and Hashemi [1] as:

$$\begin{cases} min \sum_{(i,j)\in A} (\tilde{c}_{ij} \otimes \tilde{x}_{ij}) \\ \text{subject to:} \\ \sum_{j:(i,j)\in A} \tilde{x}_{ij} = \tilde{a}_i;\ i = 1, \ldots, m, \\ \sum_{j:(i,j)\in A} \tilde{x}_{ij} \ominus_H \sum_{j:(j,i)\in A} \tilde{x}_{ji} = \tilde{a}_{m+i};\ i = 1, \ldots, l \\ \sum_{i:(i,j)\in A} \tilde{x}_{ij} = \tilde{b}_j;\ j = 1\ \ldots, t \\ \sum_{i:(i,j)\in A} \tilde{x}_{ij} \ominus_H \sum_{i:(j,i)\in A} \tilde{x}_{ji} = \tilde{b}_{t+j};\ j = 1, \ldots, q \\ \sum_{j:(i,j)\in A} \tilde{x}_{ij} = \sum_{j:(j,i)\in A} \tilde{x}_{ji};\ i = 1, \ldots, r \end{cases} \tag{6.1}$$

where: $\tilde{x}_{ij}$ is a non-negative fuzzy number, for all $(i, j) \in A$; A is the set of arcs (i, j) joining node i to node j.

We have now some important remarks.

Remark 10 Let $\tilde{A}_1 = (m_1, \alpha_1, \beta_1)_{LR}$ and $\tilde{A}_2 = (m_2, \alpha_2, \beta_2)_{LR}$ be two LR fuzzy numbers such that $\alpha_1 \geq \alpha_2$ and $\beta_1 \geq \beta_2$.

Then (cf. Ghatee and Hashemi [1]), there holds

$$\tilde{A}_1 \ominus_H \tilde{A}_2 = (m_1 - m_2, \alpha_1 - \alpha_2, \beta_1 - \beta_2)_{LR}$$

Remark 11 Let $\tilde{A}_1 = (m_1, n_1, \alpha_1, \beta_1)_{LR}$ and $\tilde{A}_2 = (m_2, n_2, \alpha_2, \beta_2)_{LR}$ be two LR flat fuzzy numbers such that $\alpha_1 \geq \alpha_2$ and $\beta_1 \geq \beta_2$.

Then (cf. Ghatee abd Hashemi [2]), we have

$$\tilde{A}_1 \ominus_H \tilde{A}_2 = (m_1 - m_2, n_1 - n_2, \alpha_1 - \alpha_2, \beta_1 - \beta_2)_{LR}$$

Remark 12 If

$$\sum_{i=1}^{m} \tilde{a}_i \oplus \sum_{i=1}^{l} \tilde{a}_{m+i} = \sum_{j=1}^{t} \tilde{b}_j \oplus \sum_{j=1}^{q} \tilde{b}_{t+j}$$

then the fully fuzzy transshipment problem is said to be a *balanced fully fuzzy transshipment problem* otherwise it is called an *unbalanced fully fuzzy transshipment problem.*

6.2 Outline of the Ghatee and Hashemi Method

A method for finding a fuzzy optimal solution of the balanced fully fuzzy transshipment problems was originally proposed by Ghatee and Hashemi [1] Later, it was applied to solve some real world problems, notably for: planning a bus network (cf. Ghatee and Hasemi [2]), a petroleum storage and distribution network (cf. Ghatee and Hashemi [3, 4]), transportation of hazardous materials (cf. Ghatee, Hashemi,

Zarepisheh, and Khorram [5], as well as to a more general problem of network design under uncertainty (cf. Ghatee and Hashemi [3]).

The consecutive steps of the source Ghatee and Hashemi's [1] method can be presented as:

Step 1 We assume:

$\tilde{c}_{ij}=(m'_{ij},\alpha'_{ij},\beta'_{ij})_{LR}$, $\tilde{x}_{ij}=(m_{ij},\alpha_{ij},\beta_{ij})_{LR}$, $\tilde{a}_i=(m_i,\alpha_i,\beta_i)_{LR}$, $\tilde{a}_{m+i}=(m_{m+i},\alpha_{m+i},\beta_{m+i})_{LR}$, $\tilde{b}_j=(m'_j,\alpha'_j,\beta'_j)_{LR}$, $\tilde{b}_{t+j}=(m_{t+j},\alpha_{t+j},\beta_{t+j})_{LR}$ and the fuzzy linear programming problem (6.1) can be written as:

$$\begin{cases} \min \sum\limits_{(i,j)\in A} (m'_{ij},\alpha'_{ij},\beta'_{ij})_{LR}\otimes(m_{ij},\alpha_{ij},\beta_{ij})_{LR}) \\ \text{subject to:} \\ \sum\limits_{j:(i,j)\in A}(m_{ij},\alpha_{ij},\beta_{ij})_{LR}=(m_i,\alpha_i,\beta_i)_{LR};\ i=1,\ldots,m \\ \sum\limits_{j:(i,j)\in A}(m_{ij},\alpha_{ij},\beta_{ij})_{LR}\ominus_H \sum\limits_{j:(j,i)\in A}(m_{ji},\alpha_{ji},\beta_{ji})_{LR}= \\ =(m_{m+i},\alpha_{m+i},\beta_{m+i})_{LR};\ i=1,\ldots,l \\ \sum\limits_{i:(i,j)\in A}(m_{ij},\alpha_{ij},\beta_{ij})_{LR}=(m'_j,\alpha'_j,\beta'_j)_{LR};\ j=1,\ldots,t \\ \sum\limits_{i:(i,j)\in A}(m_{ij},\alpha_{ij},\beta_{ij})_{LR}\ominus_H \sum\limits_{i:(j,i)\in A}(m_{ji},\alpha_{ji},\beta_{ji})_{LR}= \\ =(m_{t+j},\alpha_{t+j},\beta_{t+j})_{LR};\ j=1\ldots,q \\ \sum\limits_{j:(i,j)\in A}(m_{ij},\alpha_{ij},\beta_{ij})_{LR}=\sum\limits_{j:(j,i)\in A}(m_{ji},\alpha_{ji},\beta_{ji})_{LR};\ i=1\ldots,r, \end{cases} \tag{6.2}$$

where $(m_{ij},\alpha_{ij},\beta_{ij})_{LR}$ is a non-negative LR fuzzy number, for all $(i,j)\in A$.

Step 2 Using the arithmetic operations on the LR fuzzy numbers defined in Sect. 5.1, and assuming that

$$\sum_{(i,j)\in A}((m'_{ij},\alpha'_{ij},\beta'_{ij})_{LR}\otimes(m_{ij},\alpha_{ij},\beta_{ij})_{LR})=(m_0,\alpha_0,\beta_0)_{LR},$$

the fuzzy linear programming problem obtained in Step 1, i.e. (6.2), can be written as:

$$\begin{cases} \min(m_0,\alpha_0,\beta_0)_{LR} \\ \text{subject to:} \\ (\sum\limits_{j:(i,j)\in A} m_{ij},\sum\limits_{j:(i,j)\in A}\alpha_{ij},\sum\limits_{j:(i,j)\in A}\beta_{ij})_{LR}=(m_i,\alpha_i,\beta_i)_{LR};\ i=1,\ldots,m \\ (\sum\limits_{j:(i,j)\in A} m_{ij},\sum\limits_{j:(i,j)\in A}\alpha_{ij},\sum\limits_{j:(i,j)\in A}\beta_{ij})_{LR}\ominus_H \\ \ominus_H(\sum\limits_{j:(j,i)\in A} m_{ji},\sum\limits_{j:(j,i)\in A}\alpha_{ji},\sum\limits_{j:(j,i)\in A}\beta_{ji})_{LR};\ (m_{m+i},\alpha_{m+i},\beta_{m+i})_{LR},\ i=1,\ldots,l \\ (\sum\limits_{i:(i,j)\in A} m_{ij},\sum\limits_{i:(i,j)\in A}\alpha_{ij},\sum\limits_{i:(i,j)\in A}\beta_{ij})_{LR}=(m'_j,\alpha'_j,\beta'_j)_{LR},\ j=1,\ldots,t \\ (\sum\limits_{i:(i,j)\in A} m_{ij},\sum\limits_{i:(i,j)\in A}\alpha_{ij},\sum\limits_{i:(i,j)\in A}\beta_{ij})_{LR}\ominus_H \\ \ominus_H(\sum\limits_{i:(j,i)\in A} m_{ji},\sum\limits_{i:(j,i)\in A}\alpha_{ji},\sum\limits_{i:(j,i)\in A}\beta_{ji})_{LR}=(m_{t+j},\alpha_{t+j},\beta_{t+j})_{LR};\ j=1,\ldots,q \\ (\sum\limits_{j:(i,j)\in A} m_{ij},\sum\limits_{j:(i,j)\in A}\alpha_{ij},\sum\limits_{j:(i,j)\in A}\beta_{ij})_{LR}= \\ =(\sum\limits_{j:(j,i)\in A} m_{ji},\sum\limits_{j:(j,i)\in A}\alpha_{ji},\sum\limits_{j:(j,i)\in A}\beta_{ji})_{LR};\ i=1,\ldots,r \end{cases} \tag{6.3}$$

where $(m_{ij},\alpha_{ij},\beta_{ij})_{LR}$ is a non-negative LR fuzzy number, for all $(i,j)\in A$.

Step 3 Using Definitions 18, 19, and Remark 10, the fuzzy linear programming problem obtained in Step 2, i.e. (6.3), can be converted into the following fuzzy linear programming problem:

$$\begin{cases} min(m_0, \alpha_0, \beta_0)_{LR} \\ \text{subject to:} \\ \sum\limits_{j:(i,j)\in A} m_{ij} = m_i;\ i = 1, \ldots, m \\ \sum\limits_{j:(i,j)\in A} \alpha_{ij} = \alpha_i;\ i = 1, \ldots, m \\ \sum\limits_{j:(i,j)\in A} \beta_{ij} = \beta_i;\ i = 1, \ldots, m \\ \sum\limits_{j:(i,j)\in A} m_{ij} - \sum\limits_{j:(j,i)\in A} m_{ji} = m_{m+i};\ i = 1, \ldots, l \\ \sum\limits_{j:(i,j)\in A} \alpha_{ij} - \sum\limits_{j:(j,i)\in A} \alpha_{ji} = \alpha_{m+i};\ i = 1, \ldots, l \\ \sum\limits_{j:(i,j)\in A} \beta_{ij} - \sum\limits_{j:(j,i)\in A} \beta_{ji} = \beta_{m+i};\ i = 1, \ldots, l \\ \sum\limits_{i:(i,j)\in A} m_{ij} = m'_j;\ j = 1, \ldots, t \\ \sum\limits_{i:(i,j)\in A} \alpha_{ij} = \alpha'_j;\ j = 1, \ldots, t \\ \sum\limits_{i:(i,j)\in A} \beta_{ij} = \beta'_j;\ j = 1, \ldots, t \\ \sum\limits_{i:(i,j)\in A} m_{ij} - \sum\limits_{i:(j,i)\in A} m_{ji} = m_{t+j};\ j = 1, \ldots, q \\ \sum\limits_{i:(i,j)\in A} \alpha_{ij} - \sum\limits_{i:(j,i)\in A} \alpha_{ji} = \alpha_{t+j};\ j = 1, \ldots, q \\ \sum\limits_{i:(i,j)\in A} \beta_{ij} - \sum\limits_{i:(j,i)\in A} \beta_{ji} = \beta_{t+j};\ j = 1, \ldots, q \\ \sum\limits_{j:(i,j)\in A} m_{ij} = \sum\limits_{j:(j,i)\in A} m_{ji};\ i = 1, \ldots, r \\ \sum\limits_{j:(i,j)\in A} \alpha_{ij} = \sum\limits_{j:(j,i)\in A} \alpha_{ji};\ i = 1, \ldots, r \\ \sum\limits_{j:(i,j)\in A} \beta_{ij} = \sum\limits_{j:(j,i)\in A} \beta_{ji};\ i = 1, \ldots, r \end{cases} \tag{6.4}$$

where $m_{ij} - \alpha_{ij}, \alpha_{ij}, \beta_{ij} \geq 0$, for all $(i, j) \in A$,

Step 4 The solution of the fuzzy optimal solution of the fuzzy linear programming problem obtained in Step 3, i.e. (6.4), can be obtained by solving the following crisp linear programming problem:

$$
\begin{cases}
min(km_0 + l\alpha_0 + r\beta_0)\\
\text{subject to:}\\
\sum_{j:(i,j)\in A} m_{ij} = m_i;\ i = 1,\dots,m\\
\sum_{j:(i,j)\in A} \alpha_{ij} = \alpha_i;\ i = 1,\dots,m\\
\sum_{j:(i,j)\in A} \beta_{ij} = \beta_i;\ i = 1,\dots,m\\
\sum_{j:(i,j)\in A} m_{ij} - \sum_{j:(j,i)\in A} m_{ji} = m_{m+i};\ i = 1,\dots,l\\
\sum_{j:(i,j)\in A} \alpha_{ij} - \sum_{j:(j,i)\in A} \alpha_{ji} = \alpha_{m+i};\ i = 1,\dots,l\\
\sum_{j:(i,j)\in A} \beta_{ij} - \sum_{j:(j,i)\in A} \beta_{ji} = \beta_{m+i};\ i = 1,\dots,l\\
\sum_{i:(i,j)\in A} m_{ij} = m'_j;\ j = 1,\dots,t\\
\sum_{i:(i,j)\in A} \alpha_{ij} = \alpha'_j;\ j = 1,\dots,t\\
\sum_{i:(i,j)\in A} \beta_{ij} = \beta'_j;\ j = 1,\dots,t\\
\sum_{i:(i,j)\in A} m_{ij} - \sum_{i:(j,i)\in A} m_{ji} = m_{t+j};\ j = 1,\dots,q\\
\sum_{i:(i,j)\in A} \alpha_{ij} - \sum_{i:(j,i)\in A} \alpha_{ji} = \alpha_{t+j};\ j = 1,\dots,q\\
\sum_{i:(i,j)\in A} \beta_{ij} - \sum_{i:(j,i)\in A} \beta_{ji} = \beta_{t+j};\ j = 1,\dots,q\\
\sum_{j:(i,j)\in A} m_{ij} = \sum_{j:(j,i)\in A} m_{ji};\ i = 1,\dots,r\\
\sum_{j:(i,j)\in A} \alpha_{ij} = \sum_{j:(j,i)\in A} \alpha_{ji};\ i = 1,\dots,r\\
\sum_{j:(i,j)\in A} \beta_{ij} = \sum_{j:(j,i)\in A} \beta_{ji};\ i = 1,\dots,r
\end{cases}
\tag{6.5}
$$

that is, with the constraints of the problem (6.4), and where: $k = q_1\vartheta^{n_1}, l = q_2\vartheta^{n_2}$, $r = q_3\vartheta^{n_3}$, and $q_1, q_2, q_3 \in \mathbb{Q}^+$, $n_1 \neq n_2 \neq n_3$ are non-negative integers and ϑ is a positive real number.

Step 5 Solve the crisp linear programming problem, obtained in Step 4, i.e. (6.5) to find the optimal solution $\{m_{ij}, \alpha_{ij}, \beta_{ij}\}$.

Step 6 Introduce the values of $m_{ij}, \alpha_{ij}, \beta_{ij}$ obtained in Step 5 into

$$\tilde{x}_{ij} = (m_{ij}, \alpha_{ij}, \beta_{ij})_{LR}$$

to find the fuzzy optimal solution $\{\tilde{x}_{ij}\}$.

Step 7 Introduce the values of $\tilde{x}_{ij}$ obtained in Step 6 into

$$\sum_{(i,j)\in A} (\tilde{c}_{ij} \otimes \tilde{x}_{ij})$$

to find the minimum total fuzzy transportation cost sought.

Finally, some important remark is due:

Remark 13 The authors of the source method, cf. Ghatee and Hashemi [1], use the multiplication proposed by Wagenknecht and Hampel [7] which is defined as follows: if $\tilde{A}_1 = (m, \alpha, \beta)_{LR}$ and $\tilde{A}_2 = (m_1, \alpha_1, \beta_1)_{LR}$ are two non-negative LR fuzzy numbers, then

$$\tilde{A}_1 \otimes \tilde{A}_2 \simeq (mm_1, m\alpha_1 + m_1\alpha - \kappa_1\alpha\alpha_1, m\beta_1 + m_1\beta + \kappa_2\beta\beta_1)_{LR} \tag{6.6}$$

where:

$$\kappa_1 = \frac{\int_0^1 [L^{-1}(t)]^3 dt}{\int_0^1 [L^{-1}(t)]^2 dt}$$

and

$$\kappa_2 = \frac{\int_0^1 [R^{-1}(t)]^3 dt}{\int_0^1 [R^{-1}(t)]^2 dt}$$

It can be shown that the above multiplication (6.6) is an approximation of the multiplication presented in Sect. 5.1, (5.1).

6.3 On Some Limitations of the Existing Methods

In this section, the limitations of the Ghatee and Hashemi's [1] method, which has been considered as a point of departure and presented in Sect. 6.2, as well as the improved methods proposed in the previous chapters, can be briefly summarized as follows:

1. The Ghatee and Hashemi [1] method is proposed for solving the balanced fully fuzzy transshipment problems. Since, the balanced fully fuzzy transportation problems are a special type of fully fuzzy transshipment problems, then Ghatee and Hashemi's [1] method can also be used to find a fuzzy optimal solution of the balanced fully fuzzy transportation problems. However, Ghatee and Hashemi's [1] method can neither be used to find a fuzzy optimal solution of the unbalanced fully fuzzy transportation problems or to find a fuzzy optimal solution of the unbalanced fully fuzzy transshipment problems. For instance, the balanced fully fuzzy transportation problem presented in Example 18, can be solved by Ghatee and Hashemi's [1] method but the unbalanced fully fuzzy transportation problems presented in Examples 14, and 17, can not be solved by using this method. Similarly, the existing balanced fully fuzzy transshipment problem (cf. Example 17, Ghatee and Hashemi's [2]) can be solved by the Ghatee and Hashemi's [1] method in question. On the other hand, the unbalanced fully fuzzy transshipment problem to be illustrated in Example 19, can not be solved by using Ghatee and Hashemi's [1] method.

The following examples will briefly outline the essence of the particular problems mentioned above.

Example 18 In a company there are two sources S_1 and S_2 and three destinations D_1, D_2 and D_3. The fuzzy cost for transporting one unit quantity of the product from the ith source to jth destination is $\tilde{c}_{ij}$, $i, j = 1, 2$, where:

$$[\tilde{c}_{ij}]_{2\times 3} = \begin{pmatrix} (20, 30, 10, 10)_{LR} & (60, 70, 10, 20)_{LR} & (90, 110, 10, 10)_{LR} \\ (70, 80, 10, 10)_{LR} & (80, 100, 10, 20)_{LR} & (30, 50, 10, 10)_{LR} \end{pmatrix}$$

The fuzzy availabilities of the product at source s_1 and source s_2 are given as the following LR fuzzy numbers, respectively: $(90, 100, 20, 30)_{LR}$ and $(60, 70, 20, 30)_{LR}$ and the fuzzy demand of the product at destination d_1 and destination d_2 are given as the following LR fuzzy numbers, respectively: first, second and third destinations are $(40, 50, 10, 20)_{LR}$, $(30, 40, 10, 10)_{LR}$ and $(80, 80, 20, 30)_{LR}$ respectively, where, $L(x) = R(x) = \max\{0, 1 - x\}$.

The company wants to determine the fuzzy quantity of the product that should be transported from each source to each destination so that the total fuzzy transportation cost be minimized.

Example 19 Suppose that we have a network with five nodes in which there is one purely source node (1), one source node (2), one transition node (3) and two purely destination nodes (4) and (5).

The fuzzy cost for transporting one unit quantity of the product from one node to another is shown on the respective arcs. The fuzzy availabilities and the fuzzy demands are represented by the following LR flat fuzzy numbers:

- Fuzzy availabilities:

$$\tilde{a}_1 = (40, 40, 10, 20)_{LR},$$
$$\tilde{a}_2 = (30, 40, 10, 30)_{LR}.$$

- Fuzzy demands:

$$\tilde{b}_4 = (20, 30, 10, 10)_{LR},$$
$$\tilde{b}_5 = (60, 70, 30, 10)_{LR},$$

where $L(x) = R(x) = \max\{0, 1 - x^4\}$.

We seek the fuzzy optimum shipping schedule.

Notice that the methods proposed in the previous chapters can only be used to find a fuzzy optimal solution of the fully fuzzy transportation problems. However, since the fully fuzzy transshipment problems are generalization of the fully fuzzy

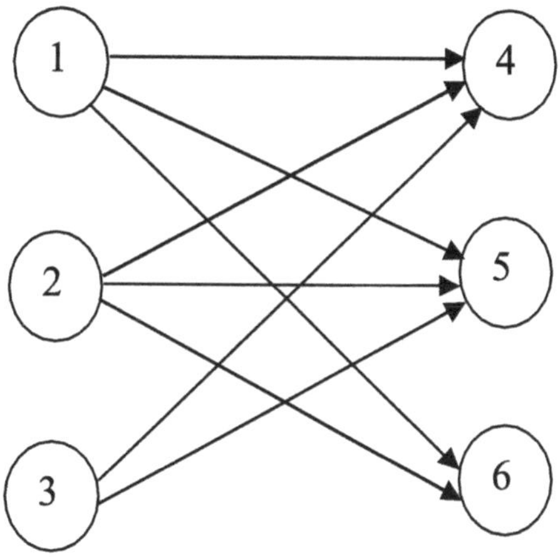

Fig. 6.1 A network representing different shipping routes for Example 18

transportation problems, then the methods, proposed in the previous chapters can not be used for solving the fully fuzzy transshipment problems. Therefore, we have to resort to other methods that will be presented in the sequel (Fig. 6.1).

6.4 New Methods

As we have indicated in the previous section (Sect. 3.3), both Ghatee and Hashemi's [1] method as well as other methods for solving the (balanced) fully fuzzy transshipment problems with all the parameters represented by LR fuzy numbers that have been outlined in Sect. 6.3 have some limitations. We will therefore in this section present two new methods to solve such problems, that is to find a fuzzy optimal solution the fully fuzzy transshipment problems in which all the parameters are represented by LR flat fuzzy numbers. Moreover, we will present advantages of these new methods as compared to the well known Ghatee and Hashemi's [1] method and methods proposed in the previous chapters. To be more specific, we will present the new methods in their versions similar to what we have adopted in previous chapters, that is, one based on a fuzzy linear programming formulation and one based on the tabular representation.

6.4.1 New Method Based on the Fuzzy Linear Programming Formulation

In this section, a new method for finding a fuzzy optimal solution of the fully fuzzy transshipment problems in which all the parameters are represented by LR flat fuzzy is presented, based on a fuzzy linear programming formulation.

The new method can be presented via the following steps:

Step 1
Find the total fuzzy availability

$$\sum_{i=1}^{m} \tilde{a}_i \oplus \sum_{i=1}^{l} \tilde{a}_{m+i}$$

and the total fuzzy demand

$$\sum_{j=1}^{t} \tilde{b}_j \oplus \sum_{j=1}^{q} \tilde{b}_{t+j}.$$

Let

$$\sum_{i=1}^{m} \tilde{a}_i \oplus \sum_{i=1}^{l} \tilde{a}_{m+i} = (m, n, \alpha, \beta)_{LR}$$

and

$$\sum_{j=1}^{t} \tilde{b}_j \oplus \sum_{j=1}^{q} \tilde{b}_{t+j} = (m', n', \alpha', \beta')_{LR}.$$

Use Definition 18, to check if the problem is balanced or not, that is, if

$$\sum_{i=1}^{m} \tilde{a}_i \oplus \sum_{i=1}^{l} \tilde{a}_{m+i} = \sum_{j=1}^{t} \tilde{b}_j \oplus \sum_{j=1}^{q} \tilde{b}_{t+j} \quad (6.7)$$

or

$$\sum_{i=1}^{m} \tilde{a}_i \oplus \sum_{i=1}^{l} \tilde{a}_{m+i} \neq \sum_{j=1}^{t} \tilde{b}_j \oplus \sum_{j=1}^{q} \tilde{b}_{t+j} \quad (6.8)$$

And then:

Case (1) If the problem is balanced, i.e., when (6.7) holds, then Go to Step 2.
Case (2) If the problem is unbalanced, i.e. when (6.8) holds, then convert the unbalanced problem into the balanced problem as follows:

Case (2a) If:

$$m - \alpha \leq m' - \alpha',$$
$$\alpha \leq \alpha',$$
$$n - m \leq n' - m',$$
$$\beta \leq \beta',$$

then introduce a dummy purely source node with the fuzzy availability

$$(m' - m, n' - n, \alpha' - \alpha, \beta' - \beta)_{LR}.$$

Assume the fuzzy cost for transporting one unit quantity of the product from the introduced dummy purely source node to all purely destination nodes and all intermediate nodes as the zero LR flat fuzzy number. Go to Step 2.
Case (2b)
If

$$\begin{aligned}
&m - \alpha \geq m' - \alpha', \\
&\alpha \geq \alpha', \\
&n - m \geq n' - m', \\
&\beta \geq \beta',
\end{aligned}$$

then introduce a dummy purely destination node with the fuzzy demand

$$(m - m', n - n', \alpha - \alpha', \beta - \beta')_{LR}.$$

Assume the fuzzy cost for transporting one unit quantity of the product from all purely source nodes and intermediate nodes to the introduced dummy purely destination node as the zero LR flat fuzzy number. Go to Step 2.
Case (2c)
If neither Case (a) nor Case (b) is satisfied, then introduce a dummy purely source node with the fuzzy availability given as the following LR fuzzy number

$$\begin{aligned}
&(\max\{0, (m' - \alpha') - (m - \alpha)\} + \max\{0, (\alpha' - \alpha)\}, \\
&\max\{0, (m' - \alpha') - (m - \alpha)\} + \max\{0, (\alpha' - \alpha)\}+ \\
&\max\{0, (n' - m') - (n - m)\}, \max\{0, (\alpha' - \alpha)\}, \\
&\max\{0, (\beta' - \beta)\})_{LR}
\end{aligned}$$

and also introduce a dummy purely destination node with the fuzzy demand given by the following LR fuzzy number

$$\begin{aligned}
&(\max\{0, (m - \alpha) - (m' - \alpha')\} + \max\{0, (\alpha - \alpha')\}, \\
&\max\{0, (m - \alpha) - (m' - \alpha')\} + \max\{0, (\alpha - \alpha')\}+ \\
&\max\{0, (n - m) - (n' - m')\}, \max\{0, (\alpha - \alpha')\}, \\
&\max\{0, (\beta - \beta')\})_{LR}
\end{aligned}$$

Assume the fuzzy cost for transporting one unit quantity of the product from the introduced dummy purely source node to all intermediate nodes, existing purely destination nodes and introduced dummy purely destination node as the zero LR flat fuzzy number. Moreover, similarly, assume the fuzzy cost for transporting one unit quantity of the product from all intermediate nodes, existing purely source nodes and introduced dummy purely source node to the introduced dummy purely destination node as the zero LR flat fuzzy number. Go to Step 2.

Step 2
Formulate the balanced fully fuzzy transshipment problem, obtained in Step 1, in the form of the fuzzy linear programming problem given as Sect. 6.4.1, that is,

$$\begin{cases} min \sum\limits_{(i,j)\in A} (\tilde{c}_{ij} \otimes \tilde{x}_{ij}) \\ \text{subject to:} \\ \sum\limits_{j:(i,j)\in A} \tilde{x}_{ij} = \tilde{a}_i;\, i = 1, \ldots, m, \\ \sum\limits_{j:(i,j)\in A} \tilde{x}_{ij} \ominus_H \sum\limits_{j:(j,i)\in A} \tilde{x}_{ji} = \tilde{a}_{m+i};\, i = 1, \ldots, l \\ \sum\limits_{i:(i,j)\in A} \tilde{x}_{ij} = \tilde{b}_j;\, j = 1 \ldots, t \\ \sum\limits_{i:(i,j)\in A} \tilde{x}_{ij} \ominus_H \sum\limits_{i:(j,i)\in A} \tilde{x}_{ji} = \tilde{b}_{t+j};\, j = 1, \ldots, q \\ \sum\limits_{j:(i,j)\in A} \tilde{x}_{ij} = \sum\limits_{j:(j,i)\in A} \tilde{x}_{ji};\, i = 1, \ldots, r \end{cases}$$

where: $\tilde{x}_{ij}$ is a non-negative fuzzy number, for all $(i, j) \in A$; A is the set of arcs (i, j) joining node i to node j.
Step 3
By assuming

$$\begin{aligned} \tilde{c}_{ij} &= (m'_{ij}, n'_{ij}, \alpha'_{ij}, \beta'_{ij})_{LR}, \\ \tilde{x}_{ij} &= (m_{ij}, n_{ij}, \alpha_{ij}, \beta_{ij})_{LR}, \\ \tilde{a}_i &= (m_i, n_i, \alpha_i, \beta_i)_{LR}, \\ \tilde{a}_{m+i} &= (m_{m+i}, n_{m+i}, \alpha_{m+i}, \beta_{m+i})_{LR}, \\ \tilde{b}_j &= (m'_j, n'_j, \alpha'_j, \beta'_j)_{LR}, \\ \tilde{b}_{t+j} &= (m_{t+j}, n_{t+j}, \alpha_{t+j}, \beta_{t+j})_{LR}, \end{aligned}$$

the fuzzy linear programming problem Sect. 6.4.1 can be written as:

$$\begin{cases} \min \sum\limits_{(i,j)\in A} \left((m'_{ij}, n'_{ij}, \alpha'_{ij}, \beta'_{ij})_{LR} \otimes (m_{ij}, n_{ij}, \alpha_{ij}, \beta_{ij})_{LR}\right) \\ \text{subject to:} \\ \sum\limits_{j:(i,j)\in A} (m_{ij}, n_{ij}, \alpha_{ij}, \beta_{ij})_{LR} = (m_i, n_i, \alpha_i, \beta_i)_{LR};\, i = 1, \ldots, m \\ \sum\limits_{j:(i,j)\in A} (m_{ij}, n_{ij}, \alpha_{ij}, \beta_{ij})_{LR} \ominus_H \\ \ominus_H \sum\limits_{j:(j,i)\in A} (m_{ji}, n_{ji}, \alpha_{ji}, \beta_{ji})_{LR} = (m_{m+i}, n_{m+i}, \alpha_{m+i}, \beta_{m+i})_{LR};\, i = 1, \ldots, l \\ \sum\limits_{i:(i,j)\in A} (m_{ij}, n_{ij}, \alpha_{ij}, \beta_{ij})_{LR} = (m'_j, n'_j, \alpha'_j, \beta'_j)_{LR};\, j = 1, \ldots, t \\ \sum\limits_{i:(i,j)\in A} (m_{ij}, n_{ij}, \alpha_{ij}, \beta_{ij})_{LR} \ominus_H \sum\limits_{i:(j,i)\in A} (m_{ji}, n_{ji}, \alpha_{ji}, \beta_{ji})_{LR} = \\ = (m_{t+j}, n_{t+j}, \alpha_{t+j}, \beta_{t+j})_{LR};\, j = 1, \ldots, q \\ \sum\limits_{j:(i,j)\in A} (m_{ij}, n_{ij}, \alpha_{ij}, \beta_{ij})_{LR} = \\ = \sum\limits_{j:(j,i)\in A} (m_{ji}, n_{ji}, \alpha_{ji}, \beta_{ji})_{LR};\, i = 1, \ldots, r \end{cases} \tag{6.9}$$

where $(m_{ij}, n_{ij}, \alpha_{ij}, \beta_{ij})_{LR}$ is a non-negative LR flat fuzzy number, for all $(i, j) \in A$.

Step 4 By using the arithmetic operations on LR flat fuzzy numbers defined in Sect. 4.3, and by assuming

$$\sum_{(i,j)\in A} \left(m'_{ij}, n'_{ij}, \alpha'_{ij}, \beta'_{ij})_{LR} \quad \otimes (m_{ij}, n_{ij}, \alpha_{ij}, \beta_{ij})_{LR}\right) = (m_0, n_0, \alpha_0, \beta_0)_{LR},$$

the fuzzy linear programming problem, obtained in Step 3, i.e. (6.9), can be written as the following fuzzy linear programming problem:

$$\begin{cases} \min(m_0, n_0, \alpha_0, \beta_0)_{LR} \\ \text{subject to:} \\ (\sum\limits_{j:(i,j)\in A} m_{ij}, \sum\limits_{j:(i,j)\in A} n_{ij}, \sum\limits_{j:(i,j)\in A} \alpha_{ij}, \sum\limits_{j:(i,j)\in A} \beta_{ij})_{LR} = \\ = (m_i, n_i, \alpha_i, \beta_i)_{LR};\ i = 1, \ldots, m \\ (\sum\limits_{j:(i,j)\in A} m_{ij}, \sum\limits_{j:(i,j)\in A} n_{ij}, \sum\limits_{j:(i,j)\in A} \alpha_{ij}, \sum\limits_{j:(i,j)\in A} \beta_{ij})_{LR} \ominus_H \\ \ominus_H (\sum\limits_{j:(j,i)\in A} m_{ji}, \sum\limits_{j:(j,i)\in A} n_{ji}, \\ \sum\limits_{j:(j,i)\in A} \alpha_{ji}, \sum\limits_{j:(j,i)\in A} \beta_{ji})_{LR} = \\ = (m_{m+i}, n_{m+i}, \alpha_{m+i}, \beta_{m+i})_{LR};\ i = 1, \ldots, l \\ (\sum\limits_{i:(i,j)\in A} m_{ij}, \sum\limits_{i:(i,j)\in A} n_{ij}, \sum\limits_{i:(i,j)\in A} \alpha_{ij}, \sum\limits_{i:(i,j)\in A} \beta_{ij})_{LR} = \\ = (m'_j, n'_j, \alpha'_j, \beta'_j)_{LR};\ j = 1, \ldots, t \\ (\sum\limits_{i:(i,j)\in A} m_{ij}, \sum\limits_{i:(i,j)\in A} n_{ij}, \sum\limits_{i:(i,j)\in A} \alpha_{ij}, \sum\limits_{i:(i,j)\in A} \beta_{ij})_{LR} \ominus_H \\ \ominus_H (\sum\limits_{i:(j,i)\in A} m_{ji}, \sum\limits_{i:(j,i)\in A} n_{ji}, \sum\limits_{i:(j,i)\in A} \alpha_{ji}, \\ \sum\limits_{i:(j,i)\in A} \beta_{ji})_{LR} = (m_{t+j}, n_{t+j}, \alpha_{t+j}, \beta_{t+j})_{LR};\ j = 1, \ldots, q \\ (\sum\limits_{j:(i,j)\in A} m_{ij}, \sum\limits_{j:(i,j)\in A} n_{ij}, \sum\limits_{j:(i,j)\in A} \alpha_{ij}, \sum\limits_{j:(i,j)\in A} \beta_{ij})_{LR} = \\ (\sum\limits_{j:(j,i)\in A} m_{ji}, \sum\limits_{j:(j,i)\in A} n_{ji}, \sum\limits_{j:(j,i)\in A} \alpha_{ji}, \sum\limits_{j:(j,i)\in A} \beta_{ji})_{LR};\ i = 1, \ldots, r \end{cases} \tag{6.10}$$

where $(m_{ij}, n_{ij}, \alpha_{ij}, \beta_{ij})_{LR}$ is a non-negative LR flat fuzzy number, for all $(i, j) \in A$,

Step 5

By using Definitions 18, 19, and Remark 11, the fuzzy linear programming problem, obtained in Step 4, (6.10), can be converted into the following fuzzy linear programming problem

$$\left\{\begin{array}{l}
\min(m_0, n_0, \alpha_0, \beta_0)_{LR} \\
\text{subject to:} \\
\sum\limits_{j:(i,j)\in A} m_{ij} = m_i; i = 1, \ldots, m \\
\sum\limits_{j:(i,j)\in A} n_{ij} = n_i; i = 1, \ldots, m \\
\sum\limits_{j:(i,j)\in A} \alpha_{ij} = \alpha_i; i = 1, \ldots, m \\
\sum\limits_{j:(i,j)\in A} \beta_{ij} = \beta_i; i = 1, \ldots, m \\
\sum\limits_{j:(i,j)\in A} m_{ij} - \sum\limits_{j:(j,i)\in A} m_{ji} = m_{m+i}; i = 1, \ldots, l \\
\sum\limits_{j:(i,j)\in A} n_{ij} - \sum\limits_{j:(j,i)\in A} n_{ji} = n_{m+i}; i = 1, \ldots, l \\
\sum\limits_{j:(i,j)\in A} \alpha_{ij} - \sum\limits_{j:(j,i)\in A} \alpha_{ji} = \alpha_{m+i}; i = 1, \ldots, l \\
\sum\limits_{j:(i,j)\in A} \beta_{ij} - \sum\limits_{j:(j,i)\in A} \beta_{ji} = \beta_{m+i}; i = 1, \ldots, l \\
\sum\limits_{i:(i,j)\in A} m_{ij} = m'_j; j = 1, \ldots, t \\
\sum\limits_{i:(i,j)\in A} n_{ij} = n'_j; j = 1, \ldots, t \\
\sum\limits_{i:(i,j)\in A} \alpha_{ij} = \alpha'_j; j = 1, \ldots, t \\
\sum\limits_{i:(i,j)\in A} \beta_{ij} = \beta'_j; j = 1, \ldots, t \\
\sum\limits_{i:(i,j)\in A} m_{ij} - \sum\limits_{i:(j,i)\in A} m_{ji} = m_{t+j} : j = 1, \ldots, q \\
\sum\limits_{i:(i,j)\in A} n_{ij} - \sum\limits_{i:(j,i)\in A} n_{ji} = n_{t+j}; j = 1, \ldots, q \\
\sum\limits_{i:(i,j)\in A} \alpha_{ij} - \sum\limits_{i:(j,i)\in A} \alpha_{ji} = \alpha_{t+j}; j = 1, \ldots, q \\
\sum\limits_{i:(i,j)\in A} \beta_{ij} - \sum\limits_{i:(j,i)\in A} \beta_{ji} = \beta_{t+j}; j = 1, \ldots, q \\
\sum\limits_{j:(i,j)\in A} m_{ij} = \sum\limits_{j:(j,i)\in A} m_{ji}; i = 1, \ldots, r \\
\sum\limits_{j:(i,j)\in A} n_{ij} = \sum\limits_{j:(j,i)\in A} n_{ji}; i = 1, \ldots, r \\
\sum\limits_{j:(i,j)\in A} \alpha_{ij} = \sum\limits_{j:(j,i)\in A} \alpha_{ji}; i = 1, \ldots, r \\
\sum\limits_{j:(i,j)\in A} \beta_{ij} = \sum\limits_{j:(j,i)\in A} \beta_{ji}; i = 1, \ldots, r
\end{array}\right. \tag{6.11}$$

where: $m_{ij} - \alpha_{ij}, n_{ij} - m_{ij}, \alpha_{ij}, \beta_{ij} \geq 0$, for all $(i, j) \in A$,

Step 6 As it has been shown in Step 6 of the method proposed in Sect. 4.8.1, a fuzzy optimal solution of fuzzy linear programming problem (6.11), can be obtained by solving the following crisp linear programming problem:

$$\begin{cases} \min \Re(m_0, n_0, \alpha_0, \beta_0)_{LR} \\ \text{subject to:} \\ \sum\limits_{j:(i,j)\in A} m_{ij} = m_i;\ i = 1, \ldots, m \\ \sum\limits_{j:(i,j)\in A} n_{ij} = n_i;\ i = 1, \ldots, m \\ \sum\limits_{j:(i,j)\in A} \alpha_{ij} = \alpha_i;\ i = 1, \ldots, m \\ \sum\limits_{j:(i,j)\in A} \beta_{ij} = \beta_i;\ i = 1, \ldots, m \\ \sum\limits_{j:(i,j)\in A} m_{ij} - \sum\limits_{j:(j,i)\in A} m_{ji} = m_{m+i};\ i = 1, \ldots, l \\ \sum\limits_{j:(i,j)\in A} n_{ij} - \sum\limits_{j:(j,i)\in A} n_{ji} = n_{m+i};\ i = 1, \ldots, l \\ \sum\limits_{j:(i,j)\in A} \alpha_{ij} - \sum\limits_{j:(j,i)\in A} \alpha_{ji} = \alpha_{m+i};\ i = 1, \ldots, l \\ \sum\limits_{j:(i,j)\in A} \beta_{ij} - \sum\limits_{j:(j,i)\in A} \beta_{ji} = \beta_{m+i};\ i = 1, \ldots, l \\ \sum\limits_{i:(i,j)\in A} m_{ij} = m'_j;\ j = 1, \ldots, t \\ \sum\limits_{i:(i,j)\in A} n_{ij} = n'_j;\ j = 1, \ldots, t \\ \sum\limits_{i:(i,j)\in A} \alpha_{ij} = \alpha'_j;\ j = 1, \ldots, t \\ \sum\limits_{i:(i,j)\in A} \beta_{ij} = \beta'_j;\ j = 1, \ldots, t \\ \sum\limits_{i:(i,j)\in A} m_{ij} - \sum\limits_{i:(j,i)\in A} m_{ji} = m_{t+j} :\ j = 1, \ldots, q \\ \sum\limits_{i:(i,j)\in A} n_{ij} - \sum\limits_{i:(j,i)\in A} n_{ji} = n_{t+j};\ j = 1, \ldots, q \\ \sum\limits_{i:(i,j)\in A} \alpha_{ij} - \sum\limits_{i:(j,i)\in A} \alpha_{ji} = \alpha_{t+j};\ j = 1, \ldots, q \\ \sum\limits_{i:(i,j)\in A} \beta_{ij} - \sum\limits_{i:(j,i)\in A} \beta_{ji} = \beta_{t+j};\ j = 1, \ldots, q \\ \sum\limits_{j:(i,j)\in A} m_{ij} = \sum\limits_{j:(j,i)\in A} m_{ji};\ i = 1, \ldots, r \\ \sum\limits_{j:(i,j)\in A} n_{ij} = \sum\limits_{j:(j,i)\in A} n_{ji};\ i = 1, \ldots, r \\ \sum\limits_{j:(i,j)\in A} \alpha_{ij} = \sum\limits_{j:(j,i)\in A} \alpha_{ji};\ i = 1, \ldots, r \\ \sum\limits_{j:(i,j)\in A} \beta_{ij} = \sum\limits_{j:(j,i)\in A} \beta_{ji};\ i = 1, \ldots, r \end{cases} \tag{6.12}$$

Step 7

By using the known formula from Liou and Wang [6]:

$$\Re(m_0, n_0, \alpha_0, \beta_0)_{LR} = \\ = \frac{1}{2}\left(\int_0^1 (m_0 - \alpha_0 L^{-1}(\lambda))d\lambda + \int_0^1 (n_0 + \beta_0 R^{-1}(\lambda))d\lambda\right), \tag{6.13}$$

the crisp linear programming problem, obtained in Step 6, i.e. (6.12), can be written as:

$$\left\{\begin{array}{l}\min \frac{1}{2}(\int_0^1 (m_0 - \alpha_0 L^{-1}(\lambda))d\lambda + \int_0^1 (n_0 + \beta_0 R^{-1}(\lambda))d\lambda) \\ \text{subject to:} \\ \sum\limits_{j:(i,j)\in A} m_{ij} = m_i;\ i = 1,\ldots,m \\ \sum\limits_{j:(i,j)\in A} n_{ij} = n_i;\ i = 1,\ldots,m \\ \sum\limits_{j:(i,j)\in A} \alpha_{ij} = \alpha_i;\ i = 1,\ldots,m \\ \sum\limits_{j:(i,j)\in A} \beta_{ij} = \beta_i;\ i = 1,\ldots,m \\ \sum\limits_{j:(i,j)\in A} m_{ij} - \sum\limits_{j:(j,i)\in A} m_{ji} = m_{m+i};\ i = 1,\ldots,l \\ \sum\limits_{j:(i,j)\in A} n_{ij} - \sum\limits_{j:(j,i)\in A} n_{ji} = n_{m+i};\ i = 1,\ldots,l \\ \sum\limits_{j:(i,j)\in A} \alpha_{ij} - \sum\limits_{j:(j,i)\in A} \alpha_{ji} = \alpha_{m+i};\ i = 1,\ldots,l \\ \sum\limits_{j:(i,j)\in A} \beta_{ij} - \sum\limits_{j:(j,i)\in A} \beta_{ji} = \beta_{m+i};\ i = 1,\ldots,l \\ \sum\limits_{i:(i,j)\in A} m_{ij} = m'_j;\ j = 1,\ldots,t \\ \sum\limits_{i:(i,j)\in A} n_{ij} = n'_j;\ j = 1,\ldots,t \\ \sum\limits_{i:(i,j)\in A} \alpha_{ij} = \alpha'_j;\ j = 1,\ldots,t \\ \sum\limits_{i:(i,j)\in A} \beta_{ij} = \beta'_j;\ j = 1,\ldots,t \\ \sum\limits_{i:(i,j)\in A} m_{ij} - \sum\limits_{i:(j,i)\in A} m_{ji} = m_{t+j} : j = 1,\ldots,q \\ \sum\limits_{i:(i,j)\in A} n_{ij} - \sum\limits_{i:(j,i)\in A} n_{ji} = n_{t+j};\ j = 1,\ldots,q \\ \sum\limits_{i:(i,j)\in A} \alpha_{ij} - \sum\limits_{i:(j,i)\in A} \alpha_{ji} = \alpha_{t+j};\ j = 1,\ldots,q \\ \sum\limits_{i:(i,j)\in A} \beta_{ij} - \sum\limits_{i:(j,i)\in A} \beta_{ji} = \beta_{t+j};\ j = 1,\ldots,q \\ \sum\limits_{j:(i,j)\in A} m_{ij} = \sum\limits_{j:(j,i)\in A} m_{ji};\ i = 1,\ldots,r \\ \sum\limits_{j:(i,j)\in A} n_{ij} = \sum\limits_{j:(j,i)\in A} n_{ji};\ i = 1,\ldots,r \\ \sum\limits_{j:(i,j)\in A} \alpha_{ij} = \sum\limits_{j:(j,i)\in A} \alpha_{ji};\ i = 1,\ldots,r \\ \sum\limits_{j:(i,j)\in A} \beta_{ij} = \sum\limits_{j:(j,i)\in A} \beta_{ji};\ i = 1,\ldots,r \end{array}\right. \tag{6.14}$$

Step 8
Solve the crisp linear programming problem, obtained in Step 7, i.e. (6.14), to find an optimal solution $\{m_{ij}, n_{ij}, \alpha_{ij}, \beta_{ij}\}$.
Step 9
Introduce the values of $m_{ij}, n_{ij}, \alpha_{ij}, \beta_{ij}$ into

$$\tilde{x}_{ij} = (m_{ij}, n_{ij}, \alpha_{ij}, \beta_{ij})_{LR}$$

to find a fuzzy optimal solution $\{\tilde{x}_{ij}\}$.

Step 10 Introduce the values of $\tilde{x}_{ij}$, obtained from Step 9, into

$$\sum_{(i,j)\in A} (\tilde{c}_{ij} \otimes \tilde{x}_{ij}),$$

to find the minimum total fuzzy transportation cost.

Since the comparison of fuzzy numbers is of utmost importance for our purposes, the following remark is important.

Remark 14 The reference method used in our discussion, the one due to Ghatee and Hashemi [1] is based on the comparison of LR fuzzy numbers so that it is not suited for cases when the comparison of the LR flat fuzzy numbers is involved. Luckily enough, the method presented by Liou and Wang [6] can be used for comparing both the LR fuzzy numbers and LR flat fuzzy numbers. That is why in Step 6 of the proposed method presented above the latter method, i.e. Liou and Wang [6] is used for comparing the LR flat fuzzy numbers.

6.4.2 New Method Based on the Tabular Representation

In this section, a new method for finding a fuzzy optimal solution of the fully fuzzy transshipment problems in which all the parameters are represented by LR flat fuzzy is presented, based on a tabular representation.

The new method can be presented in an algorithmic way via the following steps:

Step 1 Use Step 1 of the method, proposed in Sect. 6.4.1, to obtain a balanced fully fuzzy transshipment problem. For the convenience of the reader, we will repeat this step:

Find the total fuzzy availability

$$\sum_{i=1}^{m} \tilde{a}_i \oplus \sum_{i=1}^{l} \tilde{a}_{m+i}$$

and the total fuzzy demand

$$\sum_{j=1}^{t} \tilde{b}_j \oplus \sum_{j=1}^{q} \tilde{b}_{t+j}.$$

Let

$$\sum_{i=1}^{m} \tilde{a}_i \oplus \sum_{i=1}^{l} \tilde{a}_{m+i} = (m, n, \alpha, \beta)_{LR}$$

and

$$\sum_{j=1}^{t} \tilde{b}_j \oplus \sum_{j=1}^{q} \tilde{b}_{t+j} = (m', n', \alpha', \beta')_{LR}.$$

Use Definition 18, to check if the problem is balanced or not, that is, if

$$\sum_{i=1}^{m} \tilde{a}_i \oplus \sum_{i=1}^{l} \tilde{a}_{m+i} = \sum_{j=1}^{t} \tilde{b}_j \oplus \sum_{j=1}^{q} \tilde{b}_{t+j} \quad (6.15)$$

or

$$\sum_{i=1}^{m} \tilde{a}_i \oplus \sum_{i=1}^{l} \tilde{a}_{m+i} \neq \sum_{j=1}^{t} \tilde{b}_j \oplus \sum_{j=1}^{q} \tilde{b}_{t+j} \quad (6.16)$$

And then:

Case (1) If the problem is balanced, i.e., when (6.15) holds, then Go to Step 2.
Case (2) If the problem is unbalanced, i.e. when (6.16) holds, then convert the unbalanced problem into the balanced problem as follows:

Case (2a)
If:

$$m - \alpha \leq m' - \alpha',$$
$$\alpha \leq \alpha',$$
$$n - m \leq n' - m',$$
$$\beta \leq \beta',$$

then introduce a dummy purely source node with the fuzzy availability

$$(m' - m, n' - n, \alpha' - \alpha, \beta' - \beta)_{LR}.$$

Assume the fuzzy cost for transporting one unit quantity of the product from the introduced dummy purely source node to all purely destination nodes and all intermediate nodes as the zero LR flat fuzzy number. Go to Step 2.

Case (2b)
If

$$m - \alpha \geq m' - \alpha',$$
$$\alpha \geq \alpha',$$
$$n - m \geq n' - m',$$
$$\beta \geq \beta',$$

then introduce a dummy purely destination node with the fuzzy demand

$$(m - m', n - n', \alpha - \alpha', \beta - \beta')_{LR}.$$

Assume the fuzzy cost for transporting one unit quantity of the product from all purely source nodes and intermediate nodes to the introduced dummy purely destination node as the zero LR flat fuzzy number. Go to Step 2.
Case (2c)
If neither Case (a) nor Case (b) is satisfied, then introduce a dummy purely source node with the fuzzy availability given as the following LR fuzzy number

$$\begin{aligned}&(\max\{0, (m' - \alpha') - (m - \alpha)\} + \max\{0, (\alpha' - \alpha)\},\\ &\max\{0, (m' - \alpha') - (m - \alpha)\} + \max\{0, (\alpha' - \alpha)\}+\\ &\max\{0, (n' - m') - (n - m)\}, \max\{0, (\alpha' - \alpha)\},\\ &\max\{0, (\beta' - \beta)\})_{LR}\end{aligned}$$

and also introduce a dummy purely destination node with the fuzzy demand given by the following LR fuzzy number

$$\begin{aligned}&(\max\{0, (m - \alpha) - (m' - \alpha')\} + \max\{0, (\alpha - \alpha')\},\\ &\max\{0, (m - \alpha) - (m' - \alpha')\} + \max\{0, (\alpha - \alpha')\}+\\ &\max\{0, (n - m) - (n' - m')\}, \max\{0, (\alpha - \alpha')\},\\ &\max\{0, (\beta - \beta')\})_{LR}\end{aligned}$$

Assume the fuzzy cost for transporting one unit quantity of the product from the introduced dummy purely source node to all intermediate nodes, existing purely destination nodes and introduced dummy purely destination node as the zero LR flat fuzzy number. Moreover, similarly, assume the fuzzy cost for transporting one unit quantity of the product from all intermediate nodes, existing purely source nodes and introduced dummy purely source node to the introduced dummy purely destination node as the zero LR flat fuzzy number. Go to Step 2.

Step 2 Represent the balanced fully fuzzy transshipment problem, obtained above in Step 1, in the tabular form as shown in Table 6.1.
Step 3 Convert the balanced fully fuzzy transshipment problem represented by Table 6.1, into the balanced fully fuzzy transportation problem as follows: add some amount of fuzzy buffer stock

$$\tilde{P} = \sum_{i=1}^{m} \tilde{a}_i \oplus \sum_{i=1}^{l} \tilde{a}_{m+i} = (P_1, P_2, \alpha_p, \beta_p)_{LR}$$

or

$$\tilde{P} = \sum_{j=1}^{t} \tilde{b}_j \oplus \sum_{j=1}^{q} \tilde{b}_{t+j} = (P_1, P_2, \alpha_p, \beta_p)_{LR}$$

into the fuzzy availability and fuzzy demand corresponding to intermediate nodes of Table 6.1.

Therefore, after the above addition of the fuzzy buffer stock $\tilde{P}$, the new fuzzy availabilities corresponding to the ith intermediate nodes: N_{S_i}, N_{T_i} and N_{D_i} are, respectively:

$$\tilde{a}'_{m+i} = \tilde{a}_{m+i} \oplus \tilde{P} = (m''_{m+i}, n''_{m+i}, \alpha''_{m+i}, \beta''_{m+i})_{LR},$$

and the new fuzzy demands corresponding to the jth intermediate nodes N_{D_j}, N_{T_j} and N_{S_j} are

$$\tilde{b}'_{t+j} = \tilde{b}_{t+j} \oplus \tilde{P} = (m''_{t+j}, n''_{t+j}, \alpha''_{t+j}, \beta''_{t+j})_{LR},$$

and Table 6.1 in transformed into Table 6.2.

Step 4 Split Table 6.2 into four crisp transportation problem tables, Tables 6.3, 6.4, 6.5 and 6.6, respectively.

The cost for transporting one unit quantity of the product from the ith node to the jth node in these tables, i.e Tables 6.3, 6.4, 6.5 and 6.6, are represented by η_{ij}, ρ_{ij}, δ_{ij} and ξ_{ij}, respectively, where:

$$\eta_{ij} =$$
$$= \frac{1}{2}((m'_{ij} + n'_{ij}) - \alpha'_{ij} \int_0^1 L^{-1}(\lambda)d\lambda + \beta'_{ij} \int_0^1 R^{-1}(\lambda)d\lambda);$$
$$i = 1, \ldots, m + l + q + r;\ j = 1, 2, \ldots, t + q + r + l;$$

$$\rho_{ij} =$$
$$= \frac{1}{2}((m'_{ij} + n'_{ij}) - m'_{ij} \int_0^1 L^{-1}(\lambda)d\lambda + \beta'_{ij} \int_0^1 R^{-1}(\lambda)d\lambda);$$
$$i = 1, \ldots, m + l + q + r;\ j = 1, \ldots, t + q + r + l;$$

$$\delta_{ij} =$$
$$= \frac{1}{2}(n'_{ij} + \beta'_{ij} \int_0^1 R^{-1}(\lambda)d\lambda);$$
$$i = 1, \ldots, m + l + q + r;\ j = 1, \ldots, t + q + r + l,$$

and

$$\xi_{ij} =$$
$$= \frac{1}{2}((n'_{ij} + \beta'_{ij}) \int_0^1 R^{-1}(\lambda)d\lambda);$$
$$i = 1, \ldots, m + l + q + r;\ j = 1, \ldots, t + q + r + l.$$

Step 5 Solve the crisp transportation problems, shown by Tables 6.3, 6.4, 6.5 and 6.6, to find the optimal solution:

$$\begin{aligned}&\{m_{ij}-\alpha_{ij}\},\\&\{\alpha_{ij}\};\\&\{n_{ij}-m_{ij}\};\\&\{\beta_{ij}\}\end{aligned}$$

respectively.

Step 6

Solve the equations, shown in Step 5, to find the values of $m_{ij}, n_{ij}, \alpha_{ij}$ and β_{ij}.

Step 7 Find the fuzzy optimal solution $\{\tilde{x}_{ij}\}$ by introducing the values of $m_{ij}, n_{ij}, \alpha_{ij}, \beta_{ij}$ obtained in Step 6 into

$$\tilde{x}_{ij}=(m_{ij}, n_{ij}, \alpha_{ij}, \beta_{ij})_{LR}.$$

Step 8 Find the minimum total fuzzy transportation cost by putting the values of $\tilde{x}_{ij}$ obtained in Step 7 into

$$\sum_{(i,j)\in A}(\tilde{c}_{ij}\otimes\tilde{x}_{ij}).$$

6.4.3 Advantages of the New Methods

In this section we will briefly discuss the advantages of the methods proposed in this chapter over, first, the well known and often employed method by Ghatee and Hashemi [1] as well as other method presented in the previous chapters.

These advantages can briefly be subsumed as follows:

1. The Ghatee and Hashemi [1] method can be employed for solving both the balanced fully fuzzy transshipment problems and the balanced fully fuzzy transportation problems. However, unfortunatly, it can not be used for solving the unbalanced fully fuzzy transportation problems or the unbalanced fully fuzzy transshipment problems which, in turn, can be solved by the new method proposed in this chapter.
2. The methods, proposed in the previous chapters, can be used to find a fuzzy optimal solution of the fully fuzzy transportation problems but can not be used for solving the fully fuzzy transshipment problems. Since the fully fuzzy transportation problems are a special type of the fully fuzzy transshipment problems, then the methods proposed in this chapter can also be used to find a fuzzy optimal solution of the fully fuzzy transportation problems.

Table 6.1 Tabular representation of the balanced fully fuzzy transshipment problem

	N_{PD_1}	⋯	N_{PD_t}	N_{D_1}	⋯	N_{D_q}	N_{T_1}	⋯	N_{T_r}	N_{S_1}	⋯	N_{S_l}	
N_{PS_1}	$\tilde{c}_{11}$	⋯	$\tilde{c}_{1t}$	$\tilde{c}_{1(t+1)}$	⋯	$\tilde{c}_{1(t+q)}$	$\tilde{c}_{1(t+q+1)}$	⋯	$\tilde{c}_{1(t+q+r)}$	$\tilde{c}_{1(t+q+r+1)}$	⋯	$\tilde{c}_{1(t+q+r+l)}$	$\tilde{a}_1$
⋮	⋮	⋮	⋮	⋮	⋮	⋮	⋮	⋮	⋮	⋮	⋮	⋮	⋮
N_{PS_m}	$\tilde{c}_{m1}$	⋯	$\tilde{c}_{mt}$	$\tilde{c}_{m(t+1)}$	⋯	$\tilde{c}_{m(t+q)}$	$\tilde{c}_{m(t+q+1)}$	⋯	$\tilde{c}_{m(t+q+r)}$	$\tilde{c}_{m(t+q+1)}$	⋯	$\tilde{c}_{m(t+q+r+l)}$	$\tilde{a}_m$
N_{S_1}	$\tilde{c}_{(m+1)1}$	⋮	$\tilde{c}_{(m+1)t}$	$\tilde{c}_{(m+1)(t+1)}$	⋯	$\tilde{c}_{(m+1)(t+q)}$	$\tilde{c}_{(m+1)(t+q+1)}$	⋯	$\tilde{c}_{(m+1)(t+q+r)}$	$\tilde{c}_{(m+1)(t+q+r+1)}$	⋯	$\tilde{c}_{(m+1)(t+q+r+l)}$	$\tilde{a}_{m+1}$
⋮	⋮	⋮	⋮	⋮	⋮	⋮	⋮	⋮	⋮	⋮	⋮	⋮	⋮
N_{S_l}	$\tilde{c}_{(m+l)1}$	⋮	$\tilde{c}_{(m+l)t}$	$\tilde{c}_{(m+l)(t+1)}$	⋯	$\tilde{c}_{(m+l)(t+q)}$	$\tilde{c}_{(m+l)(t+q+1)}$	⋯	$\tilde{c}_{(m+l)(t+q+r)}$	$\tilde{c}_{(m+l)(t+q+r+1)}$	⋯	$\tilde{c}_{(m+l)(t+q+r+l)}$	$\tilde{a}_{m+l}$
N_{D_1}	$\tilde{c}_{(m+l+1)1}$	⋯	$\tilde{c}_{(m+l+1)t}$	$\tilde{c}_{(m+l+1)(t+1)}$	⋯	$\tilde{c}_{(m+l+1)(t+q)}$	$\tilde{c}_{(m+l+1)(t+q+1)}$	⋯	$\tilde{c}_{(m+l+1)(t+q+r)}$	$\tilde{c}_{(m+l+1)(t+q+r+1)}$	⋯	$\tilde{c}_{(m+l+1)(t+q+r+l)}$	–
⋮	⋮	⋮	⋮	⋮	⋮	⋮	⋮	⋮	⋮	⋮	⋮	⋮	–
N_{D_q}	$\tilde{c}_{(m+l+q)1}$	⋯	$\tilde{c}_{(m+l+q)t}$	$\tilde{c}_{(m+l+q)(t+1)}$	⋯	$\tilde{c}_{(m+l+q)(t+q)}$	$\tilde{c}_{(m+l+q)(t+q+1)}$	⋯	$\tilde{c}_{(m+l+q)(t+q+r)}$	$\tilde{c}_{(m+l+q)(t+q+r+1)}$	⋯	$\tilde{c}_{(m+l+q)(t+q+r+l)}$	–
N_{T_1}	$\tilde{c}_{(m+l+q+1)1}$	⋯	$\tilde{c}_{(m+l+q+1)t}$	$\tilde{c}_{(m+l+q+1)(t+1)}$	⋯	$\tilde{c}_{(m+l+q+1)(t+q)}$	$\tilde{c}_{(m+l+q+1)(t+q+1)}$	⋯	$\tilde{c}_{(m+l+q+1)(t+q+r)}$	$\tilde{c}_{(m+l+q+1)(t+q+r+1)}$	⋯	$\tilde{c}_{(m+l+q+1)(t+q+r+l)}$	–
⋮	⋮	⋮	⋮	⋮	⋮	⋮	⋮	⋮	⋮	⋮	⋮	⋮	–
N_{T_r}	$\tilde{c}_{(m+l+q+r)1}$	⋯	$\tilde{c}_{(m+l+q+r)t}$	$\tilde{c}_{(m+l+q+r)(t+1)}$	⋯	$\tilde{c}_{(m+l+q+r)(t+q)}$	$\tilde{c}_{(m+l+q+r)(t+q+r)}$	⋯	$\tilde{c}_{(m+l+q+r)(t+q+1)}$	$\tilde{c}_{(m+l+q+r)(t+q+r+1)}$	⋯	$\tilde{c}_{(m+l+q+r)(t+q+r+l)}$	–
	$\tilde{b}_1$	⋯	$\tilde{b}_t$	$\tilde{b}_{t+1}$	⋯	$\tilde{b}_{t+q}$	–	–	–	–	–	–	

$$\tilde{c}_{ij} = \begin{cases} (0,0,0,0)_{LR} & ;\ \text{If the product is supplied from some intermediate node to same intermediate node} \\ & \text{or} \\ & \text{If the product is supplied from some dummy purely source node to any intermediate node or} \\ & \text{to any purely destination node} \\ & \text{or} \\ & \text{If the product is supplied from any purely source node to dummy purely destination node} \\ & \text{or} \\ & \text{If the product is supplied from any intermediate node to dummy purely destination node} \\ (M,M,0,0)_{LR} & ;\ \text{If the product can not be directly supplied from } i\text{th node to } j\text{th node} \\ & \text{or If the product can not be supplied from } i\text{th node to } j\text{th node} \\ (m'_{ij},n'_{ij},\alpha'_{ij},\beta'_{ij})_{LR} & ;\ \text{If the product can be directly supplied from } i\text{thnode to } j\text{th node} \end{cases}$$

Table 6.2 Tabular representation of the balanced fully fuzzy transportation problem obtained by adding fuzzy buffer stock

	N_{PD_1}	⋯	N_{PD_t}	N_{D_1}	⋯	N_{D_q}	N_{T_1}	⋯	N_{T_r}	N_{S_1}	⋯	N_{S_l}	
N_{PS_1}	$\tilde{c}_{11}$	⋯	$\tilde{c}_{1t}$	$\tilde{c}_{1(t+1)}$	⋯	$\tilde{c}_{1(t+q)}$	$\tilde{c}_{1(t+q+1)}$	⋯	$\tilde{c}_{1(t+q+r)}$	$\tilde{c}_{1(t+q+r+1)}$	⋯	$\tilde{c}_{1(t+q+r+l)}$	$\tilde{a}_1$
⋮	⋮	⋮	⋮	⋮	⋮	⋮	⋮	⋮	⋮	⋮	⋮	⋮	⋮
N_{PS_m}	$\tilde{c}_{m1}$	⋯	$\tilde{c}_{mt}$	$\tilde{c}_{m(t+1)}$	⋯	$\tilde{c}_{m(t+q)}$	$\tilde{c}_{m(t+q+1)}$	⋯	$\tilde{c}_{m(t+q+r)}$	$\tilde{c}_{m(t+q+1)}$	⋯	$\tilde{c}_{m(t+q+r+l)}$	$\tilde{a}_m$
N_{S_1}	$\tilde{c}_{(m+1)1}$	⋮	$\tilde{c}_{(m+1)t}$	$\tilde{c}_{(m+1)(t+1)}$	⋯	$\tilde{c}_{(m+1)(t+q)}$	$\tilde{c}_{(m+1)(t+q+1)}$	⋯	$\tilde{c}_{(m+1)(t+q+r)}$	$\tilde{c}_{(m+1)(t+q+r+1)}$	⋯	$\tilde{c}_{(m+1)(t+q+r+l)}$	$\tilde{a}'_{m+1}$
⋮	⋮	⋮	⋮	⋮	⋮	⋮	⋮	⋮	⋮	⋮	⋮	⋮	⋮
N_{S_l}	$\tilde{c}_{(m+l)1}$	⋮	$\tilde{c}_{(m+l)t}$	$\tilde{c}_{(m+l)(t+1)}$	⋯	$\tilde{c}_{(m+l)(t+q)}$	$\tilde{c}_{(m+l)(t+q+1)}$	⋯	$\tilde{c}_{(m+l)(t+q+r)}$	$\tilde{c}_{(m+l)(t+q+r+1)}$	⋯	$\tilde{c}_{(m+l)(t+q+r+l)}$	$\tilde{a}'_{m+l}$
N_{D_1}	$\tilde{c}_{(m+l+1)1}$	⋯	$\tilde{c}_{(m+l+1)t}$	$\tilde{c}_{(m+l+1)(t+1)}$	⋯	$\tilde{c}_{(m+l+1)(t+q)}$	$\tilde{c}_{(m+l+1)(t+q+1)}$	⋯	$\tilde{c}_{(m+l+1)(t+q+r)}$	$\tilde{c}_{(m+l+1)(t+q+r+1)}$	⋯	$\tilde{c}_{(m+l+1)(t+q+r+l)}$	$\widetilde{P}$
⋮	⋮	⋮	⋮	⋮	⋮	⋮	⋮	⋮	⋮	⋮	⋮	⋮	⋮
N_{D_q}	$\tilde{c}_{(m+l+q)1}$	⋯	$\tilde{c}_{(m+l+q)t}$	$\tilde{c}_{(m+l+q)(t+1)}$	⋯	$\tilde{c}_{(m+l+q)(t+q)}$	$\tilde{c}_{(m+l+q)(t+q+1)}$	⋯	$\tilde{c}_{(m+l+q)(t+q+r)}$	$\tilde{c}_{(m+l+q)(t+q+r+1)}$	⋯	$\tilde{c}_{(m+l+q)(t+q+r+l)}$	$\widetilde{P}$
N_{T_1}	$\tilde{c}_{(m+l+q+1)1}$	⋯	$\tilde{c}_{(m+l+q+1)t}$	$\tilde{c}_{(m+l+q+1)(t+1)}$	⋯	$\tilde{c}_{(m+l+q+1)(t+q)}$	$\tilde{c}_{(m+l+q+1)(t+q+1)}$	⋯	$\tilde{c}_{(m+l+q+1)(t+q+r)}$	$\tilde{c}_{(m+l+q+1)(t+q+r+1)}$	⋯	$\tilde{c}_{(m+l+q+1)(t+q+r+l)}$	$\widetilde{P}$
⋮	⋮	⋮	⋮	⋮	⋮	⋮	⋮	⋮	⋮	⋮	⋮	⋮	⋮
N_{T_r}	$\tilde{c}_{(m+l+q+r)1}$	⋯	$\tilde{c}_{(m+l+q+r)t}$	$\tilde{c}_{(m+l+q+r)(t+1)}$	⋯	$\tilde{c}_{(m+l+q+r)(t+q)}$	$\tilde{c}_{(m+l+q+r)(t+q+r)}$	⋯	$\tilde{c}_{(m+l+q+r)(t+q+1)}$	$\tilde{c}_{(m+l+q+r)(t+q+r+1)}$	⋯	$\tilde{c}_{(m+l+q+r)(t+q+r+l)}$	$\widetilde{P}$
	$\widetilde{b}_1$	⋯	$\widetilde{b}_t$	$\widetilde{b}'_{t+1}$	⋯	$\widetilde{b}'_{t+q}$	$\widetilde{P}$	⋯	$\widetilde{P}$	$\widetilde{P}$	⋯	$\widetilde{P}$	

Table 6.3 Tabular representation of the first crisp transportation table

	N_{PD_1}	$\cdots$	N_{PD_t}	N_{D_1}	$\cdots$	N_{D_q}	N_{T_1}	$\cdots$	N_{T_r}	N_{S_1}	$\cdots$	N_{S_l}	
N_{PS_1}	η_{11}	$\cdots$	η_{1t}	$\eta_{1(t+1)}$	$\cdots$	$\eta_{1(t+q)}$	$\eta_{1(t+q+1)}$	$\cdots$	$\eta_{1(t+q+r)}$	$\eta_{1(t+q+r+1)}$	$\cdots$	$\eta_{1(t+q+r+l)}$	$m_1-\alpha_1$
$\vdots$	$\vdots$	$\vdots$	$\vdots$	$\vdots$	$\vdots$	$\vdots$	$\vdots$	$\vdots$	$\vdots$	$\vdots$	$\vdots$	$\vdots$	$\vdots$
N_{PS_m}	η_{m1}	$\cdots$	η_{mt}	$\eta_{m(t+1)}$	$\cdots$	$\eta_{m(t+q)}$	$\eta_{m(t+q+1)}$	$\cdots$	$\eta_{m(t+q+r)}$	$\eta_{m(t+q+1)}$	$\cdots$	$\eta_{m(t+q+r+l)}$	m_m- α_m
N_{S_1}	$\eta_{(m+1)1}$	$\vdots$	$\eta_{(m+1)t}$	$\eta_{(m+1)}$ $(t+1)$	$\cdots$	$\eta_{(m+1)}$ $(t+q)$	$\eta_{(m+1)}$ $(t+q+1)$	$\cdots$	$\eta_{(m+1)}$ $(t+q+r)$	$\eta_{(m+1)}$ $(t+q+r+1)$	$\cdots$	$\eta_{(m+1)}$ $(t+q+r+l)$	$m''_{m+1}-$ α''_{m+1}
$\vdots$	$\vdots$	$\vdots$	$\vdots$	$\vdots$	$\vdots$	$\vdots$	$\vdots$	$\vdots$	$\vdots$	$\vdots$	$\vdots$	$\vdots$	$\vdots$
N_{S_l}	$\eta_{(m+l)1}$	$\vdots$	$\eta_{(m+l)t}$	$\eta_{(m+l)}$ $(t+1)$	$\cdots$	$\eta_{(m+l)}$ $(t+q)$	$\eta_{(m+l)}$ $(t+q+1)$	$\cdots$	$\eta_{(m+l)}$ $(t+q+r)$	$\eta_{(m+l)}$ $(t+q+r+1)$	$\cdots$	$\eta_{(m+l)}$ $(t+q+r+l)$	$m''_{m+l}-$ α''_{m+l}
N_{D_1}	$\eta_{(m+l+1)1}$	$\cdots$	$\eta_{(m+l+1)t}$	$\eta_{(m+l+1)}$ $(t+1)$	$\cdots$	$\eta_{(m+l+1)}$ $(t+q)$	$\eta_{(m+l+1)}$ $(t+q+1)$	$\cdots$	$\eta_{(m+l+1)}$ $(t+q+r)$	$\eta_{(m+l+1)}$ $(t+q+r+1)$	$\cdots$	$\eta_{(m+l+1)}$ $(t+q+r+l)$	$P_1-\alpha p$
$\vdots$	$\vdots$	$\vdots$	$\vdots$	$\vdots$	$\vdots$	$\vdots$	$\vdots$	$\vdots$	$\vdots$	$\vdots$	$\vdots$	$\vdots$	$\vdots$
N_{D_q}	$\eta_{(m+l+q)1}$	$\cdots$	$\eta_{(m+l+q)t}$	$\eta_{(m+l+q)}$ $(t+1)$	$\cdots$	$\eta_{(m+l+q)}$ $(t+q)$	$\eta_{(m+l+q)}$ $(t+q+1)$	$\cdots$	$\eta_{(m+l+q)}$ $(t+q+r)$	$\eta_{(m+l+q)}$ $(t+q+r+1)$	$\cdots$	$\eta_{(m+l+q)}$ $(t+q+r+l)$	$P_1-\alpha p$
N_{T_1}	$\eta_{(m+l+q+1)1}$	$\cdots$	$\eta_{(m+l+q+1)t}$	$\eta_{(m+l+q+1)}$ $(t+1)$	$\cdots$	$\eta_{(m+l+q+1)}$ $(t+q)$	$\eta_{(m+l+q+1)}$ $(t+q+1)$	$\cdots$	$\eta_{(m+l+q+1)}$ $(t+q+r)$	$\eta_{(m+l+q+1)}$ $(t+q+r+1)$	$\cdots$	$\eta_{(m+l+q+1)}$ $(t+q+r+l)$	$P_1-\alpha p$
$\vdots$	$\vdots$	$\vdots$	$\vdots$	$\vdots$	$\vdots$	$\vdots$	$\vdots$	$\vdots$	$\vdots$	$\vdots$	$\vdots$	$\vdots$	$\vdots$
N_{T_r}	$\eta_{(m+l+q+r)1}$	$\cdots$	$\eta_{(m+l+q+r)t}$	$\eta_{(m+l+q+r)}$ $(t+1)$	$\cdots$	$\eta_{(m+l+q+r)}$ $(t+q)$	$\eta_{(m+l+q+r)}$ $(t+q+r)$	$\cdots$	$\eta_{(m+l+q+r)}$ $(t+q+1)$	$\eta_{(m+l+q+r)}$ $(t+q+r+1)$	$\cdots$	$\eta_{(m+l+q+r)}$ $(t+q+r+l)$	$P_1-\alpha p$
	$m'_1-\alpha'_1$	$\cdots$	$m'_t-\alpha'_t$	$m''_{t+1}-\alpha''_{t+1}$	$\cdots$	$m''_{t+q}-\alpha''_{t+q}$	$P_1-\alpha p$	$\cdots$	$P_1-\alpha p$	$P_1-\alpha p$	$\cdots$	$P_1-\alpha p$	

Table 6.4 Tabular representation of the second crisp transportation table

	N_{PD_1}	$\cdots$	N_{PD_t}	N_{D_1}	$\cdots$	N_{D_q}	N_{T_1}	$\cdots$	N_{T_r}	N_{S_1}	$\cdots$	N_{S_l}	
N_{PS_1}	ρ_{11}	$\cdots$	ρ_{1t}	$\rho_{1(t+1)}$	$\cdots$	$\rho_{1(t+q)}$	$\rho_{1(t+q+1)}$	$\cdots$	$\rho_{1(t+q+r)}$	$\rho_{1(t+q+r+1)}$	$\cdots$	$\rho_{1(t+q+r+l)}$	α_1
$\vdots$	$\vdots$	$\vdots$	$\vdots$	$\vdots$	$\vdots$	$\vdots$	$\vdots$	$\vdots$	$\vdots$	$\vdots$	$\vdots$	$\vdots$	$\vdots$
N_{PS_m}	ρ_{m1}	$\cdots$	ρ_{mt}	$\rho_{m(t+1)}$	$\cdots$	$\rho_{m(t+q)}$	$\rho_{m(t+q+1)}$	$\cdots$	$\rho_{m(t+q+r)}$	$\rho_{m(t+q+1)}$	$\cdots$	$\rho_{m(t+q+r+l)}$	α_m
N_{S_1}	$\rho_{(m+1)1}$	$\vdots$	$\rho_{(m+1)t}$	$\rho_{(m+1)(t+1)}$	$\cdots$	$\rho_{(m+1)(t+q)}$	$\rho_{(m+1)(t+q+1)}$	$\cdots$	$\rho_{(m+1)(t+q+r)}$	$\rho_{(m+1)(t+q+r+1)}$	$\cdots$	$\rho_{(m+1)(t+q+r+l)}$	α''_{m+1}
$\vdots$	$\vdots$	$\vdots$	$\vdots$	$\vdots$	$\vdots$	$\vdots$	$\vdots$	$\vdots$	$\vdots$	$\vdots$	$\vdots$	$\vdots$	$\vdots$
N_{S_l}	$\rho_{(m+l)1}$	$\vdots$	$\rho_{(m+l)t}$	$\rho_{(m+l)(t+1)}$	$\cdots$	$\rho_{(m+l)(t+q)}$	$\rho_{(m+l)(t+q+1)}$	$\cdots$	$\rho_{(m+l)(t+q+r)}$	$\rho_{(m+l)(t+q+r+1)}$	$\cdots$	$\rho_{(m+l)(t+q+r+l)}$	α''_{m+l}
N_{D_1}	$\rho_{(m+l+1)1}$	$\cdots$	$\rho_{(m+l+1)t}$	$\rho_{(m+l+1)(t+1)}$	$\cdots$	$\rho_{(m+l+1)(t+q)}$	$\rho_{(m+l+1)(t+q+1)}$	$\cdots$	$\rho_{(m+l+1)(t+q+r)}$	$\rho_{(m+l+1)(t+q+r+1)}$	$\cdots$	$\rho_{(m+l+1)(t+q+r+l)}$	α_p
$\vdots$	$\vdots$	$\vdots$	$\vdots$	$\vdots$	$\vdots$	$\vdots$	$\vdots$	$\vdots$	$\vdots$	$\vdots$	$\vdots$	$\vdots$	$\vdots$
N_{D_q}	$\rho_{(m+l+q)1}$	$\cdots$	$\rho_{(m+l+q)t}$	$\rho_{(m+l+q)(t+1)}$	$\cdots$	$\rho_{(m+l+q)(t+q)}$	$\rho_{(m+l+q)(t+q+1)}$	$\cdots$	$\rho_{(m+l+q)(t+q+r)}$	$\rho_{(m+l+q)(t+q+r+1)}$	$\cdots$	$\rho_{(m+l+q)(t+q+r+l)}$	α_p
N_{T_1}	$\rho_{(m+l+q+1)1}$	$\cdots$	$\rho_{(m+l+q+1)t}$	$\rho_{(m+l+q+1)(t+1)}$	$\cdots$	$\rho_{(m+l+q+1)(t+q)}$	$\rho_{(m+l+q+1)(t+q+1)}$	$\cdots$	$\rho_{(m+l+q+1)(t+q+r)}$	$\rho_{(m+l+q+1)(t+q+r+1)}$	$\cdots$	$\rho_{(m+l+q+1)(t+q+r+l)}$	α_p
$\vdots$	$\vdots$	$\vdots$	$\vdots$	$\vdots$	$\vdots$	$\vdots$	$\vdots$	$\vdots$	$\vdots$	$\vdots$	$\vdots$	$\vdots$	$\vdots$
N_{T_r}	$\rho_{(m+l+q+r)1}$	$\cdots$	$\rho_{(m+l+q+r)t}$	$\rho_{(m+l+q+r)(t+1)}$	$\cdots$	$\rho_{(m+l+q+r)(t+q)}$	$\rho_{(m+l+q+r)(t+q+r)}$	$\cdots$	$\rho_{(m+l+q+r)(t+q+1)}$	$\rho_{(m+l+q+r)(t+q+r+1)}$	$\cdots$	$\rho_{(m+l+q+r)(t+q+r+l)}$	α_p
	α'_1	$\cdots$	α'_t	α''_{t+1}	$\cdots$	α''_{t+q}	α_p	$\cdots$	α_p	α_p	$\cdots$	α_p	

Table 6.5 Tabular representation of the third crisp transportation table

	N_{PD_1}	$\cdots$	N_{PD_t}	N_{D_1}	$\cdots$	N_{D_q}	N_{T_1}	$\cdots$	N_{T_r}	N_{S_1}	$\cdots$	N_{S_l}	
N_{PS_1}	δ_{11}	$\cdots$	δ_{1t}	$\delta_{1(t+1)}$	$\cdots$	$\delta_{1(t+q)}$	$\delta_{1(t+q+1)}$	$\cdots$	$\delta_{1(t+q+r)}$	$\delta_{1(t+q+r+1)}$	$\cdots$	$\delta_{1(t+q+r+l)}$	$n_1 - m_1$
$\vdots$	$\vdots$	$\vdots$	$\vdots$	$\vdots$	$\vdots$	$\vdots$	$\vdots$	$\vdots$	$\vdots$	$\vdots$	$\vdots$	$\vdots$	$\vdots$
N_{PS_m}	δ_{m1}	$\cdots$	δ_{mt}	$\delta_{m(t+1)}$	$\cdots$	$\delta_{m(t+q)}$	$\delta_{m(t+q+1)}$	$\cdots$	$\delta_{m(t+q+r)}$	$\delta_{m(t+q+1)}$	$\cdots$	$\delta_{m(t+q+r+l)}$	$n_m - m_m$
N_{S_1}	$\delta_{(m+1)1}$	$\vdots$	$\delta_{(m+1)t}$	$\delta_{(m+1)(t+1)}$	$\cdots$	$\delta_{(m+1)(t+q)}$	$\delta_{(m+1)(t+q+1)}$	$\cdots$	$\delta_{(m+1)(t+q+r)}$	$\delta_{(m+1)(t+q+r+1)}$	$\cdots$	$\delta_{(m+1)(t+q+r+l)}$	$n''_{m+1} - m''_{m+1}$
$\vdots$	$\vdots$	$\vdots$	$\vdots$	$\vdots$	$\vdots$	$\vdots$	$\vdots$	$\vdots$	$\vdots$	$\vdots$	$\vdots$	$\vdots$	$\vdots$
N_{S_l}	$\delta_{(m+l)1}$	$\vdots$	$\delta_{(m+l)t}$	$\delta_{(m+l)(t+1)}$	$\cdots$	$\delta_{(m+l)(t+q)}$	$\delta_{(m+l)(t+q+1)}$	$\cdots$	$\delta_{(m+l)(t+q+r)}$	$\delta_{(m+l)(t+q+r+1)}$	$\cdots$	$\delta_{(m+l)(t+q+r+l)}$	$n''_{m+l} - m''_{m+l}$
N_{D_1}	$\delta_{(m+l+1)1}$	$\cdots$	$\delta_{(m+l+1)t}$	$\delta_{(m+l+1)(t+1)}$	$\cdots$	$\delta_{(m+l+1)(t+q)}$	$\delta_{(m+l+1)(t+q+1)}$	$\cdots$	$\delta_{(m+l+1)(t+q+r)}$	$\delta_{(m+l+1)(t+q+r+1)}$	$\cdots$	$\delta_{(m+l+1)(t+q+r+l)}$	$P_2 - P_1$
$\vdots$	$\vdots$	$\vdots$	$\vdots$	$\vdots$	$\vdots$	$\vdots$	$\vdots$	$\vdots$	$\vdots$	$\vdots$	$\vdots$	$\vdots$	$\vdots$
N_{D_q}	$\delta_{(m+l+q)1}$	$\cdots$	$\delta_{(m+l+q)t}$	$\delta_{(m+l+q)(t+1)}$	$\cdots$	$\delta_{(m+l+q)(t+q)}$	$\delta_{(m+l+q)(t+q+1)}$	$\cdots$	$\delta_{(m+l+q)(t+q+r)}$	$\delta_{(m+l+q)(t+q+r+1)}$	$\cdots$	$\delta_{(m+l+q)(t+q+r+l)}$	$P_2 - P_1$
N_{T_1}	$\delta_{(m+l+q+1)1}$	$\cdots$	$\delta_{(m+l+q+1)t}$	$\delta_{(m+l+q+1)(t+1)}$	$\cdots$	$\delta_{(m+l+q+1)(t+q)}$	$\delta_{(m+l+q+1)(t+q+1)}$	$\cdots$	$\delta_{(m+l+q+1)(t+q+r)}$	$\delta_{(m+l+q+1)(t+q+r+1)}$	$\cdots$	$\delta_{(m+l+q+1)(t+q+r+l)}$	$P_2 - P_1$
$\vdots$	$\vdots$	$\vdots$	$\vdots$	$\vdots$	$\vdots$	$\vdots$	$\vdots$	$\vdots$	$\vdots$	$\vdots$	$\vdots$	$\vdots$	$\vdots$
N_{T_r}	$\delta_{(m+l+q+r)1}$	$\cdots$	$\delta_{(m+l+q+r)t}$	$\delta_{(m+l+q+r)(t+1)}$	$\cdots$	$\delta_{(m+l+q+r)(t+q)}$	$\delta_{(m+l+q+r)(t+q+r)}$	$\cdots$	$\delta_{(m+l+q+r)(t+q+1)}$	$\delta_{(m+l+q+r)(t+q+r+1)}$	$\cdots$	$\delta_{(m+l+q+r)(t+q+r+l)}$	$P_2 - P_1$
	$n'_1 - m'_1$	$\cdots$	$n'_t - m'_t$	$n''_{t+1} - m''_{t+1}$	$\cdots$	$n''_{t+q} - m''_{t+q}$	$P_2 - P_1$	$\cdots$	$P_2 - P_1$	$P_2 - P_1$	$\cdots$	$P_2 - P_1$	

Table 6.6 Tabular representation of the fourth crisp transportation table

	N_{PD_1}	$\cdots$	N_{PD_t}	N_{D_1}	$\cdots$	N_{D_q}	N_{T_1}	$\cdots$	N_{T_r}	N_{S_1}	$\cdots$	N_{S_l}	
N_{PS_1}	ξ_{11}	$\cdots$	ξ_{1t}	$\xi_{1(t+1)}$	$\cdots$	$\xi_{1(t+q)}$	$\xi_{1(t+q+1)}$	$\cdots$	$\xi_{1(t+q+r)}$	$\xi_{1(t+q+r+1)}$	$\cdots$	$\xi_{1(t+q+r+l)}$	β_1
$\vdots$	$\vdots$	$\vdots$	$\vdots$	$\vdots$	$\vdots$	$\vdots$	$\vdots$	$\vdots$	$\vdots$	$\vdots$	$\vdots$	$\vdots$	$\vdots$
N_{PS_m}	ξ_{m1}	$\cdots$	ξ_{mt}	$\xi_{m(t+1)}$	$\cdots$	$\xi_{m(t+q)}$	$\xi_{m(t+q+1)}$	$\cdots$	$\xi_{m(t+q+r)}$	$\xi_{m(t+q+1)}$	$\cdots$	$\xi_{m(t+q+r+l)}$	β_m
N_{S_1}	$\xi_{(m+1)1}$	$\vdots$	$\xi_{(m+1)t}$	$\xi_{(m+1)(t+1)}$	$\cdots$	$\xi_{(m+1)(t+q)}$	$\xi_{(m+1)(t+q+1)}$	$\cdots$	$\xi_{(m+1)(t+q+r)}$	$\xi_{(m+1)(t+q+r+1)}$	$\cdots$	$\xi_{(m+1)(t+q+r+l)}$	β''_{m+1}
$\vdots$	$\vdots$	$\vdots$	$\vdots$	$\vdots$	$\vdots$	$\vdots$	$\vdots$	$\vdots$	$\vdots$	$\vdots$	$\vdots$	$\vdots$	$\vdots$
N_{S_l}	$\xi_{(m+l)1}$	$\vdots$	$\xi_{(m+l)t}$	$\xi_{(m+l)(t+1)}$	$\cdots$	$\xi_{(m+l)(t+q)}$	$\xi_{(m+l)(t+q+1)}$	$\cdots$	$\xi_{(m+l)(t+q+r)}$	$\xi_{(m+l)(t+q+r+1)}$	$\cdots$	$\xi_{(m+l)(t+q+r+l)}$	β''_{m+l}
N_{D_1}	$\xi_{(m+l+1)1}$	$\cdots$	$\xi_{(m+l+1)t}$	$\xi_{(m+l+1)(t+1)}$	$\cdots$	$\xi_{(m+l+1)(t+q)}$	$\xi_{(m+l+1)(t+q+1)}$	$\cdots$	$\xi_{(m+l+1)(t+q+r)}$	$\xi_{(m+l+1)(t+q+r+1)}$	$\cdots$	$\xi_{(m+l+1)(t+q+r+l)}$	β_p
$\vdots$	$\vdots$	$\vdots$	$\vdots$	$\vdots$	$\vdots$	$\vdots$	$\vdots$	$\vdots$	$\vdots$	$\vdots$	$\vdots$	$\vdots$	$\vdots$
N_{D_q}	$\xi_{(m+l+q)1}$	$\cdots$	$\xi_{(m+l+q)t}$	$\xi_{(m+l+q)(t+1)}$	$\cdots$	$\xi_{(m+l+q)(t+q)}$	$\xi_{(m+l+q)(t+q+1)}$	$\cdots$	$\xi_{(m+l+q)(t+q+r)}$	$\xi_{(m+l+q)(t+q+r+1)}$	$\cdots$	$\xi_{(m+l+q)(t+q+r+l)}$	β_p
N_{T_1}	$\xi_{(m+l+q+1)1}$	$\cdots$	$\xi_{(m+l+q+1)t}$	$\xi_{(m+l+q+1)(t+1)}$	$\cdots$	$\xi_{(m+l+q+1)(t+q)}$	$\xi_{(m+l+q+1)(t+q+1)}$	$\cdots$	$\xi_{(m+l+q+1)(t+q+r)}$	$\xi_{(m+l+q+1)(t+q+r+1)}$	$\cdots$	$\xi_{(m+l+q+1)(t+q+r+l)}$	β_p
$\vdots$	$\vdots$	$\vdots$	$\vdots$	$\vdots$	$\vdots$	$\vdots$	$\vdots$	$\vdots$	$\vdots$	$\vdots$	$\vdots$	$\vdots$	$\vdots$
N_{T_r}	$\xi_{(m+l+q+r)1}$	$\cdots$	$\xi_{(m+l+q+r)t}$	$\xi_{(m+l+q+r)(t+1)}$	$\cdots$	$\xi_{(m+l+q+r)(t+q)}$	$\xi_{(m+l+q+r)(t+q+r)}$	$\cdots$	$\xi_{(m+l+q+r)(t+q+1)}$	$\xi_{(m+l+q+r)(t+q+r+1)}$	$\cdots$	$\xi_{(m+l+q+r)(t+q+r+l)}$	β_p
	β'_1	$\cdots$	β'_t	β''_{t+1}	$\cdots$	β''_{t+q}	β_p	$\cdots$	β_p	β_p	$\cdots$	β_p	

6.5 Illustrative Example

For illustration, and to best show the essence and advantages of the proposed methods, we will solve now the fully fuzzy transshipment problem given in Example 19. This will be done by using the two approaches presented in Sects. 6.4.1 and 6.4.2, that is, the one based on the fuzzy linear programming the one based on the tabular representation, respectively.

6.5.1 Determination of the Optimal Solution Using the Method Based on the Fuzzy Linear Programming Formulation

A fuzzy optimal solution of the fully fuzzy transshipment problem from Example 19 can be obtained by using the method based on the fuzzy linear programming formulation presented in Sect. 6.4.1 can be obtained as follows, by following the consecutive steps of the algorithm presented therein:

Step 1 We have the total fuzzy availability $= (70, 80, 20, 50)_{LR}$ and the total fuzzy demand $= (80, 100, 40, 20)_{LR}$. Since the total fuzzy availability is not equal to the total fuzzy demand, we have an unbalanced fully fuzzy transshipment problem. Therefore, due to Case (c) of Step 1, the unbalanced fuzzy transshipment problem can be converted into the balanced fully fuzzy transshipment problem by introducing a dummy purely source node 6 with the fuzzy availability $(20, 30, 20, 0)_{LR}$ and a dummy purely destination node 7 with the fuzzy demand $(10, 10, 0, 30)_{LR}$. Suppose that the fuzzy cost for transporting one unit quantity of the product from the introduced dummy purely source node (6) to the transition node (3), source node (2), existing purely destination nodes (4 and 5) and introduced dummy purely destination node 7 are the zero LR flat fuzzy number. Similarly, assume that the fuzzy cost for transporting one unit quantity of the product from the source node (2), transition node (3), existing purely source node 1 and introduced dummy purely source node 6 to the introduced dummy purely destination node (7) are the zero LR flat fuzzy number, that is

$$\tilde{c}_{17} = \tilde{c}_{27} = \tilde{c}_{37} = \tilde{c}_{62} = \tilde{c}_{63} = \tilde{c}_{64} = \tilde{c}_{65} = \tilde{c}_{67} = (0, 0, 0, 0)_{LR}.$$

Step 2 The balanced fully fuzzy transshipment problem as defined in Step 1, can be transformed into the following fuzzy linear programming problem:

$$\left\{\begin{array}{l}\min((2,4,1,2)_{LR}\otimes\tilde{x}_{12}\oplus(1,1,1,1)_{LR}\otimes\tilde{x}_{13}\oplus(0,0,0,0)_{LR}\otimes\tilde{x}_{17}\oplus\\ \oplus(1,3,1,1)_{LR}\otimes\tilde{x}_{23}\oplus(0,0,0,0)_{LR}\otimes\tilde{x}_{27}\oplus(5,7,2,2)_{LR}\otimes\tilde{x}_{34}\oplus\\ \oplus(5,6,3,1)_{LR}\otimes\tilde{x}_{35}\oplus(0,0,0,0)_{LR}\otimes\tilde{x}_{37}\oplus\oplus(0,0,0,0)_{LR}\otimes\tilde{x}_{62}\oplus\\ \oplus(0,0,0,0)_{LR}\otimes\tilde{x}_{63}\oplus(0,0,0,0)_{LR}\otimes\tilde{x}_{64}\oplus(0,0,0,0)_{LR}\otimes\tilde{x}_{65}\oplus\\ \oplus(0,0,0,0)_{LR}\otimes\tilde{x}_{67})\\ \text{subject to:}\\ \tilde{x}_{12}\oplus\tilde{x}_{13}\oplus\tilde{x}_{17}=(40,40,10,20)_{LR}\\ (\tilde{x}_{23}\oplus\tilde{x}_{27})\ominus_H(\tilde{x}_{12}\oplus\tilde{x}_{62})=(30,40,10,30)_{LR}\\ \tilde{x}_{62}\oplus\tilde{x}_{63}\oplus\tilde{x}_{64}\oplus\tilde{x}_{65}\oplus\tilde{x}_{67}=(20,30,20,0)_{LR}\\ \tilde{x}_{34}\oplus\tilde{x}_{35}\oplus\tilde{x}_{37}=\tilde{x}_{13}\oplus\tilde{x}_{23}\oplus\tilde{x}_{63}\\ \tilde{x}_{34}\oplus\tilde{x}_{64}=(20,30,10,10)_{LR}\\ \tilde{x}_{35}\oplus\tilde{x}_{65}=(60,70,30,10)_{LR}\\ \tilde{x}_{17}\oplus\tilde{x}_{27}\oplus\tilde{x}_{37}\oplus\tilde{x}_{67}=(10,10,0,30)_{LR}\end{array}\right. \tag{6.17}$$

where: $\tilde{x}_{12},\tilde{x}_{13},\tilde{x}_{17},\tilde{x}_{23},\tilde{x}_{27},\tilde{x}_{34},\tilde{x}_{35},\tilde{x}_{37},\tilde{x}_{62},\tilde{x}_{63},\tilde{x}_{64},\tilde{x}_{65},\tilde{x}_{67}$ are non-negative LR flat fuzzy numbers.

Step 3 Using Step 3 to Step 7, of the method, proposed in Sect. 5.6.1, the fuzzy linear programming problem obtained in Step 2, (6.17), can be converted into the following crisp linear programming problem:

$$\left\{\begin{array}{l}\min(\frac{1}{10}(6m_{12}+28n_{12}-4\alpha_{12}+24\beta_{12}+m_{13}+9n_{13}+8\beta_{13}+m_{23}+19n_{23}+\\ +16\beta_{23}+17m_{34}+43n_{34}-12\alpha_{34}+36\beta_{34}+13m_{35}+34n_{35}-8\alpha_{35}+28\beta_{35}))\\ \text{subject to:}\\ m_{12}+m_{13}+m_{17}=40\\ m_{23}+m_{27}-m_{12}-m_{62}=30\\ m_{62}+m_{63}+m_{64}+m_{65}+m_{67}=20\ n_{12}+n_{13}+n_{17}=40\\ n_{23}+n_{27}-n_{12}-n_{62}=40\\ n_{62}+n_{63}+n_{64}+n_{65}+n_{67}=30\\ \alpha_{12}+\alpha_{13}+\alpha_{17}=10\\ \alpha_{23}+\alpha_{27}-\alpha_{12}-\alpha_{62}=10\\ \alpha_{62}+\alpha_{63}+\alpha_{64}+\alpha_{65}+\alpha_{67}=20\\ \beta_{12}+\beta_{13}+\beta_{17}=20\\ \beta_{23}+\beta_{27}-\beta_{12}-\beta_{62}=30\\ \beta_{62}+\beta_{63}+\beta_{64}+\beta_{65}+\beta_{67}=0\\ m_{34}+m_{35}+m_{37}=m_{13}+m_{23}+m_{63}\\ m_{34}+m_{64}=20\\ m_{35}+m_{65}=60\\ n_{34}+n_{35}+n_{37}=n_{13}+n_{23}+n_{63}\\ n_{34}+n_{64}=30\\ n_{35}+n_{65}=70\\ \alpha_{34}+\alpha_{35}+\alpha_{37}=\alpha_{13}+\alpha_{23}+\alpha_{63}\\ \alpha_{34}+\alpha_{64}=10\\ \alpha_{35}+\alpha_{65}=30\\ \beta_{34}+\beta_{35}+\beta_{37}=\beta_{13}+\beta_{23}+\beta_{63}\\ \beta_{34}+\beta_{64}=10\\ \beta_{35}+\beta_{65}=10\\ m_{17}+m_{27}+m_{37}+m_{67}=10\\ n_{17}+n_{27}+n_{37}+n_{67}=10\\ \alpha_{17}+\alpha_{27}+\alpha_{37}+\alpha_{67}=0\\ \beta_{17}+\beta_{27}+\beta_{37}+\beta_{67}=30\end{array}\right. \tag{6.18}$$

where: $m_{12}-\alpha_{12},n_{12}-m_{12},m_{13}-\alpha_{13},n_{13}-m_{13},m_{23}-\alpha_{23},n_{23}-m_{23},m_{27}-\alpha_{27},\ n_{27}-m_{27},\ m_{34}-\alpha_{34},\ n_{34}-m_{34},\ m_{35}-\alpha_{35},\ n_{35}-m_{35},\ m_{62}-\beta_{62},\ n_{62}-m_{62},\ m_{63}-\alpha_{63},\ n_{63}-m_{63},\ m_{64}-\alpha_{64},\ n_{64}-m_{64},\ m_{65}-\alpha_{65},\ n_{65}-m_{65},\ m_{67}-\alpha_{67},\ n_{67}-m_{67},m_{12},n_{12},\alpha_{12},\beta_{12},m_{13},n_{13},\alpha_{13},\beta_{13},m_{23},n_{23},\alpha_{23},\beta_{23},m_{27},n_{27},$

$\alpha_{27}, \beta_{27}, m_{34}, n_{34}, \alpha_{34}, \beta_{34}, m_{35}, n_{35}, \alpha_{35}, \beta_{35}, m_{62}, n_{62}, \alpha_{62}, \beta_{62}, m_{63}, n_{63}, \alpha_{63}, \beta_{63}, m_{64}, n_{64}, \alpha_{64}, \beta_{64}, m_{65}, n_{65}, \alpha_{65}, \beta_{65}, m_{67}, n_{67}, \alpha_{67}, \beta_{67} \geq 0.$

Step 4 The optimal solution of the crisp linear programming problem from Step 3, is:

$$\begin{array}{l} m_{13} = 40\ n_{13} = 40\ \alpha_{13} = 10\ \beta_{13} = 20 \\ m_{23} = 20\ n_{23} = 30\ \alpha_{23} = 10\ \beta_{23} = 0 \\ m_{27} = 10\ n_{27} = 10\ \alpha_{27} = 0\ \ \beta_{27} = 30 \\ m_{34} = 10\ n_{34} = 10\ \alpha_{34} = 0\ \ \beta_{34} = 10 \\ m_{35} = 50\ n_{35} = 60\ \alpha_{35} = 20\ \beta_{35} = 10 \\ m_{64} = 10\ n_{64} = 20\ \alpha_{64} = 10\ \beta_{64} = 0 \\ m_{65} = 10\ n_{65} = 10\ \alpha_{65} = 10\ \beta_{65} = 0 \end{array}$$

and the remaining values of $m_{ij}, n_{ij}, \alpha_{ij}, \beta_{ij}$ are equal 0.

Step 5 By introducing the optimal values of $m_{ij}, n_{ij}, \alpha_{ij}$ and β_{ij} from Step 4 into $\tilde{x}_{ij} = (m_{ij}, n_{ij}, \alpha_{ij}, \beta_{ij})$, the fuzzy optimal solution is becomes

$$\begin{array}{l} \tilde{x}_{13} = (40, 40, 10, 20)_{LR}\ \tilde{x}_{23} = (20, 30, 10, 0)_{LR}\ \ \tilde{x}_{27} = (10, 10, 0, 30)_{LR} \\ \tilde{x}_{34} = (10, 10, 0, 10)_{LR}\ \ \tilde{x}_{35} = (50, 60, 20, 10)_{LR} \\ \tilde{x}_{64} = (10, 20, 10, 0)_{LR}\ \ \tilde{x}_{65} = (10, 10, 10, 0)_{LR} \end{array}$$

and the remaining values of $\tilde{x}_{ij}$ are $(0, 0, 0, 0)_{LR}$.

Step 6 Introducing the values of Putting the values of $\tilde{x}_{12}, \tilde{x}_{13}, \tilde{x}_{17}, \tilde{x}_{23}, \tilde{x}_{27}, \tilde{x}_{34}, \tilde{x}_{35}, \tilde{x}_{37}, \tilde{x}_{62}, \tilde{x}_{63}, \tilde{x}_{64}, \tilde{x}_{65}, \tilde{x}_{67}$ obtained in Step 5 into

$$\begin{array}{l} (2, 4, 1, 2)_{LR} \otimes \tilde{x}_{12} \oplus (1, 1, 1, 1)_{LR} \otimes \tilde{x}_{13} \oplus (0, 0, 0, 0)_{LR} \otimes \tilde{x}_{17} \oplus \\ \oplus (1, 3, 1, 1)_{LR} \otimes \tilde{x}_{23} \oplus (0, 0, 0, 0)_{LR} \otimes \tilde{x}_{27} \oplus (5, 7, 2, 2)_{LR} \otimes \tilde{x}_{34} \\ \oplus (5, 6, 3, 1)_{LR} \otimes \tilde{x}_{35} \oplus (0, 0, 0, 0)_{LR} \otimes \tilde{x}_{37} \oplus (0, 0, 0, 0) \otimes \tilde{x}_{62} \oplus \\ \oplus (0, 0, 0, 0)_{LR} \otimes \tilde{x}_{63} \oplus (0, 0, 0, 0)_{LR} \otimes \tilde{x}_{64} \oplus (0, 0, 0, 0)_{LR} \otimes \tilde{x}_{65} \oplus \\ \oplus (0, 0, 0, 0)_{LR} \otimes \tilde{x}_{67}) \end{array}$$

the minimum total fuzzy transportation cost is $(360, 560, 270, 350)_{LR}$.

6.5.2 Determination of the Optimal Solution Using the Method Based on the Tabular Representation

Now we will obtain a fuzzy optimal solution of the fully fuzzy transshipment problem from Example 19 by using the method based on the tabular representation presented in Sect. 6.4.2. It will be obtained again by following the consecutive steps of the algorithm presented therein:

Step 1
As in the previous section, we again have the total fuzzy availability = $(70, 80, 20, 50)_{LR}$ and the total fuzzy demand = $(80, 100, 40, 20)_{LR}$. Since the total fuzzy

availability is not equal to the total fuzzy demand, we have an unbalanced fully fuzzy transshipment problem.

Therefore, due to Case (c) of Step 1, the unbalanced fuzzy transshipment problem can be converted into the balanced fully fuzzy transshipment problem by introducing a dummy purely source node 6 with the fuzzy availability $(20, 30, 20, 0)_{LR}$ and a dummy purely destination node 7 with the fuzzy demand $(10, 10, 0, 30)_{LR}$. Suppose that the fuzzy cost for transporting one unit quantity of the product from the introduced dummy purely source node (6) to the transition node (3), source node (2), existing purely destination nodes (4 and 5) and introduced dummy purely destination node 7 are the zero LR flat fuzzy number. Similarly, assume that the fuzzy cost for transporting one unit quantity of the product from the source node (2), transition node (3), existing purely source node 1 and introduced dummy purely source node 6 to the introduced dummy purely destination node (7) are the zero LR flat fuzzy number, that is

$$\tilde{c}_{17} = \tilde{c}_{27} = \tilde{c}_{37} = \tilde{c}_{62} = \tilde{c}_{63} = \tilde{c}_{64} = \tilde{c}_{65} = \tilde{c}_{67} = (0, 0, 0, 0)_{LR}.$$

Step 2

Using Step 2 of the method, proposed in Sect. 6.4.2, the balanced fully fuzzy transshipment problem, obtained from Step 1, can be represented in the tabular form as shown in Table 6.7

Step 3

Using Step 3 of the method proposed in Sect. 6.4.2, add some amount of fuzzy buffer stock

$$\tilde{P} = \sum \tilde{a}_i = \sum \tilde{b}_j = (90, 110, 40, 50)_{LR}$$

Table 6.7 Tabular representation of the balanced fully fuzzy transshipment problem

	2	3	4	5	7	
1	$(2, 4, 1, 2)_{LR}$	$(1, 1, 1, 1)_{LR}$	$(M, M, 0, 0)_{LR}$	$(M, M, 0, 0)_{LR}$	$(0, 0, 0, 0)_{LR}$	$(40, 40, 10, 20)_{LR}$
2	$(0, 0, 0, 0)_{LR}$	$(1, 3, 1, 1)_{LR}$	$(M, M, 0, 0)_{LR}$	$(M, M, 0, 0)_{LR}$	$(0, 0, 0, 0)_{LR}$	$(30, 40, 10, 30)_{LR}$
3	$(M, M, 0, 0)_{LR}$	$(0, 0, 0, 0)_{LR}$	$(5, 7, 2, 2)_{LR}$	$(5, 6, 3, 1)_{LR}$	$(0, 0, 0, 0)_{LR}$	–
6	$(0, 0, 0, 0)_{LR}$	$(0, 0, 0, 0)_{LR}$	$(0, 0, 0, 0)_{LR}$	$(0, 0, 0, 0)_{LR}$	$(0, 0, 0, 0)_{LR}$	$(20, 30, 20, 0)_{LR}$
	–	–	$(20, 30, 10, 10)_{LR}$	$(60, 70, 30, 10)_{LR}$	$(10, 10, 0, 30)_{LR}$	

Table 6.8 Tabular representation of the fully fuzzy transportation problem obtained by adding the fuzzy buffer stock

	2	3	4	5	7	
1	$(2, 4, 1, 2)_{LR}$	$(1, 1, 1, 1)_{LR}$	$(M, M, 0, 0)_{LR}$	$(M, M, 0, 0)_{LR}$	$(0, 0, 0, 0)_{LR}$	$(40, 40, 10, 20)_{LR}$
2	$(0, 0, 0, 0)_{LR}$	$(1, 3, 1, 1)_{LR}$	$(M, M, 0, 0)_{LR}$	$(M, M, 0, 0)_{LR}$	$(0, 0, 0, 0)_{LR}$	$(120, 150, 50, 80)_{LR}$
3	$(M, M, 0, 0)_{LR}$	$(0, 0, 0, 0)_{LR}$	$(5, 7, 2, 2)_{LR}$	$(5, 6, 3, 1)_{LR}$	$(0, 0, 0, 0)_{LR}$	$(90, 110, 40, 50)_{LR}$
6	$(0, 0, 0, 0)_{LR}$	$(0, 0, 0, 0)_{LR}$	$(0, 0, 0, 0)_{LR}$	$(0, 0, 0, 0)_{LR}$	$(0, 0, 0, 0)_{LR}$	$(20, 30, 20, 0)_{LR}$
	$(90, 110, 40, 50)_{LR}$	$(90, 110, 40, 50)_{LR}$	$(20, 30, 10, 10)_{LR}$	$(60, 70, 30, 10)_{LR}$	$(10, 10, 0, 30)_{LR}$	

in the fuzzy availability and fuzzy demand corresponding to each intermediate node of Table 6.7. As a result, Table 6.7 changes to Table 6.8.

Step 4 By using Step 4 of the method proposed in Sect. 6.4.2, Table 6.8 can be split into four crisp transportation tables: Tables 6.9, 6.10, 6.11 and 6.12.

Step 5 The optimal solution of the crisp transportation problems, shown by Tables 6.9, 6.10, 6.11 and 6.12 are

$$\begin{array}{lll} m_{13}-\alpha_{13}=30, & m_{22}-\alpha_{22}=50, & m_{23}-\alpha_{23}=10 \\ m_{27}-\alpha_{27}=10, & m_{33}-\alpha_{33}=10, & m_{34}-\alpha_{34}=10 \\ m_{35}-\alpha_{35}=50, & m_{64}-\alpha_{64}=0, & m_{65}-\alpha_{65}=0 \\ \alpha_{13}=10, & \alpha_{22}=50, & \alpha_{23}=10 \\ \alpha_{33}=20, & \alpha_{35}=20, & \alpha_{64}=10 \\ \alpha_{65}=10, & n_{13}-m_{13}=0, & n_{22}-m_{22}=20 \\ n_{23}-m_{23}=10, & n_{27}-m_{27}=0, & n_{33}-m_{33}=10 \\ n_{34}-m_{34}=0, & n_{35}-m_{35}=10, & n_{64}-m_{64}=10 \\ n_{65}-m_{65}=0 & & \end{array}$$

and, respectively,
$\beta_{13}=20, \beta_{22}=50, \beta_{27}=30, \beta_{33}=30, \beta_{34}=10, \beta_{35}=10$

Step 6

On solving the equations, obtained in Step 5, the values of $m_{ij}, n_{ij}, \alpha_{ij}$ and β_{ij} are:

$$\begin{array}{llll} m_{13}=40 & n_{13}=40 & \alpha_{13}=10 & \beta_{13}=20 \\ m_{22}=90 & n_{22}=110 & \alpha_{22}=40 & \beta_{22}=50 \\ m_{23}=20 & n_{23}=30 & \alpha_{23}=10 & \beta_{23}=0 \\ m_{27}=10 & n_{27}=10 & \alpha_{27}=0 & \beta_{27}=30 \\ m_{33}=30 & n_{33}=40 & \alpha_{33}=20 & \beta_{33}=30 \\ m_{34}=10 & n_{34}=10 & \alpha_{34}=0 & \beta_{34}=10 \\ m_{35}=50 & n_{35}=60 & \alpha_{35}=20 & \beta_{35}=10 \\ m_{64}=10 & n_{64}=20 & \alpha_{64}=10 & \beta_{64}=0 \\ m_{65}=10 & n_{65}=10 & \alpha_{65}=10 & \beta_{65}=0 \end{array}$$

and the remaining values of $m_{ij}, n_{ij}, \alpha_{ij}, \beta_{ij}$ are equal 0.

Step 7 By introducing the values of $m_{ij}, n_{ij}, \alpha_{ij}$ and β_{ij} into $\tilde{x}_{ij}=(m_{ij}, n_{ij}, \alpha_{ij}, \beta_{ij})_{LR}$, the fuzzy optimal solution becomes:

$$\begin{array}{lll} \tilde{x}_{13}=(40,40,10,20)_{LR} & \tilde{x}_{22}=(90,110,40,50)_{LR} & \tilde{x}_{23}=(20,30,10,0)_{LR} \\ \tilde{x}_{27}=(10,10,0,30)_{LR} & \tilde{x}_{33}=(30,40,20,30)_{LR} & \tilde{x}_{34}=(10,10,0,10)_{LR} \\ \tilde{x}_{35}=(50,60,20,10)_{LR} & \tilde{x}_{64}=(10,20,10,0)_{LR} & \tilde{x}_{65}=(10,10,10,0)_{LR} \end{array}$$

and the remaining values of $\tilde{x}_{ij}$ are $(0,0,0,0)_{LR}$.

Step 8 By introducing the values of

Table 6.9 Tabular representation of the first crisp transportation problem

	2	3	4	5	7	
1	3.4	1	M	M	0	30
2	0	2	M	M	0	70
3	M	0	6	4.7	0	50
6	0	0	0	0	0	0
	50	50	10	30	10	

Table 6.10 Tabular representation of the second crisp transportation problem

	2	3	4	5	7	
1	3	1	0.6 M	0.6 M	0	10
2	0	2	0.6 M	0.6 M	0	50
3	0.6 M	0	4.8	3.9	0	40
6	0	0	0	0	0	20
	40	40	10	30	0	

Table 6.11 Tabular representation of the third crisp transportation problem

	2	3	4	5	7	
1	2.8	0.9	0.5 M	0.5 M	0	0
2	0	1.9	0.5 M	0.5 M	0	30
3	0.5 M	0	4.3	3.4	0	20
6	0	0	0	0	0	10
	20	20	10	10	0	

$$\begin{matrix} \tilde{x}_{12} & \tilde{x}_{13} & \tilde{x}_{14} & \tilde{x}_{15} & \tilde{x}_{17} \\ \tilde{x}_{22} & \tilde{x}_{23} & \tilde{x}_{24} & \tilde{x}_{25} & \tilde{x}_{27} \\ \tilde{x}_{32} & \tilde{x}_{33} & \tilde{x}_{34} & \tilde{x}_{35} & \tilde{x}_{37} \\ \tilde{x}_{62} & \tilde{x}_{63} & \tilde{x}_{64} & \tilde{x}_{65} & \tilde{x}_{67} \end{matrix}$$

into $((2,4,1,2)_{LR} \otimes \tilde{x}_{12} \oplus (1,1,1,1)_{LR} \otimes \tilde{x}_{13} \oplus (M,M,0,0)_{LR} \otimes \tilde{x}_{14} \oplus (M,M,0,0)_{LR} \otimes \tilde{x}_{15} \oplus (0,0,0,0)_{LR} \otimes \tilde{x}_{17} \oplus (0,0,0,0)_{LR} \otimes \tilde{x}_{22} \oplus (1,3,1,1)_{LR} \otimes \tilde{x}_{23} \oplus (M,M,0,0)_{LR} \otimes \tilde{x}_{24} \oplus (M,M,0,0)_{LR} \otimes \tilde{x}_{25} \oplus (0,0,0,0)_{LR} \otimes \tilde{x}_{27} \oplus (M,M,0,0)_{LR} \otimes \tilde{x}_{32} \oplus (0,0,0,0)_{LR} \otimes \tilde{x}_{33} \oplus (5,7,2,2)_{LR} \otimes \tilde{x}_{34} \oplus (5,6,3,1)_{LR} \otimes \tilde{x}_{35} \oplus (0,0,0,0)_{LR} \otimes \tilde{x}_{37} \oplus (0,0,0,0)_{LR} \otimes \tilde{x}_{62} \oplus (0,0,0,0)_{LR} \otimes \tilde{x}_{63} \oplus (0,0,0,0)_{LR} \otimes \tilde{x}_{64} \oplus (0,0,0,0)_{LR} \otimes \tilde{x}_{65} \oplus (0,0,0,0)_{LR} \otimes \tilde{x}_{67})$
the minimum total fuzzy transportation cost is $(360, 560, 270, 350)_{LR}$ which is the same as the result obtained in Sect. 6.5.1.

Table 6.12 Tabular representation of the fourth crisp transportation problem

	2	3	4	5	7	
1	2.4	0.8	0.4 M	0.4 M	0	20
2	0	1.6	0.4 M	0.4 M	0	80
3	0.4 M	0	3.6	2.8	0	50
6	0	0	0	0	0	0
	50	50	10	10	30	

6.5.3 Interpretation of Results

For clarity and a better understanding of the new methods proposed we will now present an interpretation of the results obtained. We will interpret both the minimum total fuzzy transportation cost and the fuzzy optimal solution obtained.

By using both the proposed methods, i.e. the one using fuzzy linear programming, presented in Sect. 6.5.2, we obtain the minimum total fuzzy transportation cost equal to $(360, 560, 270, 350)_{LR}$,. This can be simply interpreted as follows:

1. the least amount of the minimum total transportation cost is 90 units,
2. the most possible amount of the minimum total transportation cost lies between 360 units and 560 units,
3. the highest amount of the minimum total transportation cost is 910 units,

so that the minimum total transportation cost will always be greater than 90 units and less than 910 units, and there is the maximum possibility that the minimum total transportation cost will be between 360 units and 560 units.

6.6 A Comparative Study

The results obtained by the known and often used method of Ghatee and Hashemi [1] and by the method proposed in this book in Chaps. 4, 5 and in this chapter are collected for comparison in Table 6.13.

The results presented in Table 6.13 can be briefly explained as follows:

1. The known and quite popular Ghatee and Hashemi's [1] method can only be used for solving the balanced fully fuzzy transportation problems and the balanced fully fuzzy transshipment problems. Since the existing fully fuzzy transshipment problem (cf. Example 3.5 from Ghatee and Hashemi [2]) and the fully fuzzy transportation problem shown in Example 18, are the balanced problems, then they can be solved by the existing Ghatee and Hashemi's [1] method mentioned above. However, the problems presented in Examples 15, 17, and 19, are unbalanced, and therefore Ghatee and Hashemi's [1] method can not be used for solving them.

Table 6.13 Results obtained by the Ghatee and Hasemi [1] method and new methods proposed in Chaps. 4, 5 and in this chapter

Example	The minimum total fuzzy transportation cost			
	Ghatee and Hasemi's [1] method [1]	Methods proposed in Chap. 4	Methods proposed in Chap. 5	Methods proposed in this chapter
Example 15	Not applicable	(2100, 2900, 3700, 4100)	$(2900, 3700, 800, 400)_{LR}$	$(2900, 3700, 800, 400)_{LR}$
Example 17	Not applicable	Not applicable	$(5800, 8400, 2800, 2900)_{LR}$	$(5800, 8400, 2800, 2900)_{LR}$
Example 18	$(4100, 6600, 2000, 2600)_{LR}$	(2100, 4100, 6600, 9200)	$(4100, 6600, 2000, 2600)_{LR}$	$(4100, 6600, 2000, 2600)_{LR}$
Example 3.5 from Ghatee and Hashemi [2]	$(1924000, 1903300, 7299800)_{LR}$	Not applicable	Not applicable	$(1924000, 1903300, 7299800)_{LR}$
Example 19	Not applicable	Not applicable	Not applicable	$(360, 560, 270, 350)_{LR}$

2. The methods proposed in Chap. 4 can only be used for solving such balanced and unbalanced fully fuzzy transportation problems in which all parameters are either represented by the triangular fuzzy numbers or the trapezoid fuzzy numbers.
   Since in the fully fuzzy transportation problems from Examples 15, and 18, all parameters are represented by the trapezoid fuzzy numbers, then these problems can be solved by the new method proposed in Chap. 4.
   However, in the fully fuzzy transportation problem shown in Example 17, the parameters are represented by the LR flat fuzzy numbers, and therefore this problem can not be solved by the methods proposed in Chap. 4.
   The problems considered in Example 3.5 from Ghatee and Hashemi [2] and Example 19, are the fully fuzzy transshipment problems which are an extension of the fully fuzzy transportation problem so that these problems can not be solved by the methods proposed in Chap. 4.
3. The new methods proposed in Chap. 5 can be used to solve such balanced and unbalanced fully fuzzy transportation problems in which all the parameters are represented by the LR flat fuzzy numbers.
   Since, in the problems discussed in Examples 15, 17, and 18, all the parameters are represented by the LR flat fuzzy numbers, then all these problems can be solved by the methods proposed in Chap. 5.
   The problem in Example from Ghatee and Hashemi [2] and the problem considered in Example 19, are the fully fuzzy transshipment problems which are extensions of the fully fuzzy transportation problems so that these problems can not be solved by the methods proposed in Chap. 5.
4. The new methods proposed in Chap. 6 can be used to solve the balanced as well as the unbalanced fully fuzzy transshipment problems. Since the fully fuzzy transportation problems are also a special type of the fully fuzzy transshipment problems, then the proposed method can also be used for solving the fully fuzzy transportation problems so that all the problems considered can be solved by the methods proposed in this chapter.

The following can be noticed:

Remark 15 Since, in the existing methods proposed by Ghatee and Hashemi [1, 2] and the methods proposed in Chaps. 4, 5 and in this chapter a different type of multiplication is used in the objective function, then different fuzzy optimal solutions are obtained. However, for the comparison of results obtained by Ghatee and Hashemi [2] and the methods proposed in Chaps. 4, 5 and in this chapter the same type of multiplication is employed for solving the fully fuzzy transportation problems and the fully fuzzy transshipment problems so that a more meaningful comparison can be made.

6.7 A Case Study

In their work, Ghatee and Hashemi [4] have claimed that if $\tilde{a}$ and $\tilde{b}$ and are two non-negative fuzzy numbers such that $\tilde{a} \neq \tilde{b}$, then, $\tilde{a} \neq \tilde{b}$ can be converted into $\tilde{a} = \tilde{b}$ via the following algorithm:

1. Find $\tilde{e} = \tilde{a} \ominus \tilde{b}$ and check whether $\tilde{e}$ is negative or positive, and then:

   Case (i) If $\tilde{e}$ is positive, then $\tilde{b} \oplus \tilde{e} = \tilde{a}$;
   Case (ii) If $\tilde{e}$ is negative, then $\tilde{a} \oplus \tilde{e'} = \tilde{b}$; where, $\tilde{e'} = \ominus_H \tilde{e}$.

However, unfortunately, it is not always possible to convert $\tilde{a} \neq \tilde{b}$ into $\tilde{a} = \tilde{b}$ by using the method described above because if $\tilde{a}$ and $\tilde{b}$ are two non-negative fuzzy numbers such that $\tilde{a} \neq \tilde{b}$, then $\tilde{e} = \tilde{a} \ominus \tilde{b}$ may be neither negative nor positive. i.e., neither $\tilde{b} \oplus \tilde{e} = \tilde{a}$ nor $\tilde{a} \oplus \tilde{e'} = \tilde{b}$. For instance, in real life fully fuzzy transshipment problems (cf. Ghatee and Hashemi [4]), to be described in Sect. 6.7, the total fuzzy availability $\tilde{a} = (1580, 49, 100)$ is not equal to the total fuzzy demand $\tilde{b} = (1498.9, 64, 59)$ so that we have to do with an unbalanced fully fuzzy transshipment problem. However, $\tilde{e} = \tilde{a} \ominus \tilde{b} = (1580 - 1498.9, 49 + 59, 100 + 64) = (81.1, 108, 164)$ is neither a negative nor a positive fuzzy number and neither $\tilde{b} \oplus \tilde{e} = \tilde{a}$ nor $\tilde{a} \oplus \tilde{e'} = \tilde{b}$.

However, since Ghatee and Hashemi [4] have used the fuzzy linear programming formulation shown in (6.9), to obtain the fuzzy optimal solution of this unbalanced real life problem without converting it into the balanced fully fuzzy transshipment problem, their results are not fully valid.

Therefore, in this section the new methods proposed in this chapter are used to find the fuzzy optimal solution of this real life problem.

We consider here a petroleum industry distribution system with a simplified version of its supply chain depicted in Fig. 6.2. In this system the crude oil is transported from production units and import terminals to refineries, export terminals and storage tanks, and then to destinations which is depicted as a block diagram in Fig. 6.3. Clearly, the cost should be minimized.

The fuzzy availability of the crude oil at different sources and destinations are shown in Table 6.14 and the fuzzy costs for transporting one unit quantity of the crude oil from different sources to different destinations are shown in Table 6.15; they are obviously represented by the LR fuzzy numbers such that $L(x) = R(x) = \max\{0, 1 - x\}$.

The fuzzy quantities of crude oil that should be transported from one node to another node as obtained by using the new methods proposed methods are shown in Table 6.16.

The results obtained can be briefly summarized as follows. We can see from Fig. 6.3 that node 5, node 7 and node 14 are not connected to node 21, node 21 and node 18, respectively. Therefore, in the fuzzy optimal solution the fuzzy quantity of the crude oil that should be transported from node 5, node 7 and node 14 to node 21, node 21 and node 18, respectively, should be represented by the zero fuzzy number. However, as it can be seen from results obtained and shown in Table 6.16 these quantities are not zero fuzzy number. One can therefore say that the obtained

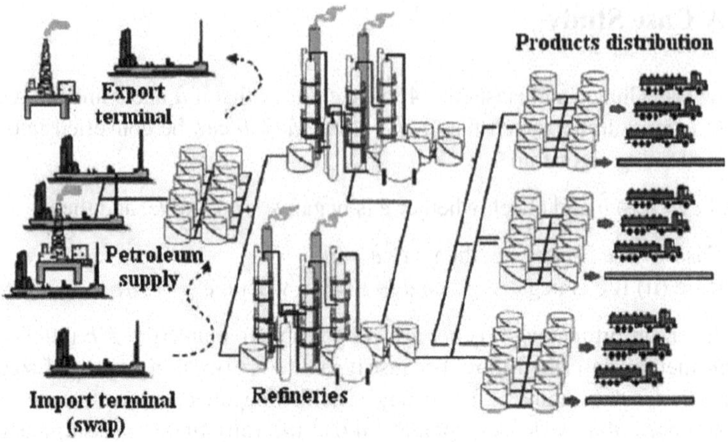

Fig. 6.2 A simplified general petroleum supply chain including the points of supply and points of demand

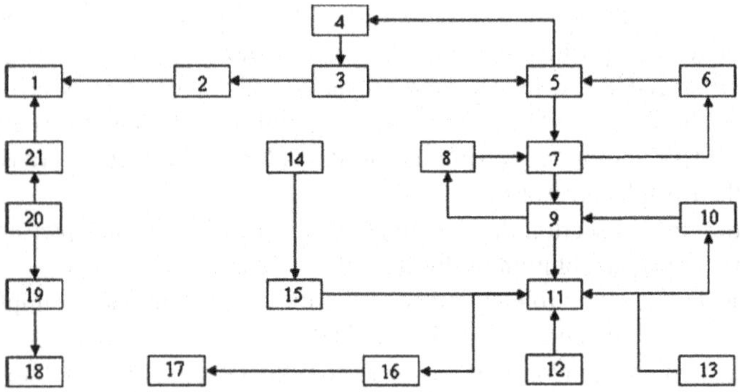

Fig. 6.3 Block diagram of the petroleum supply chain shown in Fig. 6.2

fuzzy optimal solution of the real life problem considered in this study is a "pseudo fuzzy" optimal solution.

Since, the fuzzy costs for transporting one unit quantity of the crude oil from node 5, node 7 and node 14 to node 21, node 21 and node 18, respectively, are not given in the data available (cf. Ghatee and Hashemi [4]), then by assuming this cost to be equal to M, with M being some small value, the obtained minimum total fuzzy transportation cost is $(29896400 + 7M, 26984213.51 + 7M, 9121057.2 + 2M)_{LR}$.

Then, the following interesting remarks can be made:

Remark 16 Since the real life problem considered in this section is an unbalanced problem, then to find for it a fuzzy optimal solution a purely dummy source node (22) and a purely dummy destination node (23) are introduced, as it is often done in such and similar optimization problems. In the results shown

Table 6.14 The fuzzy availability and fuzzy demand

Node	Fuzzy availability	Fuzzy demand
1	–	$(588, 10, 5)_{LR}$
2	$(0, 0, 0)_{LR}$	$(0, 0, 0)_{LR}$
3	$(200, 10, 10)_{LR}$	–
4	$(0, 0, 0)_{LR}$	$(0, 0, 0)_{LR}$
5	$(80, 4, 7)_{LR}$	–
6	$(0, 0, 0)_{LR}$	$(0, 0, 0)_{LR}$
7	$(220, 4, 12)_{LR}$	–
8	$(0, 0, 0)_{LR}$	$(0, 0, 0)_{LR}$
9	$(150, 5, 9)_{LR}$	–
10	$(0, 0, 0)_{LR}$	$(0, 0, 0)_{LR}$
11	$(0, 0, 0)_{LR}$	$(0, 0, 0)_{LR}$
12	$(220, 5, 16)_{LR}$	–
13	$(70, 2, 5)_{LR}$	–
14	$(50, 3, 5)_{LR}$	–
15	$(70, 1, 4)_{LR}$	–
16	$(0, 0, 0)_{LR}$	$(0, 0, 0)_{LR}$
17	–	$(400, 17, 20)_{LR}$
18	–	$(158.9, 3, 15)_{LR}$
19	$(0, 0, 0)_{LR}$	(0,0,0)
20	$(520, 15, 32)_{LR}$	–
21	–	$(352, 34, 19)_{LR}$

$i \longrightarrow 23$, where $i = 3, 5, 7, 9, 14$, represents the fuzzy quantity of the crude oil that should be transported from the ith node, $i = 3, 5, 7, 9, 14$, to the purely dummy destination node (23). Similarly, $22 \longrightarrow j$, where $j = 18, 21$, represents the fuzzy quantity of the crude oil that should be transported from the purely dummy source node 22 to the jth node, $j = 18, 21$.

Remark 17 If the product can not be supplied from the ith node, $i = 3, 5, 7, 9, 14$, to the jth node, $j = 18, 21$, then in the optimal solution of the transshipment problem considered the optimal quantity of the product (x_{ij}) that should be transported from the ith node to the jth node should be zero. However, if there exist any two nodes i and j such that the product can not be supplied from the ith node to the jth node, though the optimal quantity of the product (x_{ij}) is non-zero, then such an optimal solution is called a *pseudo optimal solution*. For instance, while solving the network with six nodes, including three purely source node and three purely destination nodes with the respective costs, availabilities and demands equal to:

Table 6.15 The fuzzy transportation cost

Node	Node	Fuzzy transportation cost
2	1	$(4550, 220.68, 1419.6)_{LR}$
3	2	$(4550, 22.75, 1656.2)_{LR}$
3	5	$(5000, 160, 940)_{LR}$
5	4	$(10000, 595, 2360)_{LR}$
4	3	$(10000, 830, 440)_{LR}$
5	7	$(10000, 525, 80)_{LR}$
7	6	$(10000, 610, 960)_{LR}$
6	5	$(10000, 150, 2840)_{LR}$
7	9	$(10000, 440, 920)_{LR}$
9	8	$(2500, 76.25, 940)_{LR}$
8	7	$(10000, 625, 2280)_{LR}$
9	11	$(10000, 590, 1920)_{LR}$
11	10	$(10000, 430, 1040)_{LR}$
10	9	$(10000, 515, 960)_{LR}$
13	11	$(10000, 770, 240)_{LR}$
14	15	$(10000, 895, 400)_{LR}$
15	11	$(2000, 131, 368)_{LR}$
11	16	$(5000, 127.5, 1460)_{LR}$
16	17	$(5000, 227.5, 1520)_{LR}$
12	11	$(5000, 167.5, 1520)_{LR}$
20	19	$(1600, 89.6, 96)_{LR}$
20	21	$(2500, 148.75, 700)_{LR}$
21	1	$(1600, 103.2, 89.6)_{LR}$
19	18	$(30000, 585, 7800)_{LR}$

- Costs:

$$\begin{cases} c_{14} = 2, c_{15} = 2, c_{16} = 3 \\ c_{24} = 4, c_{25} = 1, c_{26} = 2 \\ c_{34} = 1, c_{35} = 3 \end{cases}$$

- Availabilities:

$$a_1 = 10 \ a_2 = 15 \ a_3 = 40$$

- Demands:

$$b_1 = 20 \ b_2 = 15 \ b_3 = 30$$

and then the obtained optimal solution is

$$x_{16} = 10 \ x_{26} = 15 \ x_{34} = 20 \ x_{26} = 15 \ x_{36} = 5$$

and since $x_{36} \neq 0$, then the obtained optimal solution is a pseudo optimal solution.

Table 6.16 The fuzzy optimal flow between each pair of nodes in the crude oil supply chain system

Node ⟶ Node	Fuzzy flow	Node ⟶ Node	Fuzzy flow
2 ⟶ 1	$(557, 10, 5)_{LR}$	11 ⟶ 16	$(400, 17, 20)_{LR}$
3 ⟶ 2	$(557, 10, 5)_{LR}$	12 ⟶ 11	$(220, 5, 16)_{LR}$
3 ⟶ 23	$(0, 0, 5)_{LR}$	13 ⟶ 11	$(70, 2, 0)_{LR}$
4 ⟶ 3	$(357, 0, 0)_{LR}$	13 ⟶ 23	$(0, 0, 5)_{LR}$
5 ⟶ 21	$(4, 4, 0)_{LR}$	14 ⟶ 15	$(3, 3, 0)_{LR}$
5 ⟶ 4	$(357, 0, 0)_{LR}$	14 ⟶ 23	$(47, 0, 3)_{LR}$
5 ⟶ 23	$(0, 0, 7)_{LR}$	14 ⟶ 18	$(0, 0, 2)_{LR}$
6 ⟶ 5	$(281, 0, 0)_{LR}$	15 ⟶ 11	$(73, 4, 4)_{LR}$
7 ⟶ 9	$(1, 1, 0)_{LR}$	16 ⟶ 17	$(400, 17, 20)_{LR}$
7 ⟶ 6	$(281, 0, 0)_{LR}$	19 ⟶ 18	$(156, 0, 13)_{LR}$
7 ⟶ 21	$(3, 3, 0)_{LR}$	20 ⟶ 19	$(156, 0, 13)_{LR}$
7 ⟶ 23	$(0, 0, 12)_{LR}$	20 ⟶ 21	$(364, 15, 19)_{LR}$
8 ⟶ 7	$(65, 0, 0)_{LR}$	21 ⟶ 1	$(31, 0, 0)_{LR}$
9 ⟶ 8	$(65, 0, 0)_{LR}$	22 ⟶ 18	$(3, 3, 0)_{LR}$
9 ⟶ 11	$(37, 6, 0)_{LR}$	22 ⟶ 21	$(12, 12, 0)_{LR}$
9 ⟶ 23	$(49, 0, 9)_{LR}$		

6.8 Concluding Remarks

As we can see from both a formal examination of the very essence of the existing known and widely used method of Ghatee and Hashemi [1] and the numerical results obtained, all the problems which can be solved by using the Ghatee and Hashemi [1] method and the methods which have been proposed in the previous chapters can also be solved by the methods proposed in this chapter. However, there also exist some important problems which can be solved by the methods proposed in this chapter and which can not be solved by using either the traditional Ghatee and Hashemi [1] method or new methods proposed in the previous chapters. Therefore, one should recommend the use of the methods proposed in this chapter as being clearly more general ones.

References

1. M. Ghatee, S.M. Hashemi, Ranking function-based solutions of fully fuzzified minimal cost flow problem. Inf. Sci. **177**, 4271–4294 (2007)
2. M. Ghatee, S.M. Hashemi, Generalized minimal cost flow problem in fuzzy nature: an application in bus network planning problem. Appl. Math. Model. **32**, 2490–2508 (2008)
3. M. Ghatee, S.M. Hashemi, Application of fuzzy minimum cost flow problems to network design under uncertainty. Fuzzy Sets Syst. **160**, 3263–3289 (2009)

4. M. Ghatee, S.M. Hashemi, Optimal network design and storage management in petroleum distribution network under uncertainty. Eng. Appl. Artif. Intell. **22**, 796–807 (2009)
5. M. Ghatee, S.M. Hashemi, M. Zarepisheh, E. Khorram, Preemptive priority based algorithms for fuzzy minimal cost flow problem: an application in hazardous materials transportation. Comput. Ind. Eng. **57**, 341–354 (2009)
6. T.S. Liou, M.J. Wang, Ranking fuzzy number with integral values. Fuzzy Sets Syst. **50**, 247–255 (1992)
7. M. Wagenknecht, R. Hampel, V. Schneider, Computational aspects of fuzzy arithmetics based on Archimedean t-norms. Fuzzy Sets Syst. **123**, 49–62 (2001)

Chapter 7
New Methods for Solving Fully Fuzzy Solid Transportation Problems with *LR* Fuzzy Parameters

In this chapter we will analyze the well known and often employed method for finding the fuzzy optimal solution of fully fuzzy solid transportation problems proposed by Liu and Kao [5]. This method can exhibit some shortcomings and that is why we two new, improved methods are proposed in this chapter for solving fully fuzzy solid transportation problems. Moreover, to make our analysis more illustrative and complete, we will discuss the advantages of the new methods proposed over both the traditional Liu and Kao's [5] method and also methods proposed in Chaps. 4 and 5. Moreover, following what has been sone in previous chapters, we will illustrate the new method proposed on a real life fuzzy solid transportation problem.

7.1 Fuzzy Linear Programming Formulation of the Balanced Fully Fuzzy Solid Transportation Problems

According to Liu and Kao [5], the fuzzy linear programming formulation of a balanced fully fuzzy solid transportation problem with:

- p sources,
- q destinations.
- r conveyances with fuzzy availability ($\tilde{a}_i$) of the product at the ith source,
- fuzzy demand ($\tilde{b}_j$) of the product at the jth destination,
- fuzzy capacity ($\tilde{e}_k$) of the kth conveyance, i.e. the maximum quantity of the product which can be carried by the kth conveyance, and
- fuzzy cost for transporting one unit quantity of the product from the ith source to the jth destination by means of the kth conveyance, $\tilde{c}_{ijk}$,

can be written as:

A. Kaur et al., *Fuzzy Transportation and Transshipment Problems*, Studies in Fuzziness and Soft Computing 385,
https://doi.org/10.1007/978-3-030-26676-9_7

$$\begin{cases} \min \sum_{i=1}^{p} \sum_{j=1}^{q} \sum_{k=1}^{r} (\tilde{c}_{ijk} \otimes \tilde{x}_{ijk}) \\ \text{subject to:} \\ \sum_{j=1}^{q} \sum_{k=1}^{r} \tilde{x}_{ijk} = \tilde{a}_i; i = 1, 2, \ldots, p \\ \sum_{i=1}^{p} \sum_{k=1}^{r} \tilde{x}_{ijk} = \tilde{b}_j; j = 1, 2, \ldots, q \\ \sum_{i=1}^{p} \sum_{j=1}^{q} \tilde{x}_{ijk} = \tilde{e}_k; k = 1, 2, \ldots, r \\ \sum_{i=1}^{p} \tilde{a}_i = \sum_{j=1}^{q} \tilde{b}_j = \sum_{k=1}^{r} \tilde{e}_k \\ \text{where: } \tilde{x}_{ijk} \text{ is a non-negative fuzzy number.} \end{cases} \tag{7.1}$$

The following remark should be made:

Remark 18 If

$$\sum_{i=1}^{p} \tilde{a}_i = \sum_{j=1}^{q} \tilde{b}_j = \sum_{k=1}^{r} \tilde{e}_k$$

then the fully fuzzy solid transportation problem is said to be a *balanced fully fuzzy solid transportation problem*, otherwise it is called an *unbalanced fully fuzzy solid transportation problem.*

7.2 Liu and Kao's Method

Some years ago Liu and Kao [5] proposed a new method for solving the fully fuzzy solid transportation problems with p sources, q destinations and r conveyances in which the parameters are either represented by triangular fuzzy numbers or trapezoidal fuzzy numbers. This method can be presented as the following steps:

Step 1 Find the α-cuts:

$$\begin{aligned} &[(c_{ijk})^L_\alpha, (c_{ijk})^U_\alpha], \\ &[(a_i)^L_\alpha, (a_i)^U_\alpha], \\ &[(b_j)^L_\alpha, (b_j)^U_\alpha], \\ &[(e_k)^L_\alpha, (e_k)^U_\alpha] \end{aligned}$$

of $\tilde{c}_{ijk}$, $\tilde{a}_i$, $\tilde{b}_j$ and $\tilde{e}_k$ respectively.

Step 2 Verify if

$$\sum_{i=1}^{p} (a_i)^L_{\alpha=0} \geq \sum_{j=1}^{q} (b_j)^U_{\alpha=0}$$

and

$$\sum_{k=1}^{r}(e_k)_{\alpha=0}^{U} \geq \sum_{j=1}^{q}(b_j)_{\alpha=0}^{L}$$

or not, and:

Case (i) If

$$\sum_{i=1}^{p}(a_i)_{\alpha=0}^{L} \geq \sum_{j=1}^{q}(b_j)_{\alpha=0}^{U}$$

and

$$\sum_{k=1}^{r}(e_k)_{\alpha=0}^{U} \geq \sum_{j=1}^{q}(b_j)_{\alpha=0}^{L}$$

then the problem considered is *feasible* and go to Step 3.

Case (ii) If

$$\sum_{i=1}^{p}(a_i)_{\alpha=0}^{L} < \sum_{j=1}^{q}(b_j)_{\alpha=0}^{U}$$

or

$$\sum_{k=1}^{r}(e_k)_{\alpha=0}^{U} < \sum_{j=1}^{q}(b_j)_{\alpha=0}^{L}$$

then the problem considered is *infeasible*.

Step 3 Solve the following problem (7.2):

$$Z_\alpha^L = \min_{\substack{(c_{ijk})_\alpha^L \leq c_{ijk} \leq (c_{ijk})_\alpha^U \\ (a_i)_\alpha^L \leq a_i \leq (a_i)_\alpha^U \\ (b_j)_\alpha^L \leq b_j \leq (b_j)_\alpha^U \\ (e_k)_\alpha^L \leq e_k \leq (e_k)_\alpha^U \\ \forall i,j,k}} \begin{cases} \min \sum_{i=1}^{p}\sum_{j=1}^{q}\sum_{k=1}^{r} c_{ijk}x_{ijk} \\ \text{subject to:} \\ \sum_{j=1}^{q}\sum_{k=1}^{r} x_{ijk} \leq a_i;\ i = 1, 2, \ldots, p \\ \sum_{i=1}^{p}\sum_{k=1}^{r} x_{ijk} \geq b_j;\ j = 1, 2, \ldots, q \\ \sum_{i=1}^{p}\sum_{j=1}^{q} x_{ijk} \leq e_k;\ k = 1, 2, \ldots, r \\ x_{ijk} \geq 0;\ \forall i, j, k \end{cases} \tag{7.2}$$

to find:

- the left end point $(x_{ijk})_\alpha^L$ of the α-cut of the fuzzy decision variable $\tilde{x}_{ijk}$, and

- the left end point (Z^L_α) of the α-cut of the minimum total fuzzy transportation cost

$$\tilde{Z} = \sum_{i=1}^{p}\sum_{j=1}^{q}\sum_{k=1}^{r}(\tilde{c}_{ijk} \otimes \tilde{x}_{ijk})$$

corresponding to different values of $\alpha \in [0, 1]$.

Step 4 Solve the following problem (7.3):

$$Z^U_\alpha = \max_{\substack{(c_{ijk})^L_\alpha \le c_{ijk} \le (c_{ijk})^U_\alpha \\ (a_i)^L_\alpha \le a_i \le (a_i)^U_\alpha \\ (b_j)^L_\alpha \le b_j \le (b_j)^U_\alpha \\ (e_k)^L_\alpha \le e_k \le (e_k)^U_\alpha \\ \forall i,j,k}} \begin{cases} \min \sum\limits_{i=1}^{p}\sum\limits_{j=1}^{q}\sum\limits_{k=1}^{r} c_{ijk}x_{ijk} \\ \text{subject to:} \\ \sum\limits_{j=1}^{q}\sum\limits_{k=1}^{r} x_{ijk} \le a_i;\, i = 1, 2, \ldots, p \\ \sum\limits_{i=1}^{p}\sum\limits_{k=1}^{r} x_{ijk} \ge b_j;\, j = 1, 2, \ldots, q \\ \sum\limits_{i=1}^{p}\sum\limits_{j=1}^{q} x_{ijk} \le e_k;\, k = 1, 2, \ldots, r \\ x_{ijk} \ge 0;\, \forall i, j, k \end{cases} \tag{7.3}$$

to find:

- the right end point $(x_{ijk})^U_\alpha$ of the α-cut of the fuzzy decision variable $\tilde{x}_{ijk}$ and
- the right end point (Z^U_α) of the α- cut of the minimum total fuzzy transportation cost

$$\tilde{Z} = \sum_{i=1}^{p}\sum_{j=1}^{q}\sum_{k=1}^{r}(\tilde{c}_{ijk} \otimes \tilde{x}_{ijk})$$

corresponding to different values of $\alpha \in [0, 1]$.

Step 5 Use the values of

$$(x_{ijk})^L_\alpha,\, (x_{ijk})^U_\alpha,\, Z^L_\alpha,\, Z^U_\alpha,$$

obtained in Step 3 and Step 4, to find the α-cuts:

$$[(x_{ijk})^L_\alpha,\, (x_{ijk})^U_\alpha],\, [Z^L_\alpha,\, Z^U_\alpha],$$

corresponding to the optimal fuzzy quantity of the product $\tilde{x}_{ijk}$ and the minimum total fuzzy transportation cost

$$\tilde{Z} = \sum_{i=1}^{p}\sum_{j=1}^{q}\sum_{k=1}^{r}(\tilde{c}_{ijk} \otimes \tilde{x}_{ijk}).$$

7.3 Some Shortcomings of Liu and Kao's Method

In this section we will briefly indicate some shortcomings of the well known and often used Liu and Kao's [5] method considered in the previous Sect. 7.2. First, notice that the method of Liu and Kao [5] is meant to solve the fully fuzzy solid transportation problem as presented, for convenience of the reader, in the below Example 20.

Example 20 Consider a fully fuzzy solid transportation problem having two sources, three destinations and two conveyances. The fuzzy availabilities, fuzzy demands, fuzzy capacities and the fuzzy costs with $L(x) = R(x) = \max\{0, 1 - x\}$ are given as follows:

- Fuzzy availabilities:

$$\tilde{a}_1 = (80, 100, 10, 20)_{LR},$$
$$\tilde{a}_2 = (70, 70, 10, 20)_{LR}.$$

- Fuzzy demands:

$$\tilde{b}_1 = (30, 40, 20, 10)_{LR},$$
$$\tilde{b}_2 = (50, 50, 10, 10)_{LR},$$
$$\tilde{b}_3 = (40, 60, 10, 10)_{LR}.$$

- Fuzzy capacities:

$$\tilde{e}_1 = (80, 80, 10, 20)_{LR},$$
$$\tilde{e}_2 = (70, 70, 10, 20)_{LR}.$$

- Fuzzy costs:

$$\begin{aligned} \tilde{c}_{111} &= (30, 30, 10, 10)_{LR}, & c_{112} &= 70, \\ c_{121} &= 60, & c_{122} &= 60, \\ c_{131} &= 50, & c_{132} &= 30, \\ \tilde{c}_{211} &= (20, 20, 10, 10)_{LR}, & c_{212} &= 40, \\ c_{221} &= 20, & c_{222} &= 50, \\ c_{231} &= 40, & c_{232} &= 50. \end{aligned}$$

Liu and Kao [5] claimed in their paper that while solving the fully fuzzy solid transportation problem, presented in Example 20, for $\alpha = 0$, the following optimal solution is obtained $x^L_{122} = 40$ and $x^U_{122} = 10$. However, since $x^L_{122} > x^U_{122}$, then the same problem, pointed out by Liu and Kao [5] with respect to the method of Julien [3], and Parra, Bilbao and Uria [6] and presented in Chap. 4, are also valid for the Liu and Kao [5] method.

7.4 Limitations of the Methods Proposed in the Previous Chapters

Since the fully fuzzy solid transportation problem is a generalization of the fully fuzzy transportation problem and there is no link between the fully fuzzy transshipment problems and the fully fuzzy solid transportation problems, then the methods for solving the fully fuzzy transshipment problem can not be used for solving the fully fuzzy solid transportation problems, i.e. neither the existing method of Ghatee and Hashemi [1] nor the methods, proposed in previous chapters, can be used for solving the fully fuzzy solid transportation problem. Therefore, we will propose below some method to solve such a problem.

7.5 New Methods

In this section, to overcome the shortcomings of the existing method by Liu and Kao [5] which has been discussed in Sect. 7.3, and to overcome the limitations of the methods proposed in Chaps. 4 and 5, which have been discussed in Sect. 7.4, two new methods are proposed to find the fuzzy optimal solution of such a fully fuzzy solid transportation problem in which all the parameters are represented by the LR flat fuzzy numbers. Moreover, the advantages of the proposed methods over the existing method by Liu and Kao [5] and over the methods proposed in Chaps. 4 and 5 are discussed. As in the previous case of other new methods proposed, we will also present the proposed method in the form based on the fuzzy linear programming and on the tabular representation.

7.5.1 *New Method Based on the Fuzzy Linear Programming Formulation*

First, in this section we will present the new method to find the fuzzy optimal solution of such a fully fuzzy solid transportation problem in which all the parameters are represented by the LR flat fuzzy numbers using a fuzzy linear programming formulation.

The steps of the proposed method are as follows:

Step 1 Find $\sum_{i=1}^{p} \tilde{a}_i$, $\sum_{j=1}^{q} \tilde{b}_j$ and $\sum_{k=1}^{r} \tilde{e}_k$. Let

$$\sum_{i=1}^{p} \tilde{a}_i = (m, n, \alpha, \beta)_{LR}, \sum_{j=1}^{q} \tilde{b}_j = (m', n', \alpha', \beta')_{LR} \text{ and } \sum_{k=1}^{r} \tilde{e}_k = (m'', n'', \alpha'', \beta'')_{LR}.$$

Use Definition 18, to examine whether the problem considered is balanced or unbalanced:

Case (1) If the problem is balanced, i.e., $\sum_{i=1}^{p} \tilde{a}_i = \sum_{j=1}^{q} \tilde{b}_j = \sum_{k=1}^{r} \tilde{e}_k$, then go to Step 4.

Case (2) If the problem is unbalanced, i.e.

$$\sum_{i=1}^{p} \tilde{a}_i = \sum_{j=1}^{q} \tilde{b}_j \neq \sum_{k=1}^{r} \tilde{e}_k$$

or

$$\sum_{i=1}^{p} \tilde{a}_i \neq \sum_{j=1}^{q} \tilde{b}_j = \sum_{k=1}^{r} \tilde{e}_k$$

or

$$\sum_{i=1}^{p} \tilde{a}_i = \sum_{k=1}^{r} \tilde{e}_k \neq \sum_{j=1}^{q} \tilde{b}_j$$

or

$$\sum_{i=1}^{p} \tilde{a}_i \neq \sum_{j=1}^{q} \tilde{b}_j \neq \sum_{k=1}^{r} \tilde{e}_k$$

then go to Step 2.

Step 2 Check whether

$$\sum_{i=1}^{p} \tilde{a}_i = \sum_{j=1}^{q} \tilde{b}_j$$

or

$$\sum_{i=1}^{p} \tilde{a}_i \neq \sum_{j=1}^{q} \tilde{b}_j$$

and:

Case (1) If $\sum_{i=1}^{p} \tilde{a}_i = \sum_{j=1}^{q} \tilde{b}_j$ then go to Step 3.

Case (2) If $\sum_{i=1}^{p} \tilde{a}_i \neq \sum_{j=1}^{q} \tilde{b}_j$, then convert $\sum_{i=1}^{p} \tilde{a}_i \neq \sum_{j=1}^{q} \tilde{b}_j$ into $\sum_{i=1}^{u} \tilde{a}_i = \sum_{j=1}^{t} \tilde{b}_j$, where: $u = p$ or $p + 1$ and $t = q$ or $q + 1$, by using one of the following cases:

Case (2a) If $m - \alpha \leq m' - \alpha', \alpha \leq \alpha', n - m \leq n' - m'$ and $\beta \leq \beta'$, then introduce a dummy source S_{p+1} with the fuzzy availability $\tilde{a}_{p+1} = (m' - m, n' - n, \alpha' - \alpha, \beta' - \beta)_{LR}$ so that $\sum_{i=1}^{p+1} \tilde{a}_i = \sum_{j=1}^{q} \tilde{b}_j$, and go to Step 3.

Case (2b) If $m - \alpha \geq m' - \alpha', \alpha \geq \alpha', n - m \geq n' - m'$ and $\beta \geq \beta'$, then introduce a dummy destination D_{q+1} with the fuzzy demand $\tilde{b}_{q+1} = (m - m', n - n', \alpha - \alpha', \beta - \beta')_{LR}$ so that $\sum_{i=1}^{p} \tilde{a}_i = \sum_{j=1}^{q+1} \tilde{b}_j$, and go to Step 3.

Case (2c) If neither Case (2a) nor Case (2b) is satisfied, then introduce a dummy source S_{p+1} with the fuzzy availability

$$\tilde{a}_{p+1} = (\max\{0, (m' - \alpha') - (m - \alpha)\} + \max\{0, (\alpha' - \alpha)\}, \max\{0, (m' - \alpha') - (m - \alpha)\} + \max\{0, (\alpha' - \alpha)\} + \max\{0, (n' - m') - (n - m)\}, \max\{0, (\alpha' - \alpha)\}, \max\{0, (\beta' - \beta)\})_{LR}$$

and a dummy destination D_{q+1} with the fuzzy demand

$$\tilde{b}_{q+1} = (\max\{0, (m - \alpha) - (m' - \alpha')\} + \max\{0, (\alpha - \alpha')\}, \max\{0, (m - \alpha) - (m' - \alpha')\} + \max\{0, (\alpha - \alpha')\} + \max\{0, (n - m) - (n' - m')\}, \max\{0, (\alpha - \alpha')\}, \max\{0, (\beta - \beta')\})_{LR}$$

so that

$$\sum_{i=1}^{p+1} \tilde{a}_i = \sum_{j=1}^{q+1} \tilde{b}_j$$

and go to Step 3.

Step 3 Using Step 2, calculate $\sum_{i=1}^{u} \tilde{a}_i = \sum_{j=1}^{t} \tilde{b}_j$, where, $u = p$ or $p + 1$ and $t = q$ or $q + 1$. Let $\sum_{i=1}^{u} \tilde{a}_i = \sum_{j=1}^{t} \tilde{b}_j = (m_1, n_1, \alpha_1, \beta_1)_{LR}$ and $\sum_{k=1}^{r} \tilde{e}_k = (m'', n'', \alpha'', \beta'')_{LR}$.
Check if

$$\sum_{i=1}^{u} \tilde{a}_i = \sum_{j=1}^{t} \tilde{b}_j = \sum_{k=1}^{r} \tilde{e}_k$$

or

$$\sum_{i=1}^{u} \tilde{a}_i = \sum_{j=1}^{t} \tilde{b}_j \neq \sum_{k=1}^{r} \tilde{e}_k$$

and:

Case (1) If

$$\sum_{i=1}^{u} \tilde{a}_i = \sum_{j=1}^{t} \tilde{b}_j = \sum_{k=1}^{r} \tilde{e}_k$$

then go to Step 4.

Case (2) If

$$\sum_{i=1}^{u} \tilde{a}_i = \sum_{j=1}^{t} \tilde{b}_j \neq \sum_{k=1}^{r} \tilde{e}_k$$

then convert

$$\sum_{i=1}^{u} \tilde{a}_i = \sum_{j=1}^{t} \tilde{b}_j \neq \sum_{k=1}^{r} \tilde{e}_k$$

into

$$\sum_{i=1}^{m} \tilde{a}_i = \sum_{j=1}^{n} \tilde{b}_j = \sum_{k=1}^{l} \tilde{e}_k$$

where, $m = p$ or $p + 1$, $n = q$ or $q + 1$ and $l = r$ or $r + 1$, by using one of the following cases:

Case (2a) If $m_1 - \alpha_1 \leq m'' - \alpha''$, $\alpha_1 \leq \alpha''$, $n_1 - m_1 \leq n'' - m''$ and $\beta_1 \leq \beta''$, then check if in Step 2 the dummy source S_{p+1} with the fuzzy availability $\tilde{a}_{p+1}$ is introduced or not, and also check if the dummy destination D_{q+1} with the fuzzy demand $\tilde{b}_{q+1}$ is introduced or not.

Case (i) If both the dummy source S_{p+1} and the dummy destination D_{q+1} are introduced in Step 2, then increase both the fuzzy availability $\tilde{a}_{p+1}$ of the already introduced dummy source S_{p+1} and the fuzzy demand $\tilde{b}_{q+1}$ of the already introduced dummy destination D_{q+1} by the same fuzzy quantity $\tilde{a} = (m'' - m_1, n'' - n_1, \alpha'' - \alpha_1, \beta'' - \beta_1)_{LR}$, i.e. replace the fuzzy availability $\tilde{a}_{p+1}$ and te fuzzy demand $\tilde{b}_{q+1}$ of the already introduced dummy source S_{p+1} and destinations D_{q+1} by $\tilde{a}_{p+1} \oplus \tilde{a}$ and $\tilde{b}_{q+1} \oplus \tilde{a}$ so that

$$\sum_{i=1}^{p+1} \tilde{a}_i = \sum_{j=1}^{q+1} \tilde{b}_j = \sum_{k=1}^{r} \tilde{e}_k$$

and go to Step 4.

Case (ii) If, in Step 2, the dummy source S_{p+1} with the fuzzy availability $\tilde{a}_{p+1}$ is introduced but no dummy destination D_{q+1} with the fuzzy demand $\tilde{b}_{q+1}$ is introduced, then increase the fuzzy availability $\tilde{a}_{p+1}$ of the already introduced dummy source S_{p+1} by the fuzzy quantity $\tilde{a} = (m'' - m_1, n'' - n_1, \alpha'' - \alpha_1, \beta'' - \beta_1)_{LR}$, i.e. replace the fuzzy availability $\tilde{a}_{p+1}$ of the already introduced dummy source S_{p+1} by $\tilde{a}_{p+1} \oplus \tilde{a}$ and also introduce a dummy destination with the fuzzy demand

$$\tilde{b}_{q+1} = \tilde{a} = (m'' - m_1, n'' - n_1, \alpha'' - \alpha_1, \beta'' - \beta_1)_{LR}$$

so that

$$\sum_{i=1}^{p+1} \tilde{a}_i = \sum_{j=1}^{q+1} \tilde{b}_j = \sum_{k=1}^{r} \tilde{e}_k$$

and go to Step 4.

Case (iii) If the dummy destination D_{q+1} with the fuzzy demand $\tilde{b}_{q+1}$ is introduced but no dummy source S_{p+1} with the fuzzy availability $\tilde{a}_{p+1}$ is introduced, then increase the fuzzy demand $\tilde{b}_{q+1}$ of the already introduced dummy destination D_{q+1} by the fuzzy quantity $\tilde{a} = (m'' - m_1, n'' - n_1, \alpha'' - \alpha_1, \beta'' - \beta_1)_{LR}$, i.e. replace the fuzzy demand $\tilde{b}_{q+1}$ of the already introduced dummy destination D_{q+1} by $\tilde{b}_{q+1} \oplus \tilde{a}$ and also introduce a dummy source S_{p+1} with the fuzzy availability

$$\tilde{a}_{p+1} = \tilde{a} = (m'' - m_1, n'' - n_1, \alpha'' - \alpha_1, \beta'' - \beta_1)_{LR}$$

so that

$$\sum_{i=1}^{p+1} \tilde{a}_i = \sum_{j=1}^{q+1} \tilde{b}_j = \sum_{k=1}^{r} \tilde{e}_k$$

and go to Step 4.

Case (2b) If $m_1 - \alpha_1 \geq m'' - \alpha''$, $\alpha_1 \geq \alpha''$, $n_1 - m_1 \geq n'' - m''$, and $\beta_1 \geq \beta''$ then introduce a dummy conveyance E_{r+1} with the fuzzy capacity

$$\tilde{e}_{r+1} = (m_1 - m'', n_1 - n'', \alpha_1 - \alpha'', \beta_1 - \beta'')_{LR}$$

so that

$$\sum_{i=1}^{u} \tilde{a}_i = \sum_{j=1}^{t} \tilde{b}_j = \sum_{k=1}^{r+1} \tilde{e}_k$$

and go to Step 4.

Case (2c) If neither Case (2a) nor Case (2b) is satisfied, then check if in Step 2 a dummy source S_{p+1} with the fuzzy availability $\tilde{a}_{p+1}$ is introduced or not and also check if a dummy destination D_{q+1} with the fuzzy demand $\tilde{b}_{q+1}$ is introduced or not:

Case (i) If both the dummy source S_{p+1} and the dummy destination D_{q+1} are introduced, then increase both the fuzzy availability $\tilde{a}_{p+1}$ of the already introduced dummy source S_{p+1} and the fuzzy demand of the already introduced dummy destination D_{q+1} by the same fuzzy quantity

$\tilde{a} = (\max\{0, (m'' - \alpha'') - (m_1 - \alpha_1)\} + \max\{0, (\alpha'' - \alpha_1)\}, \max\{0, (m'' - \alpha'') - (m_1 - \alpha_1)\} + \max\{0, (\alpha'' - \alpha_1)\} + \max\{0, (n'' - m'') - (n_1 - m_1)\}, \max\{0, (\alpha'' - \alpha_1)\}, \max\{0, (\beta'' - \beta_1)\})_{LR}$

i.e., replace fuzzy availability $\tilde{a}_{p+1}$ and the fuzzy demand $\tilde{b}_{q+1}$ of the already introduced dummy source S_{p+1} and destinations D_{q+1} by $\tilde{a}_{p+1} \oplus \tilde{a}$ and $\tilde{b}_{q+1} \oplus \tilde{a}$, respectively.
Moreover, introduce a dummy conveyance E_{r+1} with the fuzzy capacity

$\tilde{e}_{r+1} = (\max\{0, (m_1 - \alpha_1) - (m'' - \alpha'')\} + \max\{0, (\alpha_1 - \alpha'')\}, \max\{0, (m_1 - \alpha_1) - (m'' - \alpha'')\} + \max\{0, (\alpha_1 - \alpha'')\} + \max\{0, (n_1 - m_1) - (n'' - m'')\}, \max\{0, (\alpha_1 - \alpha'')\}, \max\{0, (\beta_1 - \beta'')\})_{LR}$

so that

$$\sum_{i=1}^{p+1} \tilde{a}_i = \sum_{j=1}^{q+1} \tilde{b}_j = \sum_{k=1}^{r+1} \tilde{e}_k$$

and go to Step 4.

Case (ii) If the dummy source S_{p+1} with the fuzzy availability $\tilde{a}_{p+1}$ is introduced but no dummy destination D_{q+1} with the fuzzy demand $\tilde{b}_{q+1}$ is introduced, then increase the fuzzy availability $\tilde{a}_{p+1}$ of the already introduced dummy source S_{p+1} by the fuzzy quantity

$\tilde{a} = (\max\{0, (m'' - \alpha'') - (m_1 - \alpha_1)\} + \max\{0, (\alpha'' - \alpha_1)\}, \max\{0, (m'' - \alpha'') - (m_1 - \alpha_1)\} + \max\{0, (\alpha'' - \alpha_1)\} + \max\{0, (n'' - m'') - (n_1 - m_1)\}, \max\{0, (\alpha'' - \alpha_1)\}, \max\{0, (\beta'' - \beta_1)\})_{LR}$

i.e., replace the fuzzy availability $\tilde{a}_{p+1}$ of the already introduced dummy source S_{p+1} by $\tilde{a}_{p+1} \oplus \tilde{a}$ and also introduce a dummy destination with the fuzzy demand

$\tilde{b}_{q+1} = \tilde{a} = (\max\{0, (m'' - \alpha'') - (m_1 - \alpha_1)\} + \max\{0, (\alpha'' - \alpha_1)\}, \max\{0, (m'' - \alpha'') - (m_1 - \alpha_1)\} + \max\{0, (\alpha'' - \alpha_1)\} + \max\{0, (n'' - m'') - (n_1 - m_1)\}, \max\{0, (\alpha'' - \alpha_1)\}, \max\{0, (\beta'' - \beta_1)\})_{LR}$

and also introduce a dummy conveyance E_{r+1} with the fuzzy capacity

$\tilde{e}_{r+1} = (\max\{0, (m_1 - \alpha_1) - (m'' - \alpha'')\} + \max\{0, (\alpha_1 - \alpha'')\}, \max\{0, (m_1 - \alpha_1) - (m'' - \alpha'')\} + \max\{0, (\alpha_1 - \alpha'')\} + \max\{0, (n_1 - m_1) - (n'' - m'')\}, \max\{0, (\alpha_1 - \alpha'')\}, \max\{0, (\beta_1 - \beta'')\})_{LR}$

so that

$$\sum_{i=1}^{p+1} \tilde{a}_i = \sum_{j=1}^{q+1} \tilde{b}_j = \sum_{k=1}^{r+1} \tilde{e}_k$$

and go to Step 4.

Case (iii) If a dummy destination D_{q+1} with the fuzzy demand $\tilde{b}_{q+1}$ is introduced but no dummy source S_{p+1} with the fuzzy availability $\tilde{a}_{p+1}$ is introduced, then increase the fuzzy demand $\tilde{b}_{q+1}$ of the already introduced dummy destination D_{q+1} by the fuzzy quantity

$\tilde{a} = (\max\{0, (m'' - \alpha'') - (m_1 - \alpha_1)\} + \max\{0, (\alpha'' - \alpha_1)\}, \max\{0, (m'' - \alpha'') - (m_1 - \alpha_1)\} + \max\{0, (\alpha'' - \alpha_1)\} + \max\{0, (n'' - m'') - (n_1 - m_1)\}, \max\{0, (\alpha'' - \alpha_1)\}, \max\{0, (\beta'' - \beta_1)\})_{LR}$

i.e., replace the fuzzy demand $\tilde{b}_{q+1}$ of the already introduced dummy destination D_{q+1} by $\tilde{b}_{q+1} \oplus \tilde{a}$ and also introduce a dummy source S_{p+1} with the fuzzy availability

$\tilde{a}_{p+1} = \tilde{a} = (\max(\max\{0, (m'' - \alpha'') - (m_1 - \alpha_1)\} + \max\{0, (\alpha'' - \alpha_1)\}, \max\{0, (m'' - \alpha'') - (m_1 - \alpha_1)\} + \max\{0, (\alpha'' - \alpha_1)\} + \max\{0, (n'' - m'') - (n_1 - m_1)\}, \max\{0, (\alpha'' - \alpha_1)\}, \max\{0, (\beta'' - \beta_1)\})_{LR}$

and also, introduce a dummy conveyance E_{r+1} with the fuzzy capacity

$\tilde{e}_{r+1} = (\max\{0, (m_1 - \alpha_1) - (m'' - \alpha'')\} + \max\{0, (\alpha_1 - \alpha'')\}, \max\{0, (m_1 - \alpha_1) - (m'' - \alpha'')\} + \max\{0, (\alpha_1 - \alpha'')\} + \max\{0, (n_1 - m_1) - (n'' - m'')\}, \max\{0, (\alpha_1 - \alpha'')\}, \max\{0, (\beta_1 - \beta'')\})_{LR}$

so that

$$\sum_{i=1}^{p+1} \tilde{a}_i = \sum_{j=1}^{q+1} \tilde{b}_j = \sum_{k=1}^{r+1} \tilde{e}_k$$

and go to Step 4.

Step 4 The balanced fully fuzzy solid transportation problem, obtained by following Step 1 to Step 3, can be formulated as the fully fuzzy linear programming problem (7.4), by assuming the following fuzzy transportation costs as the zero LR flat fuzzy numbers:

(i) If it is required to add any dummy source, then assume the fuzzy cost for transporting one unit quantity of the product from the introduced dummy source to all destinations by all conveyances as the zero LR flat fuzzy number.

(ii) If it is required to add any dummy destination, then assume the fuzzy cost for transporting one unit quantity of the product from all sources to the introduced dummy destination by all conveyances as the zero LR flat fuzzy number.

(iii) If it is required to add any dummy conveyance, then assume the fuzzy cost for transporting one unit quantity of the product from the all sources to all destinations by introduced dummy conveyance as the zero LR flat fuzzy number.

Step 5 By assuming,

$$\begin{aligned}\tilde{c}_{ijk} &= (m'_{ijk}, n'_{ijk}, \alpha'_{ijk}, \beta'_{ijk})_{LR},\\ \tilde{a}_i &= (m_i, n_i, \alpha_i, \beta_i)_{LR},\\ \tilde{b}_j &= (m'_j, n'_j, \alpha'_j, \beta'_j)_{LR},\\ \tilde{e}_k &= (m''_k, n''_k, \alpha''_k, \beta''_k)_{LR},\\ \tilde{x}_{ijk} &= (m_{ijk}, n_{ijk}, \alpha_{ijk}, \beta_{ijk})_{LR},\end{aligned}$$

the fuzzy linear programming problem (7.4), can be written as:

$$\begin{cases}\min \sum_{i=1}^{m}\sum_{j=1}^{n}\sum_{k=1}^{l}\left((m'_{ijk}, n'_{ijk}, \alpha'_{ijk}, \beta'_{ijk})_{LR} \otimes (m_{ijk}, n_{ijk}, \alpha_{ijk}, \beta_{ijk})_{LR}\right)\\ \text{subject to:}\\ \sum_{j=1}^{n}\sum_{k=1}^{l}(m_{ijk}, n_{ijk}, \alpha_{ijk}, \beta_{ijk})_{LR} = (m_i, n_i, \alpha_i, \beta_i)_{LR};\ i = 1, 2, \ldots, m\\ \sum_{i=1}^{m}\sum_{k=1}^{l}(m_{ijk}, n_{ijk}, \alpha_{ijk}, \beta_{ijk})_{LR} = (m'_j, n'_j, \alpha'_j, \beta'_j)_{LR};\ j = 1, 2, \ldots, n\\ \sum_{i=1}^{m}\sum_{j=1}^{n}(m_{ijk}, n_{ijk}, \alpha_{ijk}, \beta_{ijk})_{LR} = (m''_k, n''_k, \alpha''_k, \beta''_k)_{LR};\ k = 1, 2, \ldots, l\\ \text{where: } (m_{ijk}, n_{ijk}, \alpha_{ijk}, \beta_{ijk})_{LR} \text{ is a non-negative } LR \text{ flat fuzzy number}\end{cases} \tag{7.4}$$

Step 6 Using the arithmetic operations, defined in Sect. 5.1, and assuming

$$\sum_{i=1}^{m}\sum_{j=1}^{n}\sum_{k=1}^{l}\left((m'_{ijk}, n'_{ijk}, \alpha'_{ijk}, \beta'_{ijk})_{LR} \otimes (m_{ijk}, n_{ijk}, \alpha_{ijk}, \beta_{ijk})_{LR}\right) = (m_0, n_0, \alpha_0, \beta_0)_{LR}$$

the fuzzy linear programming problem, obtained in Step 5, can be written as:

$$\begin{cases}\min (m_0, n_0, \alpha_0, \beta_0)_{LR}\\ \text{subject to:}\\ (\sum_{j=1}^{n}\sum_{k=1}^{l} m_{ijk}, \sum_{j=1}^{n}\sum_{k=1}^{l} n_{ijk}, \sum_{j=1}^{n}\sum_{k=1}^{l} \alpha_{ijk}, \sum_{j=1}^{n}\sum_{k=1}^{l} \beta_{ijk})_{LR} = (m_i, n_i, \alpha_i, \beta_i)_{LR};\ i = 1, 2, \ldots, m\\ (\sum_{i=1}^{m}\sum_{k=1}^{l} m_{ijk}, \sum_{i=1}^{m}\sum_{k=1}^{l} n_{ijk}, \sum_{i=1}^{m}\sum_{k=1}^{l} \alpha_{ijk}, \sum_{i=1}^{m}\sum_{k=1}^{l} \beta_{ijk})_{LR} = (m'_j, n'_j, \alpha'_j, \beta'_j)_{LR};\ j = 1, 2, \ldots, n\\ (\sum_{i=1}^{m}\sum_{j=1}^{n} m_{ijk}, \sum_{i=1}^{m}\sum_{j=1}^{n} n_{ijk}, \sum_{i=1}^{m}\sum_{j=1}^{n} \alpha_{ijk}, \sum_{i=1}^{m}\sum_{j=1}^{n} \beta_{ijk})_{LR} = (m''_k, n''_k, \alpha''_k, \beta''_k)_{LR};\ k = 1, 2, \ldots, l\\ \text{where: } (m_{ijk}, n_{ijk}, \alpha_{ijk}, \beta_{ijk})_{LR} \text{ is a non-negative } LR \text{ flat fuzzy number}\end{cases} \tag{7.5}$$

Step 7 Using Definitions 18 and 19, the fuzzy linear programming problem, obtained in Step 6, can be converted into the following fuzzy linear programming problem

$$\left\{\begin{array}{l}\min(m_0, n_0, \alpha_0, \beta_0)_{LR} \\ \text{subject to:} \\ \sum_{j=1}^{n}\sum_{k=1}^{l} m_{ijk} = m_i;\ i = 1, 2, \ldots, m \\ \sum_{j=1}^{n}\sum_{k=1}^{l} n_{ijk} = n_i;\ i = 1, 2, \ldots, m \\ \sum_{j=1}^{n}\sum_{k=1}^{l} \alpha_{ijk} = \alpha_i;\ i = 1, 2, \ldots, m \\ \sum_{j=1}^{n}\sum_{k=1}^{l} \beta_{ijk} = \beta_i;\ i = 1, 2, \ldots, m \\ \sum_{i=1}^{m}\sum_{k=1}^{l} m_{ijk} = m'_j;\ j = 1, 2, \ldots, n \\ \sum_{i=1}^{m}\sum_{k=1}^{l} n_{ijk} = n'_j;\ j = 1, 2, \ldots, n \\ \sum_{i=1}^{m}\sum_{k=1}^{l} \alpha_{ijk} = \alpha'_j;\ j = 1, 2, \ldots, n \\ \sum_{i=1}^{m}\sum_{k=1}^{l} \beta_{ijk} = \beta'_j;\ j = 1, 2, \ldots, n \\ \sum_{i=1}^{m}\sum_{j=1}^{n} m_{ijk} = m''_k;\ k = 1, 2, \ldots, l \\ \sum_{i=1}^{m}\sum_{j=1}^{n} n_{ijk} = n''_k;\ k = 1, 2, \ldots, l \\ \sum_{i=1}^{m}\sum_{j=1}^{n} \alpha_{ijk} = \alpha''_k;\ k = 1, 2, \ldots, l \\ \sum_{i=1}^{m}\sum_{j=1}^{n} \beta_{ijk} = \beta''_k;\ k = 1, 2, \ldots, l \\ m_{ijk} - \alpha_{ijk}, n_{ijk} - m_{ijk}, \alpha_{ijk}, \beta_{ijk} \geq 0;\ \forall\ i, j, k\end{array}\right. \tag{7.6}$$

Step 8 As discussed in Step 6, of the method proposed in Sect. 4.6.1, in Chap. 4, the fuzzy optimal solution of fuzzy linear programming problem (7.6) ($P_{5.4}$), can be obtained by solving the following crisp linear programming problem:

$$
\begin{cases}
\min \Re(m_0, n_0, \alpha_0, \beta_0)_{LR} \\
\text{subject to:} \\
\sum_{j=1}^{n}\sum_{k=1}^{l} m_{ijk} = m_i;\ i = 1, 2, \ldots, m \\
\sum_{j=1}^{n}\sum_{k=1}^{l} n_{ijk} = n_i;\ i = 1, 2, \ldots, m \\
\sum_{j=1}^{n}\sum_{k=1}^{l} \alpha_{ijk} = \alpha_i;\ i = 1, 2, \ldots, m \\
\sum_{j=1}^{n}\sum_{k=1}^{l} \beta_{ijk} = \beta_i;\ i = 1, 2, \ldots, m \\
\sum_{i=1}^{m}\sum_{k=1}^{l} m_{ijk} = m'_j;\ j = 1, 2, \ldots, n \\
\sum_{i=1}^{m}\sum_{k=1}^{l} n_{ijk} = n'_j;\ j = 1, 2, \ldots, n \\
\sum_{i=1}^{m}\sum_{k=1}^{l} \alpha_{ijk} = \alpha'_j;\ j = 1, 2, \ldots, n \\
\sum_{i=1}^{m}\sum_{k=1}^{l} \beta_{ijk} = \beta'_j;\ j = 1, 2, \ldots, n \\
\sum_{i=1}^{m}\sum_{j=1}^{n} m_{ijk} = m''_k;\ k = 1, 2, \ldots, l \\
\sum_{i=1}^{m}\sum_{j=1}^{n} n_{ijk} = n''_k;\ k = 1, 2, \ldots, l \\
\sum_{i=1}^{m}\sum_{j=1}^{n} \alpha_{ijk} = \alpha''_k;\ k = 1, 2, \ldots, l \\
\sum_{i=1}^{m}\sum_{j=1}^{n} \beta_{ijk} = \beta''_k;\ k = 1, 2, \ldots, l \\
m_{ijk} - \alpha_{ijk}, n_{ijk} - m_{ijk}, \alpha_{ijk}, \beta_{ijk} \geq 0;\ \forall\ i, j, k
\end{cases}
\tag{7.7}
$$

Step 9 Using the formula known from the paper by Liou and Wang [4]:

$$
\Re(m_0, n_0, \alpha_0, \beta_0)_{LR} = \frac{1}{2}\left(\int_0^1 (m_0 - \alpha_0 L^{-1}(\lambda))d\lambda + \int_0^1 (n_0 + \beta_0 R^{-1}(\lambda))d\lambda\right),
$$

the crisp linear programming problem, obtained in Step 8, can be written as:

$$\begin{cases} \min \frac{1}{2}(\int_0^1 (m_0 - \alpha_0 L^{-1}(\lambda))d\lambda + \int_0^1 (n_0 + \beta_0 R^{-1}(\lambda))d\lambda) \\ \text{subject to:} \\ \sum_{j=1}^{n}\sum_{k=1}^{l} m_{ijk} = m_i;\ i = 1, 2, \ldots, m \\ \sum_{j=1}^{n}\sum_{k=1}^{l} n_{ijk} = n_i;\ i = 1, 2, \ldots, m \\ \sum_{j=1}^{n}\sum_{k=1}^{l} \alpha_{ijk} = \alpha_i;\ i = 1, 2, \ldots, m \\ \sum_{j=1}^{n}\sum_{k=1}^{l} \beta_{ijk} = \beta_i;\ i = 1, 2, \ldots, m \\ \sum_{i=1}^{m}\sum_{k=1}^{l} m_{ijk} = m'_j;\ j = 1, 2, \ldots, n \\ \sum_{i=1}^{m}\sum_{k=1}^{l} n_{ijk} = n'_j;\ j = 1, 2, \ldots, n \\ \sum_{i=1}^{m}\sum_{k=1}^{l} \alpha_{ijk} = \alpha'_j;\ j = 1, 2, \ldots, n \\ \sum_{i=1}^{m}\sum_{k=1}^{l} \beta_{ijk} = \beta'_j;\ j = 1, 2, \ldots, n \\ \sum_{i=1}^{m}\sum_{j=1}^{n} m_{ijk} = m''_k;\ k = 1, 2, \ldots, l \\ \sum_{i=1}^{m}\sum_{j=1}^{n} n_{ijk} = n''_k;\ k = 1, 2, \ldots, l \\ \sum_{i=1}^{m}\sum_{j=1}^{n} \alpha_{ijk} = \alpha''_k;\ k = 1, 2, \ldots, l \\ \sum_{i=1}^{m}\sum_{j=1}^{n} \beta_{ijk} = \beta''_k;\ k = 1, 2, \ldots, l \\ m_{ijk} - \alpha_{ijk}, n_{ijk} - m_{ijk}, \alpha_{ijk}, \beta_{ijk} \geq 0;\ \forall i, j, k \end{cases} \tag{7.8}$$

Step 10 Solve the crisp linear programming problem, obtained in Step 9, to find the optimal solution $\{m_{ijk}, n_{ijk}, \alpha_{ijk}, \beta_{ijk}\}$.

Step 11 Find the fuzzy optimal solution $\{\tilde{x}_{ijk}\}$ by putting the values of m_{ijk}, n_{ijk}, α_{ijk}, β_{ijk} into $\tilde{x}_{ijk} = (m_{ijk}, n_{ijk}, \alpha_{ijk}, \beta_{ijk})_{LR}$.

Step 12 Find the minimum total fuzzy transportation cost by putting the values of $\tilde{x}_{ijk}$ into

$$\sum_{i=1}^{m}\sum_{j=1}^{n}\sum_{k=1}^{l}(\tilde{c}_{ijk} \otimes \tilde{x}_{ijk}).$$

We have some interesting remarks:

Remark 19 In Step 1 of the proposed method, to convert an unbalanced fuzzy solid transportation problem into a balanced fuzzy solid transportation problem, one should check if

$$\sum_{i=1}^{p} \tilde{a}_i \neq \sum_{j=1}^{q} \tilde{b}_j$$

or

$$\sum_{i=1}^{p} \tilde{a}_i = \sum_{j=1}^{q} \tilde{b}_j.$$

Notice that one can use here:

$$\sum_{i=1}^{p} \tilde{a}_i \text{ and } \sum_{k=1}^{r} \tilde{e}_k$$
or
$$\sum_{j=1}^{q} \tilde{b}_j \text{ and } \sum_{k=1}^{r} \tilde{e}_k$$

and the method and the solution will remain same.

Remark 20 Let $\tilde{A} = (m_{ijk}, n_{ijk}, \alpha_{ijk}, \beta_{ijk})_{LR}$ be an LR flat fuzzy number with $L(x) = R(x) = \max\{0, 1 - x\}$. Then:

$$\begin{aligned}\Re(\tilde{A}) &= \frac{1}{2}((m_{ijk} + n_{ijk}) - \alpha_{ijk} \int_0^1 L^{-1}(\lambda)d\lambda + \beta_{ijk} \int_0^1 R^{-1}(\lambda)d\lambda) \\ &= \frac{1}{4}(2m_{ijk} + 2n_{ijk} + \beta_{ijk} - \alpha_{ijk})\end{aligned}$$

7.5.2 New Method Based on the Tabular Representation

We will present now a new method section a new method for finding the fuzzy optimal solution of fully fuzzy solid transportation problems with the parameters are represented by LR flat fuzzy numbers. The method will be based on the tabular representation as shown first in Sect. 6.5.

The consecutive steps of the proposed method are as follows:

Step 1 Apply Step 1 to Step 3 of the method, proposed in Sect. 7.5.1, to obtain a balanced fully fuzzy solid transportation problem.

Step 2 Represent the balanced fully fuzzy solid transportation problem, obtained in Step 1, in the tabular form as shown in Table 7.1.

Step 3 Split Table 7.1 into four crisp transportation tables shown in: Tables 7.2, 7.3, 7.4 and 7.5. The costs for transporting one unit quantity of the product from the ith source to the jth destination by means of the kth conveyance in Tables 7.2, 7.3, 7.4 and 7.5 are represented by η_{ijk}, ρ_{ijk}, δ_{ijk} and ξ_{ijk}, respectively. where:

$$\eta_{ijk} = \frac{1}{2}((m'_{ijk} + n'_{ijk}) - \alpha'_{ijk}\int_0^1 L^{-1}(\lambda)d\lambda + \beta'_{ijk}\int_0^1 R^{-1}(\lambda)d\lambda); \quad i = 1, 2, \ldots, m;\ j = 1, 2, \ldots, n;\ k = 1, 2, \ldots, l \tag{7.9}$$

$$\rho_{ijk} = \frac{1}{2}((m'_{ijk} + n'_{ijk}) - m'_{ijk}\int_0^1 L^{-1}(\lambda)d\lambda + \beta'_{ijk}\int_0^1 R^{-1}(\lambda)d\lambda); \quad i = 1, 2, \ldots, m;\ j = 1, 2, \ldots, n;\ k = 1, 2, \ldots, l \tag{7.10}$$

$$\delta_{ijk} = \frac{1}{2}(n'_{ijk} + \beta'_{ijk}\int_0^1 R^{-1}(\lambda)d\lambda); \quad i = 1, 2, \ldots, m;\ j = 1, 2, \ldots, n;\ k = 1, 2, \ldots, l \tag{7.11}$$

$$\xi_{ijk} = \frac{1}{2}((n'_{ijk} + \beta'_{ijk})\int_0^1 R^{-1}(\lambda)d\lambda); \quad i = 1, 2, \ldots, m;\ j = 1, 2, \ldots, n;\ k = 1, 2, \ldots, l \tag{7.12}$$

Step 4 Solve the crisp solid transportation problems (cf. Haley [2]) shown by Tables 7.2, 7.3, 7.4 and 7.5, to find the optimal solutions

$$\{m_{ijk} - \alpha_{ijk}\}\ \{\alpha_{ijk}\}\ \{n_{ijk} - m_{ijk}\}\ \{\beta_{ijk}\}$$

respectively.

Step 5 Solve the equations, obtained in Step 4, to find the values of:

$$m_{ijk},\ n_{ijk},\ \alpha_{ijk},\ \beta_{ijk}$$

respectively.

Step 6 Find the fuzzy optimal solution $\{\tilde{x}_{ijk}\}$ by putting the values of m_{ijk}, n_{ijk}, α_{ijk}, β_{ijk} into

$$\tilde{x}_{ijk} = (m_{ijk}, n_{ijk}, \alpha_{ijk}, \beta_{ijk})_{LR}.$$

Step 7 Find the minimum total fuzzy transportation cost by putting the values of $\tilde{x}_{ijk}$ into

$$\sum_{i=1}^{m}\sum_{j=1}^{n}\sum_{k=1}^{l}(\tilde{c}_{ijk} \otimes \tilde{x}_{ijk}).$$

Table 7.1 Tabular representation of the balanced fully fuzzy solid transportation problem

																											Capacity ($\tilde{e}_k$)
Conveyance	E_1						E_1						$\cdots$	E_1						$\cdots$	E_1						$\tilde{e}_1$
		E_2						E_2					$\cdots$		E_2					$\cdots$		E_2					$\tilde{e}_2$
			$\cdots$						$\cdots$				$\cdots$			$\cdots$				$\cdots$			$\cdots$				$\vdots$
				E_k						E_k			$\cdots$				E_k			$\cdots$				E_k			$\tilde{e}_k$
					$\cdots$						$\cdots$		$\cdots$					$\cdots$		$\cdots$					$\cdots$		$\vdots$
						E_l						E_l	$\cdots$						E_l	$\cdots$						E_l	$\tilde{e}_l$
Destinations $\rightarrow$ Sources$\downarrow$	D_1						D_2							D_j							D_n						Availability $\tilde{a}_i$
S_1	$\tilde{c}_{111}$	$\tilde{c}_{112}$	$\cdots$	$\tilde{c}_{11k}$	$\cdots$	$\tilde{c}_{11l}$	$\tilde{c}_{121}$	$\tilde{c}_{122}$	$\cdots$	$\tilde{c}_{12k}$	$\cdots$	$\tilde{c}_{12l}$	$\cdots$	$\tilde{c}_{1j1}$	$\tilde{c}_{1j2}$	$\cdots$	$\tilde{c}_{1jk}$	$\cdots$	$\tilde{c}_{1jl}$	$\cdots$	$\tilde{c}_{1n1}$	$\tilde{c}_{1n2}$	$\cdots$	$\tilde{c}_{1nk}$	$\cdots$	$\tilde{c}_{1nl}$	$\tilde{a}_1$
S_2	$\tilde{c}_{211}$	$\tilde{c}_{212}$	$\cdots$	$\tilde{c}_{21k}$	$\cdots$	$\tilde{c}_{21l}$	$\tilde{c}_{221}$	$\tilde{c}_{222}$	$\cdots$	$\tilde{c}_{22k}$	$\cdots$	$\tilde{c}_{22l}$	$\cdots$	$\tilde{c}_{2j1}$	$\tilde{c}_{2j2}$	$\cdots$	$\tilde{c}_{2jk}$	$\cdots$	$\tilde{c}_{2jl}$	$\cdots$	$\tilde{c}_{2n1}$	$\tilde{c}_{2n2}$	$\cdots$	$\tilde{c}_{2nk}$	$\cdots$	$\tilde{c}_{2nl}$	$\tilde{a}_2$
$\vdots$	$\vdots$	$\vdots$	$\vdots$	$\vdots$	$\vdots$	$\vdots$	$\vdots$	$\vdots$	$\vdots$	$\vdots$	$\vdots$	$\vdots$	$\vdots$	$\vdots$	$\vdots$	$\vdots$	$\vdots$	$\vdots$	$\vdots$	$\vdots$	$\vdots$	$\vdots$	$\vdots$	$\vdots$	$\vdots$	$\vdots$	$\vdots$
S_i	$\tilde{c}_{i11}$	$\tilde{c}_{i12}$	$\cdots$	$\tilde{c}_{i1k}$	$\cdots$	$\tilde{c}_{i1l}$	$\tilde{c}_{i21}$	$\tilde{c}_{i22}$	$\cdots$	$\tilde{c}_{i2k}$	$\cdots$	$\tilde{c}_{i2l}$	$\cdots$	$\tilde{c}_{ij1}$	$\tilde{c}_{ij2}$	$\cdots$	$\tilde{c}_{ijk}$	$\cdots$	$\tilde{c}_{ijl}$	$\cdots$	$\tilde{c}_{in1}$	$\tilde{c}_{in2}$	$\cdots$	$\tilde{c}_{ink}$	$\cdots$	$\tilde{c}_{inl}$	$\tilde{a}_i$
$\vdots$	$\vdots$	$\vdots$	$\vdots$	$\vdots$	$\vdots$	$\vdots$	$\vdots$	$\vdots$	$\vdots$	$\vdots$	$\vdots$	$\vdots$	$\vdots$	$\vdots$	$\vdots$	$\vdots$	$\vdots$	$\vdots$	$\vdots$	$\vdots$	$\vdots$	$\vdots$	$\vdots$	$\vdots$	$\vdots$	$\vdots$	$\vdots$
S_m	$\tilde{c}_{m11}$	$\tilde{c}_{m12}$	$\cdots$	$\tilde{c}_{m1k}$	$\cdots$	$\tilde{c}_{m1l}$	$\tilde{c}_{m21}$	$\tilde{c}_{m22}$	$\cdots$	$\tilde{c}_{m2k}$	$\cdots$	$\tilde{c}_{m2l}$	$\cdots$	$\tilde{c}_{mj1}$	$\tilde{c}_{mj2}$	$\cdots$	$\tilde{c}_{mjk}$	$\cdots$	$\tilde{c}_{mjl}$	$\cdots$	$\tilde{c}_{mn1}$	$\tilde{c}_{mn2}$	$\cdots$	$\tilde{c}_{mnk}$	$\cdots$	$\tilde{c}_{mnl}$	$\tilde{a}_m$
Demand ($\tilde{b}_j$)	$\tilde{b}_1$						$\tilde{b}_2$						$\cdots$	$\tilde{b}_j$						$\cdots$	$\tilde{b}_n$						

where: $\tilde{c}_{ijk} = \begin{cases} (0,0,0,0)_{LR}; & \text{If } i\text{th source node is dummy source node} \\ (0,0,0,0)_{LR}; & \text{If } j\text{th destination node is dummy destination node} \\ (0,0,0,0)_{LR}; & \text{If } k\text{th conveyance is dummy conveyance} \\ (m'_{ijk}, n'_{ijk}, \alpha'_{ijk}, \beta'_{ijk})_{LR}; & \text{otherwise} \end{cases}$

Table 7.2 Tabular representation of the first crisp solid transportation problem

																											Capacity
Conveyance	E_1						E_1						$\cdots$	E_1						$\cdots$	E_1						$m''_1 - \alpha''_1$
		E_2						E_2					$\cdots$		E_2					$\cdots$		E_2					$m''_2 - \alpha''_2$
			$\cdots$						$\cdots$				$\cdots$			$\cdots$				$\cdots$			$\cdots$				$\vdots$
				E_k						E_k			$\cdots$				E_k			$\cdots$				E_k			$m''_k - \alpha''_k$
					$\cdots$						$\cdots$		$\cdots$					$\cdots$		$\cdots$					$\cdots$		$\vdots$
						E_l						E_l	$\cdots$						E_l	$\cdots$						E_l	$m''_l - \alpha''_l$
Destinations → Sources↓	D_1						D_2							D_j							D_n						Availability
S_1	η_{111}	η_{112}	$\cdots$	η_{11k}	$\cdots$	η_{11l}	η_{121}	η_{122}	$\cdots$	η_{12k}	$\cdots$	η_{12l}	$\cdots$	η_{1j1}	η_{1j2}	$\cdots$	η_{1jk}	$\cdots$	η_{1jl}	$\cdots$	η_{1n1}	η_{1n2}	$\cdots$	η_{1nk}	$\cdots$	η_{1nl}	$m_1 - \alpha_1$
S_2	η_{211}	η_{212}	$\cdots$	η_{21k}	$\cdots$	η_{21l}	η_{221}	η_{222}	$\cdots$	η_{22k}	$\cdots$	η_{22l}	$\cdots$	η_{2j1}	η_{2j2}	$\cdots$	η_{2jk}	$\cdots$	η_{2jl}	$\cdots$	η_{2n1}	η_{2n2}	$\cdots$	η_{2nk}	$\cdots$	η_{2nl}	$m_2 - \alpha_2$
$\vdots$	$\vdots$	$\vdots$	$\vdots$	$\vdots$	$\vdots$	$\vdots$	$\vdots$	$\vdots$	$\vdots$	$\vdots$	$\vdots$	$\vdots$	$\vdots$	$\vdots$	$\vdots$	$\vdots$	$\vdots$	$\vdots$	$\vdots$	$\vdots$	$\vdots$	$\vdots$	$\vdots$	$\vdots$	$\vdots$	$\vdots$	$\vdots$
S_i	η_{i11}	η_{i12}	$\cdots$	η_{i1k}	$\cdots$	η_{i1l}	η_{i21}	η_{i22}	$\cdots$	η_{i2k}	$\cdots$	η_{i2l}	$\cdots$	η_{ij1}	η_{ij2}	$\cdots$	η_{ijk}	$\cdots$	η_{ijl}	$\cdots$	η_{in1}	η_{in2}	$\cdots$	η_{ink}	$\cdots$	η_{inl}	$m_i - \alpha_i$
$\vdots$	$\vdots$	$\vdots$	$\vdots$	$\vdots$	$\vdots$	$\vdots$	$\vdots$	$\vdots$	$\vdots$	$\vdots$	$\vdots$	$\vdots$	$\vdots$	$\vdots$	$\vdots$	$\vdots$	$\vdots$	$\vdots$	$\vdots$	$\vdots$	$\vdots$	$\vdots$	$\vdots$	$\vdots$	$\vdots$	$\vdots$	$\vdots$
S_m	η_{m11}	η_{m12}	$\cdots$	η_{m1k}	$\cdots$	η_{m1l}	η_{m21}	η_{m22}	$\cdots$	η_{m2k}	$\cdots$	η_{m2l}	$\cdots$	η_{mj1}	η_{mj2}	$\cdots$	η_{mjk}	$\cdots$	η_{mjl}	$\cdots$	η_{mn1}	η_{mn2}	$\cdots$	η_{mnk}	$\cdots$	η_{mnl}	$m_m - \alpha_m$
Demand	$m'_1 - \alpha'_1$						$m'_2 - \alpha'_2$						$\cdots$	$m'_j - \alpha'_j$						$\cdots$	$m'_n - \alpha'_n$						

Table 7.3 Tabular representation of the second crisp solid transportation problem

																											Capacity
Conveyance	E_1						E_1						…	E_1						…	E_1						α''_1
		E_2						E_2					…		E_2					…		E_2					α''_2
			…						…				…			…				…			…				…
				E_k						E_k			…				E_k			…				E_k			α''_k
					…						…		…					…		…					…		…
						E_l						E_l	…						E_l	…						E_l	α''_l
Destinations → Sources↓	D_1						D_2							D_j							D_n						Availability
S_1	ρ_{111}	ρ_{112}	…	ρ_{11k}	…	ρ_{11l}	ρ_{121}	ρ_{122}	…	ρ_{12k}	…	ρ_{12l}	…	ρ_{1j1}	ρ_{1j2}	…	ρ_{1jk}	…	ρ_{1jl}	…	ρ_{1n1}	ρ_{1n2}	…	ρ_{1nk}	…	ρ_{1nl}	α_1
S_2	ρ_{211}	ρ_{212}	…	ρ_{21k}	…	ρ_{21l}	ρ_{221}	ρ_{222}	…	ρ_{22k}	…	ρ_{22l}	…	ρ_{2j1}	ρ_{2j2}	…	ρ_{2jk}	…	ρ_{2jl}	…	ρ_{2n1}	ρ_{2n2}	…	ρ_{2nk}	…	ρ_{2nl}	α_2
…	…	…	…	…	…	…	…	…	…	…	…	…	…	…	…	…	…	…	…	…	…	…	…	…	…	…	…
S_i	ρ_{i11}	ρ_{i12}	…	ρ_{i1k}	…	ρ_{i1l}	ρ_{i21}	ρ_{i22}	…	ρ_{i2k}	…	ρ_{i2l}	…	ρ_{ij1}	ρ_{ij2}	…	ρ_{ijk}	…	ρ_{ijl}	…	ρ_{in1}	ρ_{in2}	…	ρ_{ink}	…	ρ_{inl}	α_i
…	…	…	…	…	…	…	…	…	…	…	…	…	…	…	…	…	…	…	…	…	…	…	…	…	…	…	…
S_m	ρ_{m11}	ρ_{m12}	…	ρ_{m1k}	…	ρ_{m1l}	ρ_{m21}	ρ_{m22}	…	ρ_{m2k}	…	ρ_{m2l}	…	ρ_{mj1}	ρ_{mj2}	…	ρ_{mjk}	…	ρ_{mjl}	…	ρ_{mn1}	ρ_{mn2}	…	ρ_{mnk}	…	ρ_{mnl}	α_m
Demand	α'_1						α'_2						…	α'_j						…	α'_n						

Table 7.4 Tabular representation of the third crisp solid transportation problem

																											Capacity
Conveyance	E_1						E_1						…	E_1						…	E_1						$n''_1 - m''_1$
		E_2						E_2					…		E_2					…		E_2					$n''_2 - m''_2$
			…						…				…			…				…			…				⋮
				E_k						E_k			…				E_k			…				E_k			$n''_k - m''_k$
					…						…		…					…		…					…		⋮
						E_l						E_l	…						E_l	…						E_l	$n''_l - m''_l$
Destinations → Sources↓	D_1						D_2							D_j							D_n						Availability
S_1	δ_{111}	δ_{112}	…	δ_{11k}	…	δ_{11l}	δ_{121}	δ_{122}	…	δ_{12k}	…	δ_{12l}		δ_{1j1}	δ_{1j2}	…	δ_{1jk}	…	δ_{1jl}		δ_{1n1}	δ_{1n2}	…	δ_{1nk}	…	δ_{1nl}	$n_1 - m_1$
S_2	δ_{211}	δ_{212}	…	δ_{21k}	…	δ_{21l}	δ_{221}	δ_{222}	…	δ_{22k}	…	δ_{22l}		δ_{2j1}	δ_{2j2}	…	δ_{2jk}	…	δ_{2jl}		δ_{2n1}	δ_{2n2}	…	δ_{2nk}	…	δ_{2nl}	$n_2 - m_2$
⋮	⋮	⋮	⋮	⋮	⋮	⋮	⋮	⋮	⋮	⋮	⋮	⋮	⋮	⋮	⋮	⋮	⋮	⋮	⋮	⋮	⋮	⋮	⋮	⋮	⋮	⋮	⋮
S_i	δ_{i11}	δ_{i12}	…	δ_{i1k}	…	δ_{i1l}	δ_{i21}	δ_{i22}	…	δ_{i2k}	…	δ_{i2l}		δ_{ij1}	δ_{ij2}	…	δ_{ijk}	…	δ_{ijl}		δ_{in1}	δ_{in2}	…	δ_{ink}	…	δ_{inl}	$n_i - m_i$
⋮	⋮	⋮	⋮	⋮	⋮	⋮	⋮	⋮	⋮	⋮	⋮	⋮	⋮	⋮	⋮	⋮	⋮	⋮	⋮	⋮	⋮	⋮	⋮	⋮	⋮	⋮	⋮
S_m	δ_{m11}	δ_{m12}	…	δ_{m1k}	…	δ_{m1l}	δ_{m21}	δ_{m22}	…	δ_{m2k}	…	δ_{m2l}		δ_{mj1}	δ_{mj2}	…	δ_{mjk}	…	δ_{mjl}		δ_{mn1}	δ_{mn2}	…	δ_{mnk}	…	δ_{mnl}	$n_m - m_m$
Demand	$n'_1 - m'_1$						$n'_2 - n'_2$						…	$n'_j - n'_j$						…	$n'_n - n'_n$						

Table 7.5 Tabular representation of the fourth crisp solid transportation problem

																											Capacity
Conveyance	E_1						E_1						$\cdots$	E_1						$\cdots$	E_1						β''_1
		E_2						E_2					$\cdots$		E_2					$\cdots$		E_2					β''_2
			$\cdots$						$\cdots$				$\cdots$			$\cdots$				$\cdots$			$\cdots$				$\vdots$
				E_k						E_k			$\cdots$				E_k			$\cdots$				E_k			β''_k
					$\cdots$						$\cdots$		$\cdots$					$\cdots$		$\cdots$					$\cdots$		$\vdots$
						E_l						E_l	$\cdots$						E_l	$\cdots$						E_l	β''_l
Destinations → Sources↓	D_1						D_2							D_j							D_n						Availability
S_1	ξ_{111}	ξ_{112}	$\cdots$	ξ_{11k}	$\cdots$	ξ_{11l}	ξ_{121}	ξ_{122}	$\cdots$	ξ_{12k}	$\cdots$	ξ_{12l}		ξ_{1j1}	ξ_{1j2}	$\cdots$	ξ_{1jk}	$\cdots$	ξ_{1jl}		ξ_{1n1}	ξ_{1n2}	$\cdots$	ξ_{1nk}	$\cdots$	ξ_{1nl}	β_1
S_2	ξ_{211}	ξ_{212}	$\cdots$	ξ_{21k}	$\cdots$	ξ_{21l}	ξ_{221}	ξ_{222}	$\cdots$	ξ_{22k}	$\cdots$	ξ_{22l}		ξ_{2j1}	ξ_{2j2}	$\cdots$	ξ_{2jk}	$\cdots$	ξ_{2jl}		ξ_{2n1}	ξ_{2n2}	$\cdots$	ξ_{2nk}	$\cdots$	ξ_{2nl}	β_2
$\vdots$	$\vdots$	$\vdots$	$\vdots$	$\vdots$	$\vdots$	$\vdots$	$\vdots$	$\vdots$	$\vdots$	$\vdots$	$\vdots$	$\vdots$	$\vdots$	$\vdots$	$\vdots$	$\vdots$	$\vdots$	$\vdots$	$\vdots$	$\vdots$	$\vdots$	$\vdots$	$\vdots$	$\vdots$	$\vdots$	$\vdots$	$\vdots$
S_i	ξ_{i11}	ξ_{i12}	$\cdots$	ξ_{i1k}	$\cdots$	ξ_{i1l}	ξ_{i21}	ξ_{i22}	$\cdots$	ξ_{i2k}	$\cdots$	ξ_{i2l}		ξ_{ij1}	ξ_{ij2}	$\cdots$	ξ_{ijk}	$\cdots$	ξ_{ijl}		ξ_{in1}	ξ_{in2}	$\cdots$	ξ_{ink}	$\cdots$	ξ_{inl}	β_i
$\vdots$	$\vdots$	$\vdots$	$\vdots$	$\vdots$	$\vdots$	$\vdots$	$\vdots$	$\vdots$	$\vdots$	$\vdots$	$\vdots$	$\vdots$	$\vdots$	$\vdots$	$\vdots$	$\vdots$	$\vdots$	$\vdots$	$\vdots$	$\vdots$	$\vdots$	$\vdots$	$\vdots$	$\vdots$	$\vdots$	$\vdots$	$\vdots$
S_m	ξ_{m11}	ξ_{m12}	$\cdots$	ξ_{m1k}	$\cdots$	ξ_{m1l}	ξ_{m21}	ξ_{m22}	$\cdots$	ξ_{m2k}	$\cdots$	ξ_{m2l}		ξ_{mj1}	ξ_{mj2}	$\cdots$	ξ_{mjk}	$\cdots$	ξ_{mjl}		ξ_{mn1}	ξ_{mn2}	$\cdots$	ξ_{mnk}	$\cdots$	ξ_{mnl}	β_m
Demand	β'_1						β'_2						$\cdots$	β'_j						$\cdots$	β'_n						

The following remark is important:

Remark 21 Let $\tilde{A} = (m_{ijk}, n_{ijk}, \alpha_{ijk}, \beta_{ijk})_{LR}$ be an LR flat fuzzy number with $L(x) = R(x) = \max\{0, 1 - x\}$. Then:

$$\eta_{ijk} = \frac{1}{2}((m_{ijk} + n_{ijk}) - \alpha_{ijk}\int_0^1 L^{-1}(\lambda)d\lambda + \beta_{ijk}\int_0^1 R^{-1}(\lambda)d\lambda)$$
$$= \frac{1}{4}(2m_{ijk} + 2n_{ijk} + \beta_{ijk} - \alpha_{ijk}),$$

$$\rho_{ijk} = \frac{1}{2}((m_{ijk} + n_{ijk}) - m_{ijk}\int_0^1 L^{-1}(\lambda)d\lambda + \beta_{ijk}\int_0^1 R^{-1}(\lambda)d\lambda)$$
$$= \frac{1}{4}(m_{ijk} + 2n_{ijk} + \beta_{ijk}),$$

$$\delta_{ijk} = \frac{1}{2}(n_{ijk} + \beta_{ijk}\int_0^1 R^{-1}(\lambda)d\lambda)$$
$$= \frac{1}{4}(2n_{ijk} + \beta_{ijk})$$

and

$$\xi_{ijk} = \frac{1}{2}((n_{ijk} + \beta_{ijk})\int_0^1 R^{-1}(\lambda)d\lambda)$$
$$= \frac{1}{4}(n_{ijk} + \beta_{ijk}).$$

7.5.3 Advantages of the New Methods

In this section we will discuss the advantages of the new methods proposed in this chapter, over the existing Liu and Kao's [5] method, as well as over the new method proposed in this book in Chaps. 4 and 5.

Basically, the above can be summarized briefly as follows:

(1) Since, in the proposed methods we have the constraints $b - a \geq 0, c - b \geq 0$ and $d - c \geq 0$, due to which the constraint $d \geq a$ (or $x^U \geq x^L$) will always hold. Therefore, by using the proposed methods the shortcomings of the existing methods, mentioned above, notably Liu and Kao's [5] method, as indicated in Sect. 7.3, are overcome.

(2) The methods proposed in Chaps. 4 and 5 can be used to find the fuzzy optimal solution of the fully fuzzy transportation problems but can not be used for solving the fully fuzzy solid transportation problems. However, since the fully fuzzy transportation problems are a special type of the fully fuzzy solid transportation problems, then the methods proposed in this chapter, can also be used to find the fuzzy optimal solution of the fully fuzzy transportation problems.

7.6 Illustrative Example

Now, to illustrate the proposed methods, and show their strengths, the existing fully fuzzy solid transportation problem, presented in Example 20 will be solved.

7.6.1 Determination of the Fuzzy Optimal Solution Using the New Method Based on the Fuzzy Linear Programming Formulation

By using the proposed method, based on the fuzzy linear programming formulation presented in Sect. 7.5.2, the fuzzy optimal solution of the fully fuzzy solid transportation problem shown in Example 20, can be obtained as follows:

Step 1 Using the values of

$$\tilde{a}_1, \tilde{a}_2, \tilde{b}_1, \tilde{b}_2, \tilde{b}_3, \tilde{e}_1, \tilde{e}_2,$$

the total fuzzy availability is

$$\sum_{i=1}^{2} \tilde{a}_i = (150, 170, 20, 40)_{LR},$$

the total fuzzy demand is

$$\sum_{j=1}^{3} \tilde{b}_j = (120, 150, 40, 30)_{LR}$$

and the total fuzzy capacity is

$$\sum_{k=1}^{2} \tilde{e}_k = (150, 150, 20, 40)_{LR}.$$

Since

$$\sum_{i=1}^{2} \tilde{a}_i \neq \sum_{j=1}^{3} \tilde{b}_j \neq \sum_{k=1}^{2} \tilde{e}_k,$$

then this is an unbalanced fully fuzzy solid transportation problem.

Step 2 Comparing

$$\sum_{i=1}^{2} \tilde{a}_i = (150, 170, 20, 40)_{LR}$$

with

$$\sum_{i=1}^{p} \tilde{a}_i = (m, n, \alpha, \beta)_{LR}$$

and

$$\sum_{j=1}^{3} \tilde{b}_j = (120, 150, 40, 30)_{LR}$$

with

$$\sum_{j=1}^{q} \tilde{b}_j = (m', n', \alpha', \beta')_{LR}$$

the values of

$$m, n, \alpha, \beta, m', n', \alpha', \beta'$$

are equal to 150, 170, 20, 40, 120, 150, 40 and 30, respectively.
Since

$$\sum_{i=1}^{2} \tilde{a}_i \neq \sum_{j=1}^{3} \tilde{b}_j$$

and neither

$$m - \alpha \leq m' - \alpha',\ \alpha \leq \alpha',\ n - m \leq n' - m',\ \beta \leq \beta'$$

nor

$$m - \alpha \geq m' - \alpha',\ \alpha \geq \alpha',\ n - m \geq n' - m',\ \beta \geq \beta'$$

Therefore, as described in Step 2 (Case (2c)) of the proposed method, there is a need to introduce:

- a dummy source S_3 with the fuzzy availability $\tilde{a}_3 = (20, 30, 20, 0)_{LR}$ and
- a dummy destination D_4 with the fuzzy demand $\tilde{b}_4 = (50, 50, 0, 10)_{LR}$,

so that

$$\sum_{i=1}^{3} \tilde{a}_i = \sum_{j=1}^{4} \tilde{b}_j.$$

Step 3 Using Step 2, we obtain:

$$\sum_{i=1}^{3} \tilde{a}_i = \sum_{j=1}^{4} \tilde{b}_j = (170, 200, 40, 40)_{LR}$$

Since

$$\sum_{i=1}^{3} \tilde{a}_i = \sum_{j=1}^{4} \tilde{b}_j \neq \sum_{k=1}^{2} \tilde{e}_k$$

then go to Step 3.
By comparing

$$\sum_{i=1}^{3} \tilde{a}_i = \sum_{j=1}^{4} \tilde{b}_j = (170, 200, 40, 40)_{LR}$$

with

$$\sum_{i=1}^{3} \tilde{a}_i = \sum_{j=1}^{4} \tilde{b}_j = (m_1, n_1, \alpha_1, \beta_1)_{LR}$$

and by comparing

$$\sum_{k=1}^{2} \tilde{e}_k = (150, 150, 20, 40)_{LR}$$

with

$$\sum_{k=1}^{r} \tilde{e}_k = (m'', n'', \alpha'', \beta'')_{LR}$$

the values of $m_1, n_1, \alpha_1, \beta_1, m'', n'', \alpha''$ and β'' are obtained as 170, 200, 40, 40, 150, 150, 20 and 40 respectively.
Since, the conditions

$$\begin{aligned} m_1 &\geq m'', m_1 - \alpha_1 \geq m'' - \alpha'' \\ \alpha_1 &\geq \alpha'', n_1 - m_1 \ \geq n'' - m'' \end{aligned}$$

is satisfied as described in Step 3 (Case (2b)) of the proposed method, then there is a need to introduce a dummy conveyance E_3 with the fuzzy capacity $\tilde{e}_3$ = $(20, 50, 20, 0)_{LR}$ so that

$$\sum_{i=1}^{3} \tilde{a}_i = \sum_{j=1}^{4} \tilde{b}_j = \sum_{k=1}^{3} \tilde{e}_k.$$

Step 4 Since we have introduced:

- a dummy source S_3 with the fuzzy availability $\tilde{a}_3$,
- a dummy destination D_4 with the fuzzy demand $\tilde{b}_4$ and
- a dummy conveyance E_3 with the fuzzy capacity $\tilde{e}_3$,

then, as described in Step 4 of the proposed method, by assuming

$$\tilde{c}_{3jk} = \tilde{c}_{i4k} = \tilde{c}_{ij3} = (0, 0, 0, 0)_{LR},$$

for all $i = 1, 2, 3;\ j = 1, 2, 3, 4;\ k = 1, 2, 3$, the fuzzy linear programming formulation of the balanced fully fuzzy solid transportation problem, obtained from Step 3, can be written as:

$$\begin{cases}
\min((30, 30, 10, 10)_{LR} \otimes \tilde{x}_{111} \oplus (70, 70, 0, 0)_{LR} \otimes \tilde{x}_{112} \oplus (0, 0, 0, 0)_{LR} \otimes \\
\tilde{x}_{113} \oplus (60, 60, 0, 0)_{LR} \otimes \tilde{x}_{121} \oplus (60, 60, 0, 0)_{LR} \otimes \tilde{x}_{122} \oplus (0, 0, 0, 0)_{LR} \otimes \\
\tilde{x}_{123} \oplus (50, 50, 0, 0)_{LR} \otimes \tilde{x}_{131} \oplus (30, 30, 0, 0)_{LR} \otimes \tilde{x}_{132} \oplus (0, 0, 0, 0)_{LR} \otimes \\
\tilde{x}_{133} \oplus (0, 0, 0, 0)_{LR} \otimes \tilde{x}_{141} \oplus (0, 0, 0, 0)_{LR} \otimes \tilde{x}_{142} \oplus (0, 0, 0, 0)_{LR} \otimes \tilde{x}_{143} \\
\oplus (20, 20, 10, 10)_{LR} \otimes \tilde{x}_{211} \oplus (40, 40, 0, 0)_{LR} \otimes \tilde{x}_{212} \oplus (0, 0, 0, 0)_{LR} \otimes \tilde{x}_{213} \\
\oplus (20, 20, 0, 0)_{LR} \otimes \tilde{x}_{221} \oplus (50, 50, 0, 0)_{LR} \otimes \tilde{x}_{222} \oplus (0, 0, 0, 0)_{LR} \otimes \tilde{x}_{223} \\
\oplus (40, 40, 0, 0)_{LR} \otimes \tilde{x}_{231} \oplus (50, 50, 0, 0)_{LR} \otimes \tilde{x}_{232} \oplus (0, 0, 0, 0)_{LR} \otimes \tilde{x}_{233} \\
\oplus (0, 0, 0, 0)_{LR} \otimes \tilde{x}_{241} \oplus (0, 0, 0, 0)_{LR} \otimes \tilde{x}_{242} \oplus (0, 0, 0, 0)_{LR} \otimes \tilde{x}_{243} \oplus \\
(0, 0, 0, 0)_{LR} \otimes \tilde{x}_{311} \oplus (0, 0, 0, 0)_{LR} \otimes \tilde{x}_{312} \oplus (0, 0, 0, 0)_{LR} \otimes \tilde{x}_{313} \oplus \\
(0, 0, 0, 0)_{LR} \otimes \tilde{x}_{321} \oplus (0, 0, 0, 0)_{LR} \otimes \tilde{x}_{322} \oplus (0, 0, 0, 0)_{LR} \otimes \tilde{x}_{323} \oplus \\
(0, 0, 0, 0)_{LR} \otimes \tilde{x}_{331} \oplus (0, 0, 0, 0)_{LR} \otimes \tilde{x}_{332} \oplus (0, 0, 0, 0)_{LR} \otimes \tilde{x}_{333} \oplus \\
(0, 0, 0, 0)_{LR} \otimes \tilde{x}_{341} \oplus (0, 0, 0, 0)_{LR} \otimes \tilde{x}_{342} \oplus (0, 0, 0, 0)_{LR} \otimes \tilde{x}_{343}) \\
\text{subject to:} \\
\sum_{j=1}^{4} \sum_{k=1}^{3} \tilde{x}_{1jk} = (80, 100, 10, 20)_{LR} \\
\sum_{j=1}^{4} \sum_{k=1}^{3} \tilde{x}_{2jk} = (70, 70, 10, 20)_{LR} \\
\sum_{j=1}^{4} \sum_{k=1}^{3} \tilde{x}_{3jk} = (20, 30, 20, 0)_{LR} \\
\sum_{i=1}^{3} \sum_{k=1}^{3} \tilde{x}_{i1k} = (30, 40, 20, 10)_{LR} \\
\sum_{i=1}^{3} \sum_{k=1}^{3} \tilde{x}_{i2k} = (50, 50, 10, 10)_{LR} \\
\sum_{i=1}^{3} \sum_{k=1}^{3} \tilde{x}_{i3k} = (40, 60, 10, 10)_{LR} \\
\sum_{i=1}^{3} \sum_{k=1}^{3} \tilde{x}_{i4k} = (50, 50, 0, 10)_{LR} \\
\sum_{i=1}^{3} \sum_{j=1}^{4} \tilde{x}_{ij1} = (80, 80, 10, 20)_{LR} \\
\sum_{i=1}^{3} \sum_{j=1}^{4} \tilde{x}_{ij2} = (70, 70, 10, 20)_{LR} \\
\sum_{i=1}^{3} \sum_{j=1}^{4} \tilde{x}_{ij3} = (20, 50, 20, 0)_{LR}
\end{cases}$$

where: $\tilde{x}_{ijk}$'s are non-negative trapezoidal fuzzy numbers, for all $i = 1, 2, 3;\ j = 1, 2, 3, 4;\ k = 1, 2, 3$.

Step 5 Using Steps 7–9 of the method proposed in Sect. 7.5.2, the fuzzy linear programming problem, obtained in Step 4, can be converted into the following crisp linear programming problem:

$$
\begin{cases}
\min(\frac{1}{4}(50m_{111} + 70n_{111} - 20\alpha_{111} + 40\beta_{111} + 140m_{112} + \\
+ 140n_{112} - 70\alpha_{112} + 70\beta_{112} + 120m_{121} + 120n_{121} - 60\alpha_{121} + \\
+ 60\beta_{121} + 120m_{122} + 120n_{122} - 60\alpha_{122} + 60\beta_{122} + 100m_{131} + \\
+ 100n_{131} - 50\alpha_{131} + 50\beta_{131} + 60m_{132} + 60n_{132} - 30\alpha_{132} + \\
+ 30\beta_{132} + 30m_{211} + 50n_{211} - 10\alpha_{211} + 30\beta_{211} + 80m_{212} + \\
+ 80n_{212} - 40\alpha_{212} + 40\beta_{212} + 40m_{221} + 40n_{221} - 20\alpha_{221} + \\
+ 20\beta_{221} + 100m_{222} + 100n_{222} - 50\alpha_{222} + 50\beta_{222} + 80m_{231} + \\
+ 80n_{231} - 40\alpha_{231} + 40\beta_{231} + 100m_{232} + 100n_{232} - 50\alpha_{232} + 50\beta_{232})) \\
\text{subject to:} \\
\sum_{j=1}^{4}\sum_{k=1}^{3} m_{1jk} = 80,\ \sum_{j=1}^{4}\sum_{k=1}^{3} n_{1jk} = 100 \\
\sum_{j=1}^{4}\sum_{k=1}^{3} \alpha_{1jk} = 10,\ \sum_{j=1}^{4}\sum_{k=1}^{3} \beta_{1jk} = 20 \\
\sum_{j=1}^{4}\sum_{k=1}^{3} m_{2jk} = 70,\ \sum_{j=1}^{4}\sum_{k=1}^{3} n_{2jk} = 70 \\
\sum_{j=1}^{4}\sum_{k=1}^{3} \alpha_{2jk} = 10,\ \sum_{j=1}^{4}\sum_{k=1}^{3} \beta_{2jk} = 20 \\
\sum_{j=1}^{4}\sum_{k=1}^{3} m_{3jk} = 20,\ \sum_{j=1}^{4}\sum_{k=1}^{3} n_{3jk} = 30 \\
\sum_{j=1}^{4}\sum_{k=1}^{3} \alpha_{3jk} = 20,\ \sum_{j=1}^{4}\sum_{k=1}^{3} \beta_{3jk} = 0 \\
\sum_{i=1}^{3}\sum_{k=1}^{3} m_{i1k} = 30,\ \sum_{i=1}^{3}\sum_{k=1}^{3} n_{i1k} = 40 \\
\sum_{i=1}^{3}\sum_{k=1}^{3} \alpha_{i1k} = 20,\ \sum_{i=1}^{3}\sum_{k=1}^{3} \beta_{i1k} = 10 \\
\sum_{i=1}^{3}\sum_{k=1}^{3} m_{i2k} = 50,\ \sum_{i=1}^{3}\sum_{k=1}^{3} n_{i2k} = 50 \\
\sum_{i=1}^{3}\sum_{k=1}^{3} \alpha_{i2k} = 10,\ \sum_{i=1}^{3}\sum_{k=1}^{3} \beta_{i2k} = 10 \\
\sum_{i=1}^{3}\sum_{k=1}^{3} m_{i3k} = 40,\ \sum_{i=1}^{3}\sum_{k=1}^{3} n_{i3k} = 60 \\
\sum_{i=1}^{3}\sum_{k=1}^{3} \alpha_{i3k} = 10,\ \sum_{i=1}^{3}\sum_{k=1}^{3} \beta_{i3k} = 10 \\
\sum_{i=1}^{3}\sum_{k=1}^{3} m_{i4k} = 50,\ \sum_{i=1}^{3}\sum_{k=1}^{3} n_{i4k} = 50 \\
\sum_{i=1}^{3}\sum_{k=1}^{3} \alpha_{i4k} = 0,\ \sum_{i=1}^{3}\sum_{k=1}^{3} \beta_{i4k} = 10 \\
\sum_{i=1}^{3}\sum_{j=1}^{4} m_{ij1} = 80,\ \sum_{i=1}^{3}\sum_{j=1}^{4} n_{ij1} = 80 \\
\sum_{i=1}^{3}\sum_{j=1}^{4} \alpha_{ij1} = 10,\ \sum_{i=1}^{3}\sum_{j=1}^{4} \beta_{ij1} = 20 \\
\sum_{i=1}^{3}\sum_{j=1}^{4} m_{ij2} = 70,\ \sum_{i=1}^{3}\sum_{j=1}^{4} n_{ij2} = 70 \\
\sum_{i=1}^{3}\sum_{j=1}^{4} \alpha_{ij2} = 10,\ \sum_{i=1}^{3}\sum_{j=1}^{4} \beta_{ij2} = 20 \\
\sum_{i=1}^{3}\sum_{j=1}^{4} m_{ij3} = 20,\ \sum_{i=1}^{3}\sum_{j=1}^{4} n_{ij3} = 50 \\
\sum_{i=1}^{3}\sum_{j=1}^{4} \alpha_{ij3} = 20,\ \sum_{i=1}^{3}\sum_{j=1}^{4} \beta_{ij3} = 0
\end{cases}
$$

where: $m_{ijk} - \alpha_{ijk}$, $n_{ijk} - m_{ijk}$, α_{ijk}, $\beta_{ijk} \geq 0$, for all $i = 1, 2, 3$; $j = 1, 2, 3, 4$; $k = 1, 2, 3$

Step 6 The optimal solution of the crisp linear programming problem obtained in Step 5, is:

$$\begin{aligned}
&m_{113} = 10,\ n_{113} = 10,\ \alpha_{113} = 10,\\
&m_{132} = 30,\ n_{132} = 30,\ \beta_{132} = 10,\\
&n_{133} = 20,\ m_{141} = 10,\ n_{141} = 10,\\
&m_{142} = 30,\ n_{142} = 30,\ \beta_{142} = 10,\\
&m_{211} = 10,\ n_{211} = 10,\ \beta_{211} = 10,\\
&m_{221} = 40,\ n_{221} = 40,\ \beta_{221} = 10,\\
&m_{241} = 10,\ n_{241} = 10,\ m_{213} = 10,\\
&n_{213} = 10,\ \alpha_{213} = 10,\ m_{321} = 10,\\
&n_{321} = 10,\ \alpha_{321} = 10,\ n_{313} = 10,\\
&m_{332} = 10,\ n_{332} = 10,\ \alpha_{332} = 10,
\end{aligned}$$

and the remaining values of m_{ijk}, n_{ijk}, α_{ijk} and β_{ijk} are 0, respectively.

Step 7 Putting the values of m_{ijk}, n_{ijk}, α_{ijk} and β_{ijk} into

$$\tilde{x}_{ijk} = (m_{ijk}, n_{ijk}, \alpha_{ijk}, \beta_{ijk})_{LR}$$

the fuzzy optimal solution is obtained as

$$\begin{aligned}
&\tilde{x}_{113} = (10, 10, 10, 0)_{LR},\ \tilde{x}_{132} = (30, 30, 0, 10)_{LR},\ \tilde{x}_{133} = (0, 20, 0, 0)_{LR},\\
&\tilde{x}_{141} = (10, 10, 0, 0)_{LR},\ \tilde{x}_{142} = (30, 30, 0, 10)_{LR},\ \tilde{x}_{211} = (10, 10, 0, 10)_{LR},\\
&\tilde{x}_{221} = (40, 40, 0, 10)_{LR},\ \tilde{x}_{241} = (10, 10, 0, 0)_{LR},\ \tilde{x}_{213} = (10, 10, 10, 0)_{LR},\\
&\tilde{x}_{321} = (10, 10, 10, 0)_{LR},\ \tilde{x}_{313} = (0, 10, 0, 0)_{LR},\ \tilde{x}_{332} = (10, 10, 10, 0)_{LR},
\end{aligned}$$

and the remaining values of $\tilde{x}_{ijk}$ are 0, respectively.

Step 8 Putting the values of

$$\begin{aligned}
&\tilde{x}_{111}, \tilde{x}_{112}, \tilde{x}_{113}\\
&\tilde{x}_{121}, \tilde{x}_{122}, \tilde{x}_{123}\\
&\tilde{x}_{131}, \tilde{x}_{132}, \tilde{x}_{133}\\
&\tilde{x}_{141}, \tilde{x}_{142}, \tilde{x}_{143}\\
&\tilde{x}_{211}, \tilde{x}_{212}, \tilde{x}_{213}\\
&\tilde{x}_{221}, \tilde{x}_{222}, \tilde{x}_{223}\\
&\tilde{x}_{231}, \tilde{x}_{232}, \tilde{x}_{233}\\
&\tilde{x}_{241}, \tilde{x}_{242}, \tilde{x}_{243}\\
&\tilde{x}_{311}, \tilde{x}_{312}, \tilde{x}_{313}\\
&\tilde{x}_{321}, \tilde{x}_{322}, \tilde{x}_{323}\\
&\tilde{x}_{331}, \tilde{x}_{332}, \tilde{x}_{333}\\
&\tilde{x}_{341}, \tilde{x}_{342}, \tilde{x}_{343}
\end{aligned}$$

into

$((30, 30, 10, 10)_{LR} \otimes \tilde{x}_{111} \oplus (70, 70, 0, 0)_{LR} \otimes \tilde{x}_{112} \oplus (0, 0, 0, 0)_{LR} \otimes \tilde{x}_{113}$
$\oplus (60, 60, 0, 0)_{LR} \otimes \tilde{x}_{121} \oplus (60, 60, 0, 0)_{LR} \otimes \tilde{x}_{122} \oplus (0, 0, 0, 0)_{LR} \otimes \tilde{x}_{123}$
$\oplus (50, 50, 0, 0)_{LR} \otimes \tilde{x}_{131} \oplus (30, 30, 0, 0)_{LR} \otimes \tilde{x}_{132} \oplus (0, 0, 0, 0)_{LR} \otimes \tilde{x}_{133}$
$\oplus (0, 0, 0, 0)_{LR} \otimes \tilde{x}_{141} \oplus (0, 0, 0, 0)_{LR} \otimes \tilde{x}_{142} \oplus (0, 0, 0, 0)_{LR} \otimes \tilde{x}_{143}$
$\oplus (20, 20, 10, 10)_{LR} \otimes \tilde{x}_{211} \oplus (40, 40, 0, 0)_{LR} \otimes \tilde{x}_{212} \oplus (0, 0, 0, 0)_{LR}$
$\otimes \tilde{x}_{213} \oplus (20, 20, 0, 0)_{LR} \otimes \tilde{x}_{221} \oplus (50, 50, 0, 0)_{LR} \otimes \tilde{x}_{222} \oplus (0, 0, 0, 0)_{LR}$
$\otimes \tilde{x}_{223} \oplus (40, 40, 0, 0)_{LR} \otimes \tilde{x}_{231} \oplus (50, 50, 0, 0)_{LR} \otimes \tilde{x}_{232} \oplus (0, 0, 0, 0)_{LR}$
$\otimes \tilde{x}_{233} \oplus (0, 0, 0, 0)_{LR} \otimes \tilde{x}_{241} \oplus (0, 0, 0, 0)_{LR} \otimes \tilde{x}_{242} \oplus (0, 0, 0, 0)_{LR}$
$\otimes \tilde{x}_{243} \oplus (0, 0, 0, 0)_{LR} \otimes \tilde{x}_{311} \oplus (0, 0, 0, 0)_{LR} \otimes \tilde{x}_{312} \oplus (0, 0, 0, 0)_{LR}$
$\otimes \tilde{x}_{313} \oplus (0, 0, 0, 0)_{LR} \otimes \tilde{x}_{321} \oplus (0, 0, 0, 0)_{LR} \otimes \tilde{x}_{322} \oplus (0, 0, 0, 0)_{LR} \otimes$
$\tilde{x}_{323} \oplus (0, 0, 0, 0)_{LR} \otimes \tilde{x}_{331} \oplus (0, 0, 0, 0)_{LR} \otimes \tilde{x}_{332} \oplus (0, 0, 0, 0)_{LR} \otimes \tilde{x}_{333}$
$\oplus (0, 0, 0, 0)_{LR} \otimes \tilde{x}_{341} \oplus (0, 0, 0, 0)_{LR} \otimes \tilde{x}_{342} \oplus$
$(0, 0, 0, 0)_{LR} \otimes \tilde{x}_{343})$

we obtain the minimum total fuzzy transportation $(1900, 1900, 100, 900)_{LR}$.

7.6.2 Determination of the Fuzzy Optimal Solution Using the New Method Based on the Tabular Representation

By using the proposed method, based on the tabular representation presented in Sect. 7.5.2, the fuzzy optimal solution of the fully fuzzy solid transportation problem shown in Example 20, can be obtained as follows:

Step 1 The balanced fully fuzzy solid transportation problem, obtained in Steps 1–Step 3 in Sect. 7.6, can be represented by Table 7.6.
Step 2 Using Step 3 of the method proposed in Sect. 7.5.2, Table 7.6 can be split into four crisp solid transportation tables: Tables 7.7, 7.8, 7.9 and 7.10.
Step 3 The optimal solution of the crisp solid transportation problems represented by by Tables 7.7, 7.8, 7.9 and 7.10 is:

$m_{113} - \alpha_{113} = 0, m_{132} - \alpha_{132} = 30, m_{133} - \alpha_{133} = 0, m_{141} - \alpha_{141} = 10,$
$m_{142} - \alpha_{142} = 30, m_{211} - \alpha_{211} = 10, m_{221} - \alpha_{221} = 40, m_{242} - \alpha_{242} = 10,$
$m_{213} - \alpha_{213} = 0, m_{321} - \alpha_{321} = 0, m_{313} - \alpha_{313} = 0, m_{332} - \alpha_{332} = 0,$
$\alpha_{113} = 10, \alpha_{132} = 0, \alpha_{133} = 0, \alpha_{141} = 0, \alpha_{142} = 0, \alpha_{211} = 0, \alpha_{221} = 0,$
$\alpha_{241} = 0, \alpha_{213} = 10, \alpha_{321} = 10, \alpha_{313} = 0, \alpha_{332} = 10, n_{113} - m_{113} = 0,$
$n_{132} - m_{132} = 0, n_{133} - m_{133} = 20, n_{141} - m_{141} = 0, n_{142} - m_{142} = 0$
$n_{211} - m_{211} = 0, n_{221} - m_{221} = 0, n_{242} - m_{242} = 0, n_{213} - m_{213} = 0$
$n_{321} - m_{321} = 0, n_{313} - m_{313} = 10, n_{332} - m_{332} = 0, \beta_{132} = 10, \beta_{142} = 10,$
$\beta_{211} = 10, \beta_{221} = 10.$

Step 4 By solving the set of equations shown in the above Step 3, we obtain the following values of m_{ijk}, n_{ijk}, α_{ijk} and β_{ijk}:

Table 7.6 Tabular representation of the balanced fully fuzzy solid transportation problem

													Capacity
	E_1			E_1			E_1			E_1			$(80, 80, 10, 20)_{LR}$
		E_2			E_2			E_2			E_2		$(70, 70, 10, 20)_{LR}$
			E_3			E_3			E_3			E_3	$(20, 50, 20, 0)_{LR}$
	D_1			D_2			D_3			D_4			
S_1	$(30, 30, 10, 10)_{LR}$	$(70, 70, 0, 0)_{LR}$	$(0, 0, 0, 0)_{LR}$	$(60, 60, 0, 0)_{LR}$	$(60, 60, 0, 0)_{LR}$	$(0, 0, 0, 0)_{LR}$	$(50, 50, 0, 0)_{LR}$	$(30, 30, 0, 0)_{LR}$	$(0, 0, 0, 0)_{LR}$	$(0, 0, 0, 0)_{LR}$	$(0, 0, 0, 0)_{LR}$	$(0, 0, 0, 0)_{LR}$	$(80, 100, 10, 20)_{LR}$
S_2	$(20, 20, 10, 10)_{LR}$	$(40, 40, 0, 0)_{LR}$	$(0, 0, 0, 0)_{LR}$	$(20, 20, 0, 0)_{LR}$	$(50, 50, 0, 0)_{LR}$	$(0, 0, 0, 0)_{LR}$	$(40, 40, 0, 0)_{LR}$	$(50, 50, 0, 0)_{LR}$	$(0, 0, 0, 0)_{LR}$	$(0, 0, 0, 0)_{LR}$	$(0, 0, 0, 0)_{LR}$	$(0, 0, 0, 0)_{LR}$	$(70, 70, 10, 20)_{LR}$
S_3	$(0, 0, 0, 0)_{LR}$	$(0, 0, 0, 0)_{LR}$	$(0, 0, 0, 0)_{LR}$	$(0, 0, 0, 0)_{LR}$	$(0, 0, 0, 0)_{LR}$	$(0, 0, 0, 0)_{LR}$	$(0, 0, 0, 0)_{LR}$	$(0, 0, 0, 0)_{LR}$	$(0, 0, 0, 0)_{LR}$	$(0, 0, 0, 0)_{LR}$	$(0, 0, 0, 0)_{LR}$	$(0, 0, 0, 0)_{LR}$	$(20, 30, 20, 0)_{LR}$
	$(30, 40, 20, 10)_{LR}$			$(50, 50, 10, 10)_{LR}$			$(40, 60, 10, 10)_{LR}$			$(50, 50, 0, 10)_{LR}$			

Table 7.7 Tabular representation of first crisp solid transportation problem

													Capacity $(m''_{ijk} - \alpha''_{ijk})$
	E_1			E_1			E_1			E_1			70
		E_2			E_2			E_2			E_2		60
			E_3			E_3			E_3			E_3	0
	D_1			D_2			D_3			D_4			$(m_i - \alpha_i)$
S_1	30	70	0	60	60	0	50	30	0	0	0	0	70
S_2	20	40	0	20	50	0	40	50	0	0	0	0	60
S_3	0	0	0	0	0	0	0	0	0	0	0	0	0
$(m'_j - \alpha'_j)$	10			40			30			50			

Table 7.8 Tabular representation of second crisp solid transportation problem

													Capacity (α''_{ijk})
	E_1			E_1			E_1			E_1			10
		E_2			E_2			E_2			E_2		10
			E_3			E_3			E_3			E_3	20
	D_1			D_2			D_3			D_4			(α_i)
S_1	25	52.5	0	45	45	0	37.5	22.5	0	0	0	0	10
S_2	17.5	30	0	15	37.5	0	30	37.5	0	0	0	0	10
S_3	0	0	0	0	0	0	0	0	0	0	0	0	20
(α'_j)	20			10			10			0			

Table 7.9 Tabular representation of third crisp solid transportation problem

													Capacity $(n''_{ijk} - m''_{ijk})$
	E_1			E_1			E_1			E_1			0
		E_2			E_2			E_2			E_2		0
			E_3			E_3			E_3			E_3	30
	D_1			D_2			D_3			D_4			$(n_i - m_i)$
S_1	17.5	35	0	30	30	0	25	15	0	0	0	0	20
S_2	12.5	20	0	10	25	0	20	25	0	0	0	0	0
S_3	0	0	0	0	0	0	0	0	0	0	0	0	10
$(n'_j - m'_j)$	10			0			20			0			

Table 7.10 Tabular representation of fourth crisp solid transportation problem

													Capacity (β''_{ijk})
	E_1			E_1			E_1			E_1			20
		E_2			E_2			E_2			E_2		20
			E_3			E_3			E_3			E_3	0
	D_1			D_2			D_3			D_4			(β_i)
S_1	10	17.5	0	15	15	0	12.5	7.5	0	0	0	0	20
S_2	7.5	10	0	5	12.5	0	10	12.5	0	0	0	0	20
S_3	0	0	0	0	0	0	0	0	0	0	0	0	0
(β'_{ijk})	10			10			10			10			

$$m_{113} = 10, n_{113} = 10, \alpha_{113} = 10, \beta_{113} = 0$$
$$m_{132} = 30, n_{132} = 30, \alpha_{132} = 0, \beta_{132} = 10$$
$$m_{133} = 0, n_{133} = 20, \alpha_{133} = 0, \beta_{133} = 0$$
$$m_{141} = 10, n_{141} = 10, \alpha_{141} = 0, \beta_{141} = 0$$
$$m_{142} = 30, n_{142} = 30, \alpha_{142} = 0, \beta_{142} = 10$$
$$m_{211} = 10, n_{211} = 10, \alpha_{211} = 0, \beta_{211} = 10$$
$$m_{221} = 40, n_{221} = 40, \alpha_{221} = 0, \beta_{221} = 10$$
$$m_{241} = 10, n_{241} = 10, \alpha_{241} = 0, \beta_{241} = 0$$
$$m_{213} = 10, n_{213} = 10, \alpha_{213} = 10, \beta_{213} = 0$$
$$m_{321} = 10, n_{321} = 10, \alpha_{321} = 10, \beta_{321} = 0$$
$$m_{313} = 0, n_{313} = 10, \alpha_{313} = 0, \beta_{313} = 0$$
$$m_{332} = 10, n_{332} = 10, \alpha_{332} = 10, \beta_{332} = 0$$

and the remaining values of $m_{ijk}, n_{ijk}, \alpha_{ijk}$ and β_{ijk} are zero.

Step 5 By putting the values of $m_{ijk}, n_{ijk}, \alpha_{ijk}$ and β_{ijk} into $\tilde{x}_{ijk} = (m_{ijk}, n_{ijk}, \alpha_{ijk}, \beta_{ijk})_{LR}$, the fuzzy optimal solution is obtained as:

$$\tilde{x}_{113} = (10, 10, 10, 0)_{LR}$$
$$\tilde{x}_{132} = (30, 30, 0, 10)_{LR}$$
$$\tilde{x}_{133} = (0, 20, 0, 0)_{LR}$$
$$\tilde{x}_{141} = (10, 10, 0, 0)_{LR}$$
$$\tilde{x}_{142} = (30, 30, 0, 10)_{LR}$$
$$\tilde{x}_{211} = (10, 10, 0, 10)_{LR}$$
$$\tilde{x}_{221} = (40, 40, 0, 10)_{LR}$$
$$\tilde{x}_{241} = (10, 10, 0, 0)_{LR}$$
$$\tilde{x}_{213} = (10, 10, 10, 0)_{LR}$$
$$\tilde{x}_{321} = (10, 10, 10, 0)_{LR}$$
$$\tilde{x}_{313} = (0, 10, 0, 0)_{LR}$$
$$\tilde{x}_{332} = (10, 10, 10, 0)_{LR}$$

and the remaining values of $\tilde{x}_{ijk}$ are zero.

Step 6 by putting the values of

$$\begin{array}{c}\tilde{x}_{111}, \tilde{x}_{112}, \tilde{x}_{113}\\ \tilde{x}_{121}, \tilde{x}_{122}, \tilde{x}_{123}\\ \tilde{x}_{131}, \tilde{x}_{132}, \tilde{x}_{133}\\ \tilde{x}_{141}, \tilde{x}_{142}, \tilde{x}_{143}\\ \tilde{x}_{211}, \tilde{x}_{212}, \tilde{x}_{213}\\ \tilde{x}_{221}, \tilde{x}_{222}, \tilde{x}_{223}\\ \tilde{x}_{231}, \tilde{x}_{232}, \tilde{x}_{233}\\ \tilde{x}_{241}, \tilde{x}_{242}, \tilde{x}_{243}\\ \tilde{x}_{311}, \tilde{x}_{312}, \tilde{x}_{313}\\ \tilde{x}_{321}, \tilde{x}_{322}, \tilde{x}_{323}\\ \tilde{x}_{331}, \tilde{x}_{332}, \tilde{x}_{333}\\ \tilde{x}_{341}, \tilde{x}_{342}, \tilde{x}_{343}\end{array}$$

into

$((30, 30, 10, 10)_{LR} \otimes \tilde{x}_{111} \oplus (70, 70, 0, 0)_{LR} \otimes \tilde{x}_{112} \oplus (0, 0, 0, 0)_{LR} \otimes \tilde{x}_{113} \oplus (60, 60, 0, 0)_{LR} \otimes \tilde{x}_{121} \oplus (60, 60, 0, 0)_{LR} \otimes \tilde{x}_{122} \oplus (0, 0, 0, 0)_{LR} \otimes \tilde{x}_{123} \oplus (50, 50, 0, 0)_{LR} \otimes \tilde{x}_{131} \oplus (30, 30, 0, 0)_{LR} \otimes \tilde{x}_{132} \oplus (0, 0, 0, 0)_{LR} \otimes \tilde{x}_{133} \oplus (0, 0, 0, 0)_{LR} \otimes \tilde{x}_{141} \oplus (0, 0, 0, 0)_{LR} \otimes \tilde{x}_{142} \oplus (0, 0, 0, 0)_{LR} \otimes \tilde{x}_{143} \oplus (20, 20, 10, 10)_{LR} \otimes \tilde{x}_{211} \oplus (40, 40, 0, 0)_{LR} \otimes \tilde{x}_{212} \oplus (0, 0, 0, 0)_{LR} \otimes \tilde{x}_{213} \oplus (20, 20, 0, 0)_{LR} \otimes \tilde{x}_{221} \oplus (50, 50, 0, 0)_{LR} \otimes \tilde{x}_{222} \oplus (0, 0, 0, 0)_{LR} \otimes \tilde{x}_{223} \oplus (40, 40, 0, 0)_{LR} \otimes \tilde{x}_{231} \oplus (50, 50, 0, 0)_{LR} \otimes \tilde{x}_{232} \oplus (0, 0, 0, 0)_{LR} \otimes \tilde{x}_{233} \oplus (0, 0, 0, 0)_{LR} \otimes \tilde{x}_{241} \oplus (0, 0, 0, 0)_{LR} \otimes \tilde{x}_{242} \oplus (0, 0, 0, 0)_{LR} \otimes \tilde{x}_{243} \oplus (0, 0, 0, 0)_{LR} \otimes \tilde{x}_{311} \oplus (0, 0, 0, 0)_{LR} \otimes \tilde{x}_{312} \oplus (0, 0, 0, 0)_{LR} \otimes \tilde{x}_{313} \oplus (0, 0, 0, 0)_{LR} \otimes \tilde{x}_{321} \oplus (0, 0, 0, 0)_{LR} \otimes \tilde{x}_{322} \oplus (0, 0, 0, 0)_{LR} \otimes \tilde{x}_{323} \oplus (0, 0, 0, 0)_{LR} \otimes \tilde{x}_{331} \oplus (0, 0, 0, 0)_{LR} \otimes \tilde{x}_{332} \oplus (0, 0, 0, 0)_{LR} \otimes \tilde{x}_{333} \oplus (0, 0, 0, 0)_{LR} \otimes \tilde{x}_{341} \oplus (0, 0, 0, 0)_{LR} \otimes \tilde{x}_{342} \oplus (0, 0, 0, 0)_{LR} \otimes \tilde{x}_{343})$

the minimum total fuzzy transportation cost is $(1900, 1900, 100, 900)_{LR}$.

7.6.3 *Interpretation of Results*

The results obtained by using the method proposed, more specifically the minimum fuzzy total fuzzy transportation cost, found by solving the example employed, i.e. Example 20, solved by using the proposed method based on the use of fuzzy mathematical programming (cf. Sect. 7.5.2) and based on the tabular representation can be interpreted in an intuitively appealing and comprehensible way. Namely:

1. The least amount of the minimum total transportation cost is 1800,
2. The most possible amount of the minimum total transportation cost is 1900,
3. The greatest amount of the minimum total transportation cost is 2800,

that is, the minimum total transportation cost will be always greater than 1800 and lower than 2800, and the most possible value of the minimum total transportation cost will be 1900.

7.7 A Comparative Study

An interesting and illustrative comparison of the results obtained by the methods for solving the fully fuzzy transportation problem proposed in this chapter, In this chapter, and by the methods proposed in Chaps. 4 and 5, is shown in Table 7.11.

The results shown in Table 7.11 can be summarized as follows:

1. The methods proposed in Chap. 4, can only be used for solving such fully fuzzy transportation problems in which either all the parameters are represented by the triangular fuzzy numbers or by the trapezoidal fuzzy numbers. Similarly, the methods proposed in Chap. 5 can only be used for solving such fully fuzzy transportation problems in which all the parameters are represented by the LR flat fuzzy numbers. Since, in the fully fuzzy transportation problem presented in Example 17, all the parameters are represented by the LR flat fuzzy numbers, then the problem considered can not be solved by the method proposed in Chap. 4 but the same problem can be solved by the method proposed in Chap. 5. Moreover, since the fully fuzzy solid transportation problems are a generalization of the fully fuzzy transportation problems so, then the fully fuzzy solid transportation problem shown in Example 20, can not be solved by the methods proposed in Chaps. 4 and 5.
2. Since, the fully fuzzy transportation problems are a special type of the fully fuzzy solid transportation problems, then the methods proposed in this chapter, can also be used for solving the fully fuzzy transportation problems. That is, the method, proposed in this chapter, can be used to find the fuzzy optimal solution of all the problems considered.

Table 7.11 Results obtained by using the proposed methods for solving the fully fuzzy transportation problem

Example	Minimum total fuzzy transportation cost		
	Methods proposed in Chap. 4	Methods proposed in Chap. 5	Methods proposed in this chapter
2.2	(2100, 2900, 3500, 4200)	(2100, 2900, 3500, 4200)	(2100, 2900, 3500, 4200)
3.1	Not applicable	$(5800, 8400, 2800, 2900)_{LR}$	$(5800, 8400, 2800, 2900)_{LR}$
5.1	Not applicable	Not applicable	$(1900, 1900, 100, 900)_{LR}$

7.8 A Case Study

As we have already mentioned, Liu and Kao [5] proposed a method to find the crisp optimal solution of fully fuzzy fixed charge solid transportation problems and used it to find the crisp optimal solution of a real life fully fuzzy fixed charge solid transportation problem described in Sect. 7.8.1.

However, in Chap. 2, we have mentioned that it is better to find the fuzzy optimal solution than the crisp optimal solution. Therefore, for finding the fuzzy optimal solution of the same real life problem, assuming that there is no fixed charge, the methods proposed in this book are employed.

7.8.1 Problem Description

One of the most important and commonly used source of energy in the present day world is coal. Its important for the world economy is clearly very high. Since the use of coal is closely associated with high volume and expensive transportation from coal mines to the users, for instance power stations which may be far away, even on different continents, then the importance of devising a proper, notably most economical, transportation of coal is crucial. That is why it makes much sense to try to use for this purpose optimization tools, notably fuzzy transportation models. Obviously, due to imprecise information that is overwhelming in such problems, fuzzy transportation models can be of much use.

The illustrative and intuitively appealing problem of coal transportation to be briefly presented can be summarized as follows. Suppose that there are two coal mines to supply the coal for two cities. During the process of transportation, two kinds of transportation means are available, i.e. a train and a ship. The decision making problem consists in the determination of the transportation plan for the next month. In the beginning the decision maker needs to obtain basic data, such as the supply capacity, demand, transportation capacity, transportation cost per product unit, etc. In fact, since the transportation plan is to be made in advance, usually we do not have exact (precise) data. In such a case, the fuzzy data can be very convenient and can be obtained via expert evaluations.

In this example, the notation $\tilde{a}_i$, $\tilde{b}_j$ and $\tilde{e}_k$ is employed to denote the availability, demand and transportation capacity, respectively. The corresponding fuzzy data, with $L(x) = R(x) = \max\{0, 1 - x\}$ are listed as follows:

$\tilde{a}_1 = (28, 29, 3, 2)_{LR}$ $\tilde{a}_2 = (20, 23, 4, 2)_{LR}$ $\tilde{a}_3 = (34, 36, 2, 2)_{LR}$ $\tilde{a}_4 = (30, 32, 2, 2)_{LR}$
$\tilde{b}_1 = (13, 14, 1, 4)_{LR}$ $\tilde{b}_2 = (23, 26, 3, 1)_{LR}$ $\tilde{b}_3 = (21, 23, 2, 1)_{LR}$ $\tilde{b}_4 = (27, 29, 2, 2)_{LR}$
$\tilde{e}_1 = (45, 50, 6, 5)_{LR}$ $\tilde{e}_2 = (65, 70, 5, 5)_{LR}$

Table 7.12 Direct transportation cost of a unit amount by train

Cities→ Mines↓	1	2	3	4
1	$(5, 8, 2, 2)_{LR}$	$(8, 9, 1, 1)_{LR}$	$(17, 19, 2, 1)_{LR}$	$(15, 17, 2, 2)_{LR}$
2	$(7, 8, 2, 1)_{LR}$	$(9, 10, 3, 1)_{LR}$	$(4, 6, 1, 1)_{LR}$	$(20, 23, 4, 2)_{LR}$
3	$(8, 9, 3, 1)_{LR}$	$(15, 17, 2, 2)_{LR}$	$(6, 8, 3, 1)_{LR}$	$(8, 11, 1, 2)_{LR}$
4	$(19, 21, 1, 3)_{LR}$	$(12, 13, 3, 2)_{LR}$	$(10, 11, 3, 3)_{LR}$	$(10, 13, 1, 2)_{LR}$

Table 7.13 Direct transportation cost of a unit amount by ship

Cities→ Mines↓	1	2	3	4
1	$(9, 12, 2, 1)_{LR}$	$(9, 10, 4, 2)_{LR}$	$(12, 13, 3, 2)_{LR}$	$(25, 26, 5, 2)_{LR}$
2	$(12, 14, 2, 1)_{LR}$	$(14, 16, 2, 2)_{LR}$	$(17, 18, 6, 2)_{LR}$	$(5, 8, 2, 2)_{LR}$
3	$(10, 12, 2, 2)_{LR}$	$(23, 25, 3, 2)_{LR}$	$(25, 27, 2, 2)_{LR}$	$(8, 10, 2, 2)_{LR}$
4	$(8, 9, 2, 1)_{LR}$	$(28, 30, 2, 2)_{LR}$	$(32, 33, 2, 2)_{LR}$	$(32, 38, 2, 2)_{LR}$

The same argument applies to the transportation cost of a unit amount which practically always can not be obtained accurately in advance. It can also be modeled as a fuzzy variable and determined by expert testimonies or evaluations. For this example, the transportation cost of a unit amount is listed in Table 7.12 for train and in Table 7.13 for ship.

The problem is now to determine how much of the product (coal) is to be transported from which coal mine to which city and by which means of transportation so that the the total fuzzy transportation cost be minimized.

7.8.2 *Results*

By using the methods proposed to solve the problem outlined in the previous section, the fuzzy optimal solution $\{\tilde{x}_{ijk}\}$ determined, i.e. the fuzzy quantities of the coal that should be transported from the ith coal mine to the jth city by the kth means of transportation, which minimizes the total fuzzy transportation cost, are:

$$
\begin{aligned}
\tilde{x}_{111} &= (7, 7, 0, 0)_{LR}\\
\tilde{x}_{121} &= (16, 17, 3, 0)_{LR}\\
\tilde{x}_{231} &= (2, 3, 2, 0)_{LR}\\
\tilde{x}_{331} &= (19, 20, 0, 0)_{LR}\\
\tilde{x}_{122} &= (5, 5, 0, 0)_{LR}\\
\tilde{x}_{242} &= (18, 20, 2, 0)_{LR}\\
\tilde{x}_{342} &= (9, 9, 0, 0)_{LR}\\
\tilde{x}_{412} &= (6, 7, 1, 0)_{LR}\\
\tilde{x}_{352} &= (5, 6, 1, 0)_{LR}\\
\tilde{x}_{423} &= (2, 2, 0, 0)_{LR}\\
\tilde{x}_{452} &= (22, 23, 1, 0)_{LR}\\
\tilde{x}_{351} &= (1, 1, 1, 0)_{LR}\\
\tilde{x}_{521} &= (0, 2, 0, 0)_{LR}\\
\tilde{x}_{152} &= (0, 0, 0, 2)_{LR}\\
\tilde{x}_{213} &= (0, 0, 0, 1)_{LR}\\
\tilde{x}_{252} &= (0, 0, 0, 1)_{LR}\\
\tilde{x}_{343} &= (0, 0, 0, 2)_{LR}\\
\tilde{x}_{451} &= (0, 0, 0, 2)_{LR}\\
\tilde{x}_{511} &= (0, 0, 0, 3)_{LR}\\
\tilde{x}_{522} &= (0, 0, 0, 1)_{LR}\\
\tilde{x}_{532} &= (0, 0, 0, 1)_{LR}
\end{aligned}
$$

and the fuzzy optimal value, i.e. the minimum total fuzzy transportation cost, is $(540, 750, 214, 129)_{LR}$.

We have an important and interesting remark:

Remark 22 Since, the real life problem considered is unbalanced, then find its solution a dummy source (5), a dummy destination (5) and a dummy means of transportation (conveyance) (3) is introduced. In the results obtained, presented in Sect. 7.8.2, $\tilde{x}_{5jk}$, $\tilde{x}_{i5k}$ and $\tilde{x}_{ij3}$ represents the fuzzy quantity of the product that should be transported from the dummy source (5) to the jth destination by means of the kth means of transportation, from the ith source to the dummy destination (5) by means of the kth means of transportation, and from the ith source to the jth destination by means of the dummy means of transportation (3), respectively.

7.8.3 Interpretation of Results

In the example considered the minimum total fuzzy transportation cost is obtained as $(540, 750, 214, 129)_{LR}$ which can be interpreted as follows:

1. the least amount of the minimum total transportation cost is 326 units,
2. the most possible amount of the minimum total transportation cost lies between 540 units and 750 units, and
3. the greatest amount of the minimum total transportation cost is 879 units,

that is, the minimum total transportation cost will be always greater than 326 units and less than 879 units and the highest possibility is that the minimum total transportation cost will lie between 540 units and 750 units.

7.9 Concluding Remarks

By comparing the results obtained in the previous sections, it can be seen that the use of the new method proposed for solving the fuzzy solid transportation problems makes it possible to obtain results that do not show some deficiencies implied by the use of the traditional method by Liu and Kao [5]. Moreover, which is a very important result too, all problems which can be solved by using the methods proposed in Chaps. 4 and 5, can also be solved by the methods proposed in this chapter. There can also exist some problems which can not be solved by using the methods proposed in Chaps. 4 and 5 but can be solved by the methods proposed in this chapter. Therefore, it is better to use the methods proposed in this chapter than both the existing Liu and Kao's [5] method and the methods proposed in Chaps. 4 and 5.

References

1. M. Ghatee, S.M. Hashemi, Ranking function-based solutions of fully fuzzified minimal cost flow problem. Inf. Sci. **177**, 4271–4294 (2007)
2. K.B. Haley, The existence of a solution to the multi-index problem. Oper. Res. **16**, 471–474 (1965)
3. B. Julien, An extension to possibilistic linear programming. Fuzzy Sets Syst. **64**, 195–206 (1994)
4. T.S. Liou, M.J. Wang, Ranking fuzzy number with integral values. Fuzzy Sets Syst. **50**, 247–255 (1992)
5. S.T. Liu, C. Kao, Solving fuzzy transportation problems based on extension principle. Eur. J. Oper. Res. **153**, 661–674 (2004)
6. M.A. Parra, T.A. Bilbao, M.V.R. Uria, Solving the multiobjective possibilistic linear programming problem. Eur. J. Oper. Res. **117**, 79–90 (1999)

Chapter 8
New Methods for Solving Fully Fuzzy Solid Transshipment Problems with *LR* Flat Fuzzy Numbers

The fully fuzzy transshipment problems are obtained by introducing intermediate nodes in the fully fuzzy transportation problems while the fully fuzzy solid transportation problems are obtained by introducing additional conveyances in the fully fuzzy transportation problems. However, in real life problems both the intermediate nodes and additional conveyances occur simultaneously. Therefore, in this chapter, by combining the concepts of the fully fuzzy solid transportation problem and the fully fuzzy transshipment problem, a new type of problems, called the *fully fuzzy solid transshipment problem* is introduced and discussed. More specifically, its equivalent fuzzy linear programming formulation is presented and two new methods for finding its fuzzy optimal solution are proposed. Moreover, some advantages of the new methods proposed over both the methods proposed in the previous chapters and a known existing approach by Ghatee and Hashemi [1] are pointed out. For illustration we will solve a fully fuzzy solid transshipment problem.

8.1 New Fuzzy Linear Programming Formulation of the Balanced Fully Fuzzy Solid Transshipment Problem

In this section, a new fuzzy linear programming formulation of the balanced fully fuzzy solid transshipment problems is proposed.

Let:

- $\tilde{a}_i$ and $\tilde{a}'_i$ be the fuzzy availability of the product at the *ith* purely source node and at the *ith* source node,
- $\tilde{b}_j$ and $\tilde{b}'_j$ be the fuzzy demand of the product at the *jth* purely destination node and at the *jth* destination node,

A. Kaur et al., *Fuzzy Transportation and Transshipment Problems*, Studies in Fuzziness and Soft Computing 385,
https://doi.org/10.1007/978-3-030-26676-9_8

- $\tilde{e}_k$ be the fuzzy capacity of the *kth* conveyance (the maximum fuzzy quantity of the product which can be carried by the *kth* conveyance),
- $\tilde{c}_{ijk}$ be the fuzzy cost for transporting one unit quantity of the product from the *ith* source to the *jth* destination by means of the *kth* conveyance, and
- $\tilde{x}_{ijk}$ be the fuzzy quantity of the product that should be transported from the *ith* node to the *jth* node by means of the *kth* conveyance to minimize the total fuzzy transportation cost.

Then, the fuzzy linear programming formulation of the balanced fully fuzzy solid transshipment problem can be written as:

$$\begin{cases} \min \sum\limits_{(i,j)\in A} \sum\limits_{k\in S_C} (\tilde{c}_{ijk} \otimes \tilde{x}_{ijk}) \\ \text{subject to:} \\ \sum\limits_{j:(i,j)\in A} \sum\limits_{k\in S_C} \tilde{x}_{ijk} = \tilde{a}_i;\ i \in N_{PS} \\ \sum\limits_{j:(i,j)\in A} \sum\limits_{k\in S_C} \tilde{x}_{ijk} \ominus_H \sum\limits_{j:(j,i)\in A} \sum\limits_{k\in S_C} \tilde{x}_{jik} = \tilde{a}'_i;\ i \in N_S \\ \sum\limits_{i:(i,j)\in A} \sum\limits_{k\in S_C} \tilde{x}_{ijk} = \tilde{b}_j;\ j \in N_{PD} \\ \sum\limits_{i:(i,j)\in A} \sum\limits_{k\in S_C} \tilde{x}_{ijk} \ominus_H \sum\limits_{i:(j,i)\in A} \sum\limits_{k\in S_C} \tilde{x}_{jik} = \tilde{b}'_j;\ j \in N_D \\ \sum\limits_{j:(i,j)\in A} \sum\limits_{k\in S_C} \tilde{x}_{ijk} = \sum\limits_{j:(j,i)\in A} \sum\limits_{k\in S_C} \tilde{x}_{jik};\ i \in N_T \\ \sum\limits_{i:(i,j)\in A} \tilde{x}_{ijk} = \tilde{e}_k;\ k \in S_C \end{cases} \tag{8.1}$$

where: $\tilde{x}_{ijk}$ is a non-negative LR flat fuzzy number, $\forall\ (i, j) \in A, k \in S_C$, A is set of arcs (i, j) joining node i and node j, and S_C is the set of all available conveyances.

We have first the following important remark:

Remark 23 If

$$\sum_{i\in N_{PS}} \tilde{a}_i \oplus \sum_{i\in N_S} \tilde{a}'_i == \sum_{j\in N_{PD}} \tilde{b}_j \oplus \sum_{j\in N_D} \tilde{b}'_j = \sum_{k\in S_C} \tilde{e}_k \tag{8.2}$$

then the fully fuzzy solid transshipment problem is said to be the *balanced fully fuzzy solid transshipment problem.* Otherwise it is said to be the *unbalanced fully fuzzy solid transshipment problem.*

8.2 Limitations of the Existing Method and Methods Proposed in Previous Chapters

Obviously, the fully fuzzy solid transshipment problems are a generalization of the fully fuzzy transportation problems, the fully fuzzy solid transportation problems and the fully fuzzy transshipment problems. Therefore, the methods, proposed for solving

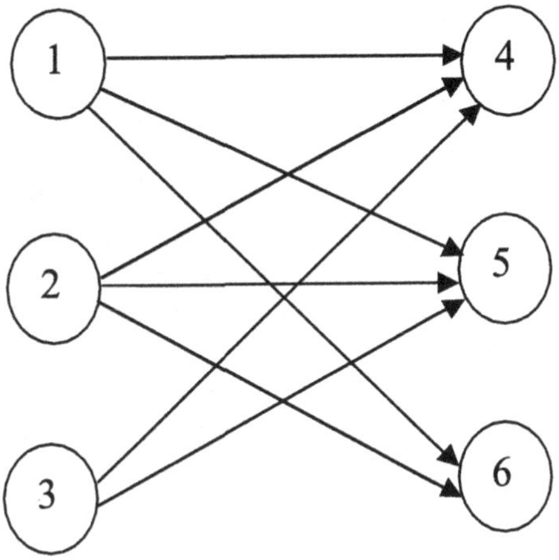

Fig. 8.1 Network representing the fully fuzzy solid transshipment problem

these problems cannot be directly used for solving the fully fuzzy solid transshipment problems, for instance, exemplified by the one shown in Example 21 presented below, either by using the known existing method by Ghatee and Hashemi [1] or the methods proposed in the previous chapters.

Example 21 Consider a network with three nodes, shown in Fig. 8.1 including one purely source node (2), one source node (1) and one purely destination node (3).

The fuzzy cost $\tilde{c}_{ijk}$, fuzzy availability $\tilde{a}_i$, fuzzy demand $\tilde{b}_j$ and the fuzzy capacity $\tilde{e}_k$ are represented by the following LR flat fuzzy numbers:

- Fuzzy costs:

$$\tilde{c}_{131} = (8, 10, 2, 2)_{LR}\ \tilde{c}_{132} = (4, 8, 3, 2)_{LR}$$
$$\tilde{c}_{211} = (8, 10, 4, 4)_{LR}\ \tilde{c}_{212} = (6, 8, 4, 4)_{LR}$$
$$\tilde{c}_{231} = (9, 12, 6, 3)_{LR}\ \tilde{c}_{232} = (3, 6, 2, 3)_{LR}$$

- Fuzzy availability:

$$\tilde{a}_1 = (60, 80, 20, 20)_{LR}\ \tilde{a}_2 = (50, 70, 20, 20)_{LR}$$

- Fuzzy demand:

$$\tilde{b}_3 = (50, 80, 30, 50)_{LR}$$

- Fuzzy capacity:

$$\tilde{e}_1 = (60, 80, 20, 10)_{LR}\ \tilde{e}_2 = (50, 70, 20, 40)_{LR}$$

where $L(x) = R(x) = \max\{0, 1 - x\}$.

The problem is to find the fuzzy optimum shipping schedule.

8.3 New Methods

In this section, to overcome some limitations of the existing known method by Ghatee and Hashemi [1], as well as the and methods proposed in the previous chapters, which are mentioned in Sect. 8.1, we will propose two new methods for finding the fuzzy optimal solution of such fully fuzzy solid transshipment problems in which all the parameters are represented by the LR flat fuzzy numbers. Needless to say that the assumption of such fuzzy numbers is well justified as they can well represent imprecision of values, can easily be derived by questioning human experts, and are relatively easy to process. We will also mention some advantages of these new methods proposed over both the Ghatee and Hashemi's [1] method and the methods presented in the previous chapters.

8.3.1 New Method Based on the Fuzzy Linear Programming Formulation

We will present here a new method for finding the fuzzy optimal solution of the fully fuzzy solid transshipment problems in which all the parameters are represented by the LR flat fuzzy numbers. It will be formulated in terms of a fuzzy linear programming problems.

The consecutive steps of the proposed method are as follows:

Step 1 Find:

$$\begin{cases} \sum\limits_{i \in N_{PS}} \tilde{a}_i \oplus \sum\limits_{i \in N_S} \tilde{a}'_i \\ \sum\limits_{j \in N_{PD}} \tilde{b}_j \oplus \sum\limits_{j \in N_D} \tilde{b}'_j \\ \sum\limits_{k \in S_C} \tilde{e}_k. \end{cases}$$

Let:

$$\begin{cases} \sum\limits_{i \in N_{PS}} \tilde{a}_i \oplus \sum\limits_{i \in N_S} \tilde{a}'_i = (m, n, \alpha, \beta)_{LR} \\ \sum\limits_{j \in N_{PD}} \tilde{b}_j \oplus \sum\limits_{j \in N_D} \tilde{b}'_j = (m', n', \alpha', \beta')_{LR} \\ \sum\limits_{k \in S_C} \tilde{e}_k = (m'', n'', \alpha'', \beta'')_{LR}. \end{cases}$$

Use Definition 18, to examine that the problem is balanced or unbalanced, and:

Case (1): If the problem is balanced, i.e.,

$$\sum_{i \in N_{PS}} \tilde{a}_i \oplus \sum_{i \in N_S} \tilde{a}'_i = \sum_{j \in N_{PD}} \tilde{b}_j \oplus \sum_{j \in N_D} \tilde{b}'_j = \sum_{k \in S_C} \tilde{e}_k,$$

then Go to Step 4;

Case (2): If the problem is unbalanced, i.e.,

$$\sum_{i\in N_{PS}} \tilde{a}_i \oplus \sum_{i\in N_S} \tilde{a}'_i = \sum_{j\in N_{PD}} \tilde{b}_j \oplus \sum_{j\in N_D} \tilde{b}'_j \neq \sum_{k\in S_C} \tilde{e}_k$$

or

$$\sum_{i\in N_{PS}} \tilde{a}_i \oplus \sum_{i\in N_S} \tilde{a}'_i \neq \sum_{j\in N_{PD}} \tilde{b}_j \oplus \sum_{j\in N_D} \tilde{b}'_j = \sum_{k\in S_C} \tilde{e}_k,$$

or

$$\sum_{i\in N_{PS}} \tilde{a}_i \oplus \sum_{i\in N_S} \tilde{a}'_i = \sum_{k\in S_C} \tilde{e}_k \neq \sum_{j\in N_{PD}} \tilde{b}_j \oplus \sum_{j\in N_D} \tilde{b}'_j,$$

or

$$\sum_{i\in N_{PS}} \tilde{a}_i \oplus \sum_{i\in N_S} c\tilde{a}'_i \neq \sum_{j\in N_{PD}} \tilde{b}_j \oplus \sum_{j\in N_D} \tilde{b}'_j \neq \sum_{k\in S_C} \tilde{e}_k,$$

then Go to Step 2.

Step 2 Check if:

$$\sum_{i\in N_{PS}} \tilde{a}_i \oplus \sum_{i\in N_S} \tilde{a}'_i = \sum_{j\in N_{PD}} \tilde{b}_j \oplus \sum_{j\in N_D} \tilde{b}'_j,$$

or

$$\sum_{i\in N_{PS}} \tilde{a}_i \oplus \sum_{i\in N_S} \tilde{a}'_i \neq \sum_{j\in N_{PD}} \tilde{b}_j \oplus \sum_{j\in N_D} \tilde{b}'_j,$$

and then:

Case (1) If

$$\sum_{i\in N_{PS}} \tilde{a}_i \oplus \sum_{i\in N_S} \tilde{a}'_i = \sum_{j\in N_{PD}} \tilde{b}_j \oplus \sum_{j\in N_D} \tilde{b}'_j$$

then Go to Step 3;

Case (2) If

$$\sum_{i\in N_{PS}} \tilde{a}_i \oplus \sum_{i\in N_S} \tilde{a}'_i \neq \sum_{j\in N_{PD}} \tilde{b}_j \oplus \sum_{j\in N_D} \tilde{b}'_j,$$

then convert

$$\sum_{i\in N_{PS}} \tilde{a}_i \oplus \sum_{i\in N_S} \tilde{a}'_i \neq \sum_{j\in N_{PD}} \tilde{b}_j \oplus \sum_{j\in N_D} \tilde{b}'_j$$

into

$$\sum_{i\in N_{PS}} \tilde{a}_i \oplus \sum_{i\in N_S} \tilde{a}'_i = \sum_{j\in N_{PD}} \tilde{b}_j \oplus \sum_{j\in N_D} \tilde{b}'_j$$

as follows:

Case (2a) If $m-\alpha \le m'-\alpha'$, $\alpha \le \alpha'$, $n-m \le n'-m'$ and $\beta \le \beta'$, then introduce a dummy purely source node with fuzzy availability $(m'-m, n'-n, \alpha'-\alpha, \beta'-\beta)_{LR}$ so that

$$\sum_{i\in N_{PS}} \tilde{a}_i \oplus \sum_{i\in N_S} \tilde{a}'_i = \sum_{j\in N_{PD}} \tilde{b}_j \oplus \sum_{j\in N_D} \tilde{b}'_j,$$

and go to Step 3;

Case (2b) If $m-\alpha \ge m'-\alpha'$, $\alpha \ge \alpha'$, $n-m \ge n'-m'$ and $\beta \ge \beta'$, then introduce a dummy purely destination node with fuzzy demand $(m-m', n-n', \alpha-\alpha', \beta-\beta')_{LR}$ so that

$$\sum_{i\in N_{PS}} \tilde{a}_i \oplus \sum_{i\in N_S} \tilde{a}'_i = \sum_{j\in N_{PD}} \tilde{b}_j \oplus \sum_{j\in N_D} \tilde{b}'_j,$$

and go to Step 3;

Case (2c) If neither Case (2a) nor Case (2b) is satisfied, then introduce a dummy purely source node with fuzzy availability

$$(\max\{0,(m'-\alpha')-(m-\alpha)\}+\max\{0,(\alpha'-\alpha)\}, \max\{0,(m'-\alpha') \\ -(m-\alpha)\}+\max\{0,(\alpha'-\alpha)\}++\max\{0,(n'-m')-(n-m)\}, \\ \max\{0,(\alpha'-\alpha)\}, \max\{0,(\beta'-\beta)\})_{LR}$$

and a dummy purely destination node with fuzzy demand

$$(\max\{0,(m-\alpha)-(m'-\alpha')\}+\max\{0,(\alpha-\alpha')\}, \max\{0,(m-\alpha) \\ -(m'-\alpha')\}++\max\{0,(\alpha-\alpha')\}+\max\{0,(n-m)-(n'-m')\}, \\ \max\{0,(\alpha-\alpha')\}, \max\{0,(\beta-\beta')\})_{LR}$$

so that

$$\sum_{i\in N_{PS}} \tilde{a}_i \oplus \sum_{i\in N_S} \tilde{a}'_i = \sum_{j\in N_{PD}} \tilde{b}_j \oplus \sum_{j\in N_D} \tilde{b}'_j,$$

and go to Step 3.

Step 3 From Step 2 we obtain

$$\sum_{i\in N_{PS}} \tilde{a}_i \oplus \sum_{i\in N_S} \tilde{a}'_i = \sum_{j\in N_{PD}} \tilde{b}_j \oplus \sum_{j\in N_D} \tilde{b}'_j$$

and let:

$$\sum_{i\in N_{PS}} \tilde{a}_i \oplus \sum_{i\in N_S} \tilde{a}'_i = \sum_{j\in N_{PD}} \tilde{b}_j \oplus \sum_{j\in N_D} \tilde{b}'_j = (m_1, n_1, \alpha_1, \beta_1)_{LR}$$

and

$$\sum_{k\in S_C} \tilde{e}_k = (m'', n'', \alpha'', \beta'')_{LR}.$$

Now check if:

$$\sum_{i\in N_{PS}} \tilde{a}_i \oplus \sum_{i\in N_S} \tilde{a}'_i = \sum_{j\in N_{PD}} \tilde{b}_j \oplus \sum_{j\in N_D} \tilde{b}'_j = \sum_{k\in S_C} \tilde{e}_k$$

or

$$\sum_{i\in N_{PS}} \tilde{a}_i \oplus \sum_{i\in N_S} \tilde{a}'_i = \sum_{j\in N_{PD}} \tilde{b}_j \oplus \sum_{j\in N_D} \tilde{b}'_j \neq \sum_{k\in S_C} \tilde{e}_k$$

and:

Case (1) If

$$\sum_{i\in N_{PS}} \tilde{a}_i \oplus \sum_{i\in N_S} \tilde{a}'_i = \sum_{j\in N_{PD}} \tilde{b}_j \oplus \sum_{j\in N_D} \tilde{b}'_j = \sum_{k\in S_C} \tilde{e}_k$$

then Go to Step 4;

Case (2) If

$$\sum_{i\in N_{PS}} \tilde{a}_i \oplus \sum_{i\in N_S} \tilde{a}'_i = \sum_{j\in N_{PD}} \tilde{b}_j \oplus \sum_{j\in N_D} \tilde{b}'_j \neq \sum_{k\in S_C} \tilde{e}_k$$

then convert

$$\sum_{i\in N_{PS}} \tilde{a}_i \oplus \sum_{i\in N_S} \tilde{a}'_i = \sum_{j\in N_{PD}} \tilde{b}_j \oplus \sum_{j\in N_D} \tilde{b}'_j \neq \sum_{k\in S_C} \tilde{e}_k$$

into

$$\sum_{i\in N_{PS}} \tilde{a}_i \oplus \sum_{i\in N_S} \tilde{a}'_i = \sum_{j\in N_{PD}} \tilde{b}_j \oplus \sum_{j\in N_D} \tilde{b}'_j = \sum_{k\in S_C} \tilde{e}_k$$

as follows:

Case (2a) If $m_1 - \alpha_1 \leq m'' - \alpha''$, $\alpha_1 \leq \alpha''$, $n_1 - m_1 \leq n'' - m''$ and $\beta_1 \leq \beta''$, then check if in Step 2 a dummy purely source node is introduced or not, and also check if a dummy purely destination node is introduced or not, and:

Case (i) If both the dummy purely source node and dummy purely destination node are introduced, then increase both the fuzzy availability of the already introduced dummy purely source node and the fuzzy demand of the already introduced dummy purely destination node by the same fuzzy quantity $(m'' - m_1, n'' - n_1, \alpha'' - \alpha_1, \beta'' - \beta_1)_{LR}$ so that

$$\sum_{i \in N_{PS}} \tilde{a}_i \oplus \sum_{i \in N_S} \tilde{a}'_i = \sum_{j \in N_{PD}} \tilde{b}_j \oplus \sum_{j \in N_D} \tilde{b}'_j = \sum_{k \in S_C} \tilde{e}_k,$$

and go to Step 4;

Case (ii) If a dummy purely source node is introduced but no dummy purely destination node is introduced, then increase the fuzzy availability of the already introduced dummy purely source node by the fuzzy quantity $(m'' - m_1, n'' - n_1, \alpha'' - \alpha_1, \beta'' - \beta_1)_{LR}$ and also introduce a dummy purely destination node with fuzzy demand $(m'' - m_1, n'' - n_1, \alpha'' - \alpha_1, \beta'' - \beta_1)_{LR}$ so that

$$\sum_{i \in N_{PS}} \tilde{a}_i \oplus \sum_{i \in N_S} \tilde{a}'_i = \sum_{j \in N_{PD}} \tilde{b}_j \oplus \sum_{j \in N_D} \tilde{b}'_j = \sum_{k \in S_C} \tilde{e}_k,$$

and go to Step 4;

Case (iii) If a dummy purely destination node is introduced but no dummy purely source node is introduced, then increase the fuzzy demand of the already introduced dummy purely destination node by the fuzzy quantity $(m'' - m_1, n'' - n_1, \alpha'' - \alpha_1, \beta'' - \beta_1)_{LR}$ and also introduce a dummy purely source node with fuzzy availability $(m'' - m_1, n'' - n_1, \alpha'' - \alpha_1, \beta'' - \beta_1)_{LR}$ so that

$$\sum_{i \in N_{PS}} \tilde{a}_i \oplus \sum_{i \in N_S} \tilde{a}'_i = \sum_{j \in N_{PD}} \tilde{b}_j \oplus \sum_{j \in N_D} \tilde{b}'_j = \sum_{k \in S_C} \tilde{e}_k$$

and go to Step 4;

Case (2b) If $m_1 - \alpha_1 \geq m'' - \alpha''$, $\alpha_1 \geq \alpha''$, $n_1 - m_1 \geq n'' - m''$ and $\beta_1 \geq \beta''$, then introduce a dummy conveyance with fuzzy capacity $(m_1 - m''$, $n_1 - n'', \alpha_1 - \alpha'', \beta_1 - \beta'')_{LR}$ so that

$$\sum_{i \in N_{PS}} \tilde{a}_i \oplus \sum_{i \in N_S} \tilde{a}'_i = \sum_{j \in N_{PD}} \tilde{b}_j \oplus \sum_{j \in N_D} \tilde{b}'_j = \sum_{k \in S_C} \tilde{e}_k$$

and go to Step 4;

Case (2c) If neither Case (2a) nor Case (2b) is satisfied, then check if in Step 2 a dummy purely source node is introduced or not and also check if a dummy purely destination node is introduced or not:

Case (i) If both the dummy purely source node and dummy purely destination node are introduced then increase both the fuzzy availability of the already introduced dummy purely source node and the fuzzy demand of the already introduced dummy purely destination node by the same fuzzy quantity

$$(\max\{0, (m'' - \alpha'') - (m_1 - \alpha_1)\} + \max\{0, (\alpha'' - \alpha_1)\}, \max\{0, (m'' - \alpha'') - (m_1 - \alpha_1)\} + \max\{0, (\alpha'' - \alpha_1)\} + + \max\{0, (n'' - m'') - (n_1 - m_1)\}, \max\{0, (\alpha'' - \alpha_1)\}, \max\{0, (\beta'' - \beta_1)\})_{LR}$$

and also introduce a dummy purely conveyance with fuzzy capacity

$$(\max\{0, (m_1-\alpha_1)-(m''-\alpha'')\}+\max\{0, (\alpha_1-\alpha'')\}, \max\{0, (m_1-\alpha_1)-(m''-\alpha'')\}+\max\{0, (\alpha_1-\alpha'')\}+ +\max\{0, (n_1-m_1)-(n''-m'')\}, \max\{0, (\alpha_1-\alpha'')\}, \max\{0, (\beta_1-\beta'')\})_{LR}$$

so that

$$\sum_{i\in N_{PS}} \tilde{a}_i \oplus \sum_{i\in N_S} \tilde{a}'_i = \sum_{j\in N_{PD}} \tilde{b}_j \oplus \sum_{j\in N_D} \tilde{b}'_j = \sum_{k\in S_C} \tilde{e}_k,$$

and go to Step 4;

Case (ii) If a dummy purely source node is introduced but no dummy purely destination node is introduced, then increase the fuzzy availability of the already introduced dummy purely source node by the fuzzy quantity

$$(\max\{0, (m''-\alpha'')-(m_1-\alpha_1)\}+\max\{0, (\alpha''-\alpha_1)\}, \max\{0, (m''-\alpha'')-(m_1-\alpha_1)\}+\max\{0, (\alpha''-\alpha_1)\}+ +\max\{0, (n''-m'')-(n_1-m_1)\}, \max\{0, (\alpha''-\alpha_1)\}, \max\{0, (\beta''-\beta_1)\})_{LR}$$

and also introduce a dummy purely destination node with fuzzy demand

$$(\max\{0, (m''-\alpha'')-(m_1-\alpha_1)\}+\max\{0, (\alpha''-\alpha_1)\}, \max\{0, (m''-\alpha'')-(m_1-\alpha_1)\}+\max\{0, (\alpha''-\alpha_1)\}+ +\max\{0, (n''-m'')-(n_1-m_1)\}, \max\{0, (\alpha''-\alpha_1)\}, \max\{0, (\beta''-\beta_1)\})_{LR}$$

and also introduce a dummy conveyance with fuzzy capacity

$$(\max\{0, (m_1-\alpha_1)-(m''-\alpha'')\}+\max\{0, (\alpha_1-\alpha'')\}, \max\{0, (m_1-\alpha_1)-(m''-\alpha'')\}+\max\{0, (\alpha_1-\alpha'')\}+ +\max\{0, (n_1-m_1)-(n''-m'')\}, \max\{0, (\alpha_1-\alpha'')\}, \max\{0, (\beta_1-\beta'')\})_{LR}$$

so that

$$\sum_{i\in N_{PS}} \tilde{a}_i \oplus \sum_{i\in N_S} \tilde{a}'_i = \sum_{j\in N_{PD}} \tilde{b}_j \oplus \sum_{j\in N_D} \tilde{b}'_j = \sum_{k\in S_C} \tilde{e}_k$$

and go to Step 4;

Case (iii) If a dummy purely destination node is introduced but no dummy purely source node is introduced, then increase the fuzzy demand of the already introduced dummy purely destination node by the fuzzy quantity

$$(\max\{0, (m''-\alpha'')-(m_1-\alpha_1)\}+\max\{0, (\alpha''-\alpha_1)\}, \max\{0, (m''-\alpha'')-(m_1-\alpha_1)\}+\max\{0, (\alpha''-\alpha_1)\}+ +\max\{0, (n''-m'')-(n_1-m_1)\}, \max\{0, (\alpha''-\alpha_1)\}, \max\{0, (\beta''-\beta_1)\})_{LR}$$

and also introduce a dummy purely source node with fuzzy availability

$(\max\{0, (m'' - \alpha'') - (m_1 - \alpha_1)\} + \max\{0, (\alpha'' - \alpha_1)\}, \max\{0, (m'' - \alpha'') - (m_1 - \alpha_1)\} + \max\{0, (\alpha'' - \alpha_1)\} + + \max\{0, (n'' - m'') - (n_1 - m_1)\}, \max\{0, (\alpha'' - \alpha_1)\}, \max\{0, (\beta'' - \beta_1)\})_{LR}$

and also, introduce a dummy conveyance with fuzzy capacity

$(\max\{0, (m_1 - \alpha_1) - (m'' - \alpha'')\} + \max\{0, (\alpha_1 - \alpha'')\}, \max\{0, (m_1 - \alpha_1) - (m'' - \alpha'')\} + \max\{0, (\alpha_1 - \alpha'')\} + \max\{0, (n_1 - m_1) - (n'' - m'')\}, \max\{0, (\alpha_1 - \alpha'')\}, \max\{0, (\beta_1 - \beta'')\})_{LR}$

so that

$$\sum_{i \in N_{PS}} \tilde{a}_i \oplus \sum_{i \in N_S} \tilde{a}'_i = \sum_{j \in N_{PD}} \tilde{b}_j \oplus \sum_{j \in N_D} \tilde{b}'_j = \sum_{k \in S_C} \tilde{e}_k$$

and go to Step 4.

Step 4 The balanced fully fuzzy solid transshipment problem obtained by using Steps 1–3 can be formulated as the fuzzy linear programming problem (8.1) by assuming the following fuzzy costs as the zero LR flat fuzzy numbers:

(i) If it is required to add any dummy purely source node, then assume the fuzzy cost for transporting one unit quantity of the product from the introduced dummy purely source node to all purely destination nodes and all intermediate nodes by any conveyance as the zero LR flat fuzzy number;
(ii) If it is required to add any dummy purely destination node, then assume the fuzzy cost for transporting one unit quantity of the product from all purely source nodes and intermediate nodes to the introduced dummy purely destination node by any conveyance as the zero LR flat fuzzy number;
(iii) If it is required to add any dummy conveyance, then assume the fuzzy cost for transporting one unit quantity of the product from all purely source nodes and intermediate nodes to all intermediate nodes and any purely destination nodes by introduced dummy conveyance as the zero LR flat fuzzy number.

Step 5 Assuming

$$\tilde{c}_{ijk} = (m'_{ijk}, n'_{ijk}, \alpha'_{ijk}, \beta'_{ijk})_{LR}$$
$$\tilde{x}_{ijk} = (m_{ijk}, n_{ijk}, \alpha_{ijk}, \beta_{ijk})_{LR},$$
$$\tilde{a}_i = (m_i, n_i, \alpha_i, \beta_i)_{LR}$$
$$\tilde{a}'_i = (m'_i, n'_i, \alpha'_i, \beta'_i)_{LR}$$
$$\tilde{b}_j = (m_j, n_j, \alpha_j, \beta_j)_{LR}$$
$$\tilde{b}'_j = (m'_j, n'_j, \alpha'_j, \beta'_j)_{LR}$$
$$\text{and } \tilde{e}_k = (m''_k, n''_k, \alpha''_k, \beta''_k)_{LR}$$

the fuzzy linear programming problem (8.1) can be written as:

$$\left\{\begin{array}{l}
\min \sum_{(i,j)\in A}\sum_{k\in S_C}\left((m'_{ijk}, n'_{ijk}, \alpha'_{ijk}, \beta'_{ijk})_{LR}\otimes(m_{ijk}, n_{ijk}, \alpha_{ijk}, \beta_{ijk})_{LR}\right)\\
\text{subject to:}\\
\sum\limits_{j:(i,j)\in A}\sum\limits_{k\in S_C}(m_{ijk}, n_{ijk}, \alpha_{ijk}, \beta_{ijk})_{LR} = (m_i, n_i, \alpha_i, \beta_i)_{LR};\ i\in N_{PS}\\
\sum\limits_{j:(i,j)\in A}\sum\limits_{k\in S_C}(m_{ijk}, n_{ijk}, \alpha_{ijk}, \beta_{ijk})_{LR}\ominus_H \sum\limits_{j:(j,i)\in A}\sum\limits_{k\in S_C}(m_{jik}, n_{jik}, \alpha_{jik}, \beta_{jik})_{LR} =\\
(m'_i, n'_i, \alpha'_i, \beta'_i)_{LR};\ i\in N_S\\
\sum\limits_{i:(i,j)\in A}\sum\limits_{k\in S_C}(m_{ijk}, n_{ijk}, \alpha_{ijk}, \beta_{ijk})_{LR} = (m_j, n_j, \alpha_j, \beta_j)_{LR};\ j\in N_{PD}\\
\sum\limits_{i:(i,j)\in A}\sum\limits_{k\in S_C}(m_{ijk}, n_{ijk}, \alpha_{ijk}, \beta_{ijk})_{LR}\ominus_H \sum\limits_{i:(j,i)\in A}\sum\limits_{k\in S_C}(m_{jik}, n_{jik}, \alpha_{jik}, \beta_{jik})_{LR} =\\
(m'_j, n'_j, \alpha'_j, \beta'_j)_{LR};\ j\in N_D\\
\sum\limits_{j:(i,j)\in A}\sum\limits_{k\in S_C}(m_{ijk}, n_{ijk}, \alpha_{ijk}, \beta_{ijk})_{LR} = \sum\limits_{j:(j,i)\in A}\sum\limits_{k\in S_C}(m_{jik}, n_{jik}, \alpha_{jik}, \beta_{jik})_{LR};\ i\in N_T\\
\sum\limits_{i:(i,j)\in A}(m_{ijk}, n_{ijk}, \alpha_{ijk}, \beta_{ijk})_{LR} = (m''_k, n''_k, \alpha''_k, \beta''_k)_{LR};\ k\in S_C,
\end{array}\right. \tag{8.3}$$

where $(m_{ijk}, n_{ijk}, \alpha_{ijk}, \beta_{ijk})_{LR}$ is non-negative LR flat fuzzy number, $\forall (i, j)\in A, k\in S_C$.

Step 6 Using the arithmetic operations on the LR flat fuzzy numbers, defined in Sect. 5.1, and assuming

$$\sum_{(i,j)\in A}\sum_{k\in S_C}(\tilde{c}_{ijk}\otimes\tilde{x}_{ijk}) = (m_0, n_0, \alpha_0, \beta_0)_{LR},$$

the fuzzy linear programming problem, obtained in Step 5, can be written as:

$$\left\{\begin{array}{l}
\max(m_0, n_0, \alpha_0, \beta_0)_{LR}\\
\text{subject to:}\\
\Big(\sum\limits_{j:(i,j)\in A}\sum\limits_{k\in S_C} m_{ijk}, \sum\limits_{j:(i,j)\in A}\sum\limits_{k\in S_C} n_{ijk}, \sum\limits_{j:(i,j)\in A}\sum\limits_{k\in S_C} \alpha_{ijk}, \sum\limits_{j:(i,j)\in A}\sum\limits_{k\in S_C} \beta_{ijk}\Big)_{LR} =\\
(m_i, n_i, \alpha_i, \beta_i)_{LR};\ i\in N_{PS}\\
\Big(\sum\limits_{j:(i,j)\in A}\sum\limits_{k\in S_C} m_{ijk}, \sum\limits_{j:(i,j)\in A}\sum\limits_{k\in S_C} n_{ijk}, \sum\limits_{j:(i,j)\in A}\sum\limits_{k\in S_C} \alpha_{ijk}, \sum\limits_{j:(i,j)\in A}\sum\limits_{k\in S_C} \beta_{ijk}\Big)_{LR}\ominus_H\\
\Big(\sum\limits_{j:(j,i)\in A}\sum\limits_{k\in S_C} m_{jik}, \sum\limits_{j:(j,i)\in A}\sum\limits_{k\in S_C} n_{jik}, \sum\limits_{j:(j,i)\in A}\sum\limits_{k\in S_C} \alpha_{jik}, \sum\limits_{j:(j,i)\in A}\sum\limits_{k\in S_C} \beta_{jik}\Big)_{LR} =\\
(m'_i, n'_i, \alpha'_i, \beta'_i)_{LR};\ i\in N_S\\
\Big(\sum\limits_{i:(i,j)\in A}\sum\limits_{k\in S_C} m_{ijk}, \sum\limits_{i:(i,j)\in A}\sum\limits_{k\in S_C} n_{ijk}, \sum\limits_{i:(i,j)\in A}\sum\limits_{k\in S_C} \alpha_{ijk}, \sum\limits_{i:(i,j)\in A}\sum\limits_{k\in S_C} \beta_{ijk}\Big)_{LR} =\\
(m_j, n_j, \alpha_j, \beta_j)_{LR};\ j\in N_{PD}\\
\Big(\sum\limits_{i:(i,j)\in A}\sum\limits_{k\in S_C} m_{ijk}, \sum\limits_{i:(i,j)\in A}\sum\limits_{k\in S_C} n_{ijk}, \sum\limits_{i:(i,j)\in A}\sum\limits_{k\in S_C} \alpha_{ijk}, \sum\limits_{i:(i,j)\in A}\sum\limits_{k\in S_C} \beta_{ijk}\Big)_{LR}\ominus_H\\
\Big(\sum\limits_{i:(j,i)\in A}\sum\limits_{k\in S_C} m_{jik}, \sum\limits_{i:(j,i)\in A}\sum\limits_{k\in S_C} n_{jik}, \sum\limits_{i:(j,i)\in A}\sum\limits_{k\in S_C} \alpha_{jik}, \sum\limits_{i:(j,i)\in A}\sum\limits_{k\in S_C} \beta_{jik}\Big)_{LR} =\\
(m'_j, n'_j, \alpha'_j, \beta'_j)_{LR};\ j\in N_D\\
\Big(\sum\limits_{j:(i,j)\in A}\sum\limits_{k\in S_C} m_{ijk}, \sum\limits_{j:(i,j)\in A}\sum\limits_{k\in S_C} n_{ijk}, \sum\limits_{j:(i,j)\in A}\sum\limits_{k\in S_C} \alpha_{ijk}, \sum\limits_{j:(i,j)\in A}\sum\limits_{k\in S_C} \beta_{ijk}\Big)_{LR} =\\
\Big(\sum\limits_{j:(j,i)\in A}\sum\limits_{k\in S_C} m_{jik}, \sum\limits_{j:(j,i)\in A}\sum\limits_{k\in S_C} n_{jik}, \sum\limits_{j:(j,i)\in A}\sum\limits_{k\in S_C} \alpha_{jik}, \sum\limits_{j:(j,i)\in A}\sum\limits_{k\in S_C} \beta_{jik}\Big)_{LR};\ i\in N_T\\
\Big(\sum\limits_{i:(i,j)\in A} m_{ijk}, \sum\limits_{i:(i,j)\in A} n_{ijk}, \sum\limits_{i:(i,j)\in A} \alpha_{ijk}, \sum\limits_{i:(i,j)\in A} \beta_{ijk}\Big)_{LR} =\\
(m''_k, n''_k, \alpha''_k, \beta''_k)_{LR};\ k\in S_C
\end{array}\right. \tag{8.4}$$

where $(m_{ijk}, n_{ijk}, \alpha_{ijk}, \beta_{ijk})_{LR}$ is a non-negative LR flat fuzzy number, $\forall\ (i, j)\in A,\ k\in S_C$;

Step 7 Using Definitions 18, 19 and 20, and p. 105, the fuzzy linear programming problem, obtained in Step 6, can be converted into the following fuzzy linear programming problem (8.5):

$$\left\{\begin{array}{l}
\min(m_0, n_0, \alpha_0, \beta_0)_{LR} \\
\text{subject to:} \\
\sum_{j:(i,j)\in A}\sum_{k\in S_C} m_{ijk} = m_i;\ i \in N_{PS} \\
\sum_{j:(i,j)\in A}\sum_{k\in S_C} n_{ijk} = n_i;\ i \in N_{PS} \\
\sum_{j:(i,j)\in A}\sum_{k\in S_C} \alpha_{ijk} = \alpha_i;\ i \in N_{PS} \\
\sum_{j:(i,j)\in A}\sum_{k\in S_C} \beta_{ijk} = \beta_i;\ i \in N_{PS} \\
\sum_{j:(i,j)\in A}\sum_{k\in S_C} m_{ijk} - \sum_{j:(j,i)\in A}\sum_{k\in S_C} m_{jik} = m'_i;\ i \in N_S \\
\sum_{j:(i,j)\in A}\sum_{k\in S_C} n_{ijk} - \sum_{j:(j,i)\in A}\sum_{k\in S_C} n_{jik} = n'_i;\ i \in N_S \\
\sum_{j:(i,j)\in A}\sum_{k\in S_C} \alpha_{ijk} - \sum_{j:(j,i)\in A}\sum_{k\in S_C} \alpha_{jik} = \alpha'_i;\ i \in N_S \\
\sum_{j:(i,j)\in A}\sum_{k\in S_C} \beta_{ijk} - \sum_{j:(j,i)\in A}\sum_{k\in S_C} \beta_{jik} = \beta'_i;\ i \in N_S \\
\sum_{i:(i,j)\in A}\sum_{k\in S_C} m_{ijk} = m_j;\ j \in N_{PD} \\
\sum_{i:(i,j)\in A}\sum_{k\in S_C} n_{ijk} = n_j;\ j \in N_{PD} \\
\sum_{i:(i,j)\in A}\sum_{k\in S_C} \alpha_{ijk} = \alpha_j;\ j \in N_{PD} \\
\sum_{i:(i,j)\in A}\sum_{k\in S_C} \beta_{ijk} = \beta_j;\ j \in N_{PD} \\
\sum_{i:(i,j)\in A}\sum_{k\in S_C} m_{ijk} - \sum_{i:(j,i)\in A}\sum_{k\in S_C} m_{jik} = m'_j;\ j \in N_D \\
\sum_{i:(i,j)\in A}\sum_{k\in S_C} n_{ijk} - \sum_{i:(j,i)\in A}\sum_{k\in S_C} n_{jik} = n'_j;\ j \in N_D \\
\sum_{i:(i,j)\in A}\sum_{k\in S_C} \alpha_{ijk} - \sum_{i:(j,i)\in A}\sum_{k\in S_C} \alpha_{jik} = \alpha'_j;\ j \in N_D \\
\sum_{i:(i,j)\in A}\sum_{k\in S_C} \beta_{ijk} - \sum_{i:(j,i)\in A}\sum_{k\in S_C} \beta_{jik} = \beta'_j;\ j \in N_D \\
\sum_{j:(i,j)\in A}\sum_{k\in S_C} m_{ijk} = \sum_{j:(j,i)\in A}\sum_{k\in S_C} m_{jik};\ i \in N_T \\
\sum_{j:(i,j)\in A}\sum_{k\in S_C} n_{ijk} = \sum_{j:(j,i)\in A}\sum_{k\in S_C} n_{jik};\ i \in N_T \\
\sum_{j:(i,j)\in A}\sum_{k\in S_C} \alpha_{ijk} = \sum_{j:(j,i)\in A}\sum_{k\in S_C} \alpha_{jik};\ i \in N_T \\
\sum_{j:(i,j)\in A}\sum_{k\in S_C} \beta_{ijk} = \sum_{j:(j,i)\in A}\sum_{k\in S_C} \beta_{jik};\ i \in N_T \\
\sum_{i:(i,j)\in A} m_{ijk} = m''_k;\ k \in S_C \\
\sum_{i:(i,j)\in A} n_{ijk} = n''_k;\ k \in S_C \\
\sum_{i:(i,j)\in A} \alpha_{ijk} = \alpha''_k;\ k \in S_C \\
\sum_{i:(i,j)\in A} \beta_{ijk} = \beta''_k;\ k \in S_C
\end{array}\right. \tag{8.5}$$

where $m_{ijk} - \alpha_{ijk}, n_{ijk} - m_{ijk}, \alpha_{ijk}, \beta_{ijk} \geq 0, \forall (i, j) \in A, k \in S_C$;

Step 8 As discussed in Step 6 of the method proposed in Sect. 4.6.1, the fuzzy optimal solution of the fuzzy linear programming problem (8.5) can be obtained by solving the following crisp linear programming problem:

$$\left\{\begin{array}{l}
\min \Re((m_0, n_0, \alpha_0, \beta_0)_{LR}) \\
\text{subject to:} \\
\sum\limits_{j:(i,j)\in A}\sum\limits_{k\in S_C} m_{ijk} = m_i;\ i \in N_{PS} \\
\sum\limits_{j:(i,j)\in A}\sum\limits_{k\in S_C} n_{ijk} = n_i;\ i \in N_{PS} \\
\sum\limits_{j:(i,j)\in A}\sum\limits_{k\in S_C} \alpha_{ijk} = \alpha_i;\ i \in N_{PS} \\
\sum\limits_{j:(i,j)\in A}\sum\limits_{k\in S_C} \beta_{ijk} = \beta_i;\ i \in N_{PS} \\
\sum\limits_{j:(i,j)\in A}\sum\limits_{k\in S_C} m_{ijk} - \sum\limits_{j:(j,i)\in A}\sum\limits_{k\in S_C} m_{jik} = m'_i;\ i \in N_S \\
\sum\limits_{j:(i,j)\in A}\sum\limits_{k\in S_C} n_{ijk} - \sum\limits_{j:(j,i)\in A}\sum\limits_{k\in S_C} n_{jik} = n'_i;\ i \in N_S \\
\sum\limits_{j:(i,j)\in A}\sum\limits_{k\in S_C} \alpha_{ijk} - \sum\limits_{j:(j,i)\in A}\sum\limits_{k\in S_C} \alpha_{jik} = \alpha'_i;\ i \in N_S \\
\sum\limits_{j:(i,j)\in A}\sum\limits_{k\in S_C} \beta_{ijk} - \sum\limits_{j:(j,i)\in A}\sum\limits_{k\in S_C} \beta_{jik} = \beta'_i;\ i \in N_S \\
\sum\limits_{i:(i,j)\in A}\sum\limits_{k\in S_C} m_{ijk} = m_j;\ j \in N_{PD} \\
\sum\limits_{i:(i,j)\in A}\sum\limits_{k\in S_C} n_{ijk} = n_j;\ j \in N_{PD} \\
\sum\limits_{i:(i,j)\in A}\sum\limits_{k\in S_C} \alpha_{ijk} = \alpha_j;\ j \in N_{PD} \\
\sum\limits_{i:(i,j)\in A}\sum\limits_{k\in S_C} \beta_{ijk} = \beta_j;\ j \in N_{PD} \\
\sum\limits_{i:(i,j)\in A}\sum\limits_{k\in S_C} m_{ijk} - \sum\limits_{i:(j,i)\in A}\sum\limits_{k\in S_C} m_{jik} = m'_j;\ j \in N_D \\
\sum\limits_{i:(i,j)\in A}\sum\limits_{k\in S_C} n_{ijk} - \sum\limits_{i:(j,i)\in A}\sum\limits_{k\in S_C} n_{jik} = n'_j;\ j \in N_D \\
\sum\limits_{i:(i,j)\in A}\sum\limits_{k\in S_C} \alpha_{ijk} - \sum\limits_{i:(j,i)\in A}\sum\limits_{k\in S_C} \alpha_{jik} = \alpha'_j;\ j \in N_D \\
\sum\limits_{i:(i,j)\in A}\sum\limits_{k\in S_C} \beta_{ijk} - \sum\limits_{i:(j,i)\in A}\sum\limits_{k\in S_C} \beta_{jik} = \beta'_j;\ j \in N_D \\
\sum\limits_{j:(i,j)\in A}\sum\limits_{k\in S_C} m_{ijk} = \sum\limits_{j:(j,i)\in A}\sum\limits_{k\in S_C} m_{jik};\ i \in N_T \\
\sum\limits_{j:(i,j)\in A}\sum\limits_{k\in S_C} n_{ijk} = \sum\limits_{j:(j,i)\in A}\sum\limits_{k\in S_C} n_{jik};\ i \in N_T \\
\sum\limits_{j:(i,j)\in A}\sum\limits_{k\in S_C} \alpha_{ijk} = \sum\limits_{j:(j,i)\in A}\sum\limits_{k\in S_C} \alpha_{jik};\ i \in N_T \\
\sum\limits_{j:(i,j)\in A}\sum\limits_{k\in S_C} \beta_{ijk} = \sum\limits_{j:(j,i)\in A}\sum\limits_{k\in S_C} \beta_{jik};\ i \in N_T \\
\sum\limits_{i:(i,j)\in A} m_{ijk} = m''_k;\ k \in S_C \\
\sum\limits_{i:(i,j)\in A} n_{ijk} = n''_k;\ k \in S_C \\
\sum\limits_{i:(i,j)\in A} \alpha_{ijk} = \alpha''_k;\ k \in S_C \\
\sum\limits_{i:(i,j)\in A} \beta_{ijk} = \beta''_k;\ k \in S_C
\end{array}\right. \quad (8.6)$$

where $m_{ijk} - \alpha_{ijk}, n_{ijk} - m_{ijk}, \alpha_{ijk}, \beta_{ijk} \geq 0, \forall (i, j) \in A, k \in S_C$.

Step 9 Using the well known formula

$$\Re(m_0, n_0, \alpha_0, \beta_0)_{LR} = \frac{1}{2}(\int_0^1 (m_0 - \alpha_0 L^{-1}(\lambda))d\lambda + \int_0^1 (n_0 + \beta_0 R^{-1}(\lambda))d\lambda)$$

the crisp linear programming problem (8.5) obtained in Step 8, can be written as:

$$\begin{cases}
\min \frac{1}{2}(\int_0^1 (m_0 - \alpha_0 L^{-1}(\lambda))d\lambda + \int_0^1 (n_0 + \beta_0 R^{-1}(\lambda))d\lambda) \\
\text{subject to:} \\
\sum_{j:(i,j)\in A} \sum_{k\in S_C} m_{ijk} = m_i;\ i \in N_{PS} \\
\sum_{j:(i,j)\in A} \sum_{k\in S_C} n_{ijk} = n_i;\ i \in N_{PS} \\
\sum_{j:(i,j)\in A} \sum_{k\in S_C} \alpha_{ijk} = \alpha_i;\ i \in N_{PS} \\
\sum_{j:(i,j)\in A} \sum_{k\in S_C} \beta_{ijk} = \beta_i;\ i \in N_{PS} \\
\sum_{j:(i,j)\in A} \sum_{k\in S_C} m_{ijk} - \sum_{j:(j,i)\in A} \sum_{k\in S_C} m_{jik} = m'_i;\ i \in N_S \\
\sum_{j:(i,j)\in A} \sum_{k\in S_C} n_{ijk} - \sum_{j:(j,i)\in A} \sum_{k\in S_C} n_{jik} = n'_i;\ i \in N_S \\
\sum_{j:(i,j)\in A} \sum_{k\in S_C} \alpha_{ijk} - \sum_{j:(j,i)\in A} \sum_{k\in S_C} \alpha_{jik} = \alpha'_i;\ i \in N_S \\
\sum_{j:(i,j)\in A} \sum_{k\in S_C} \beta_{ijk} - \sum_{j:(j,i)\in A} \sum_{k\in S_C} \beta_{jik} = \beta'_i;\ i \in N_S \\
\sum_{i:(i,j)\in A} \sum_{k\in S_C} m_{ijk} = m_j;\ j \in N_{PD} \\
\sum_{i:(i,j)\in A} \sum_{k\in S_C} n_{ijk} = n_j;\ j \in N_{PD} \\
\sum_{i:(i,j)\in A} \sum_{k\in S_C} \alpha_{ijk} = \alpha_j;\ j \in N_{PD} \\
\sum_{i:(i,j)\in A} \sum_{k\in S_C} \beta_{ijk} = \beta_j;\ j \in N_{PD} \\
\sum_{i:(i,j)\in A} \sum_{k\in S_C} m_{ijk} - \sum_{i:(j,i)\in A} \sum_{k\in S_C} m_{jik} = m'_j;\ j \in N_D \\
\sum_{i:(i,j)\in A} \sum_{k\in S_C} n_{ijk} - \sum_{i:(j,i)\in A} \sum_{k\in S_C} n_{jik} = n'_j;\ j \in N_D \\
\sum_{i:(i,j)\in A} \sum_{k\in S_C} \alpha_{ijk} - \sum_{i:(j,i)\in A} \sum_{k\in S_C} \alpha_{jik} = \alpha'_j;\ j \in N_D \\
\sum_{i:(i,j)\in A} \sum_{k\in S_C} \beta_{ijk} - \sum_{i:(j,i)\in A} \sum_{k\in S_C} \beta_{jik} = \beta'_j;\ j \in N_D \\
\sum_{j:(i,j)\in A} \sum_{k\in S_C} m_{ijk} = \sum_{j:(j,i)\in A} \sum_{k\in S_C} m_{jik};\ i \in N_T \\
\sum_{j:(i,j)\in A} \sum_{k\in S_C} n_{ijk} = \sum_{j:(j,i)\in A} \sum_{k\in S_C} n_{jik};\ i \in N_T \\
\sum_{j:(i,j)\in A} \sum_{k\in S_C} \alpha_{ijk} = \sum_{j:(j,i)\in A} \sum_{k\in S_C} \alpha_{jik};\ i \in N_T \\
\sum_{j:(i,j)\in A} \sum_{k\in S_C} \beta_{ijk} = \sum_{j:(j,i)\in A} \sum_{k\in S_C} \beta_{jik};\ i \in N_T \\
\sum_{i:(i,j)\in A} m_{ijk} = m''_k;\ k \in S_C \\
\sum_{i:(i,j)\in A} n_{ijk} = n''_k;\ k \in S_C \\
\sum_{i:(i,j)\in A} \alpha_{ijk} = \alpha''_k;\ k \in S_C \\
\sum_{i:(i,j)\in A} \beta_{ijk} = \beta''_k;\ k \in S_C
\end{cases} \tag{8.7}$$

where $m_{ijk} - \alpha_{ijk}, n_{ijk} - m_{ijk}, \alpha_{ijk}, \beta_{ijk} \geq 0, \forall\ (i, j) \in A,\ k \in S_C$;

Step 10 Solve the crisp linear programming problem (8.5) obtained in Step 9, to find the optimal solution $\{m_{ijk}, n_{ijk}, \alpha_{ijk}, \beta_{ijk}\}$.

Step 11 Put the values of $m_{ijk}, n_{ijk}, \alpha_{ijk}, \beta_{ijk}$ into $\tilde{x}_{ijk} = (m_{ijk}, n_{ijk}, \alpha_{ijk}, \beta_{ijk})_{LR}$ to find the fuzzy optimal solution $\{\tilde{x}_{ijk}\}$.

Step 12 Put the values of $\tilde{x}_{ijk}$, obtained from Step 11, into

$$\sum_{(i,j)\in A}\sum_{k\in S_C}(\tilde{c}_{ijk} \otimes \tilde{x}_{ijk})$$

to find the minimum total fuzzy transportation cost.

8.3.2 New Method Based on the Tabular Representation

In this section we will present a new method for finding the fuzzy optimal solution of the fully fuzzy solid transshipment problem (cf. (8.1), in which the parameters are represented by the LR flat fuzzy numbers. This method will be shown using the tabular representation.

The consecutive steps of the proposed method are as follows:

Step 1 Follow Steps 1–3 of the method proposed in Sect. 8.3.2, to obtain a balanced fully fuzzy solid transshipment problem.

Step 2 Convert the balanced fully fuzzy solid transshipment problem obtained in Step 1 into the balanced fully fuzzy solid transportation problem as follows: increase the fuzzy availability and the fuzzy demand corresponding to the intermediate nodes by the fuzzy quantity

$$\tilde{P} = \sum_{i\in N_{PS}} \tilde{a}_i \oplus \sum_{i\in N_S} \tilde{a}'_i$$

or

$$\sum_{j\in N_{PD}} \tilde{b}_j \oplus \sum_{j\in N_D} \tilde{b}'_j$$

or

$$\sum_{k\in S_C} \tilde{e}_k$$

and, moreover, check if in Step 1 a dummy conveyance is introduced or not, and:

Case (i) If a dummy conveyance is already introduced, then increase the fuzzy capacity of the already introduced dummy conveyance by the fuzzy quantity $K\tilde{P}$, where K is the number of intermediate nodes;

Case (ii) If no dummy conveyance is introduced, then introduce a dummy conveyance with the fuzzy capacity $K\tilde{P}$, where K is the number of intermediate nodes.

Step 3 Let in the balanced fully fuzzy solid transportation problem, obtained in Step 2, the number of purely source nodes, source nodes, transshipment nodes, purely destination nodes, destination nodes and conveyances be m, l, r, t, q and s, respectively. Moreover, let the fuzzy availability of the product at the *ith* purely source node (N_{PS_i}), the fuzzy availability of the product at the *ith* source node (N_{S_i}), the fuzzy demand of the product at the *jth* purely destination node (N_{PD_j}), the fuzzy demand of the product at the *jth* destination node (N_{D_j}), the fuzzy capacity of the *kth* conveyance (E_k) and the fuzzy cost for transporting one unit quantity of the product from the *ith* source to the *jth* destination by means of the *kth* conveyance be denoted by:$\tilde{a}_i$, $\tilde{a'}_{m+i}$, $\tilde{b}_j$, $\tilde{b'}_{t+j}$, $\tilde{e}_k$ and $\tilde{c}_{ijk}$, respectively. This can be represented by Table 8.1.

Step 4 Split Table 8.1 into four crisp solid transportation tables, i.e., Tables 8.2, 8.3, 8.4 and 8.5, respectively.

The costs for transporting one unit quantity of the product from the *ith* node to the *jth* node by means of the *kth* conveyance given in Tables 8.2, 8.3, 8.4 and 8.5 are represented by η_{ijk}, ρ_{ijk}, δ_{ijk} and ξ_{ijk}, respectively, where:

$$\eta_{ijk} = \frac{1}{2}((m'_{ijk} + n'_{ijk}) - \alpha'_{ijk}\int_0^1 L^{-1}(\lambda)d\lambda + \beta'_{ijk}\int_0^1 R^{-1}(\lambda)d\lambda),$$

$$i = 1, 2, \ldots, m + l + q + r;\ j = 1, 2, \ldots, t + q + r + l;\ k = 1, 2, \ldots, s;$$

$$\rho_{ijk} = \frac{1}{2}((m'_{ijk} + n'_{ijk}) - m'_{ijk}\int_0^1 L^{-1}(\lambda)d\lambda + \beta'_{ijk}\int_0^1 R^{-1}(\lambda)d\lambda),$$

$$i = 1, 2, \ldots, m + l + q + r;\ j = 1, 2, \ldots, t + q + r + l;\ k = 1, 2, \ldots, s;$$

$$\delta_{ijk} = \frac{1}{2}(n'_{ijk} + \beta'_{ijk}\int_0^1 R^{-1}(\lambda)d\lambda),$$

$$i = 1, 2, \ldots, m + l + q + r;\ j = 1, 2, \ldots, t + q + r + l;\ k = 1, 2, \ldots, s;$$

$$\xi_{ijk} = \frac{1}{2}((n'_{ijk} + \beta'_{ijk})\int_0^1 R^{-1}(\lambda)d\lambda),$$

$$i = 1, 2, \ldots, m + l + q + r;\ j = 1, 2, \ldots, t + q + r + l;\ k = 1, 2, \ldots, s;$$

Step 5 Solve the crisp solid transportation problems shown by Tables 8.2, 8.3, 8.4 and 8.5 to find the optimal solution $\{m_{ijk} - \alpha_{ijk}\}$, $\{\alpha_{ijk}\}$, $\{n_{ijk} - m_{ijk}\}$ and $\{\beta_{ijk}\}$, respectively.

Step 6 Solve the equations obtained in Step 5, to find the values of m_{ijk}, n_{ijk}, α_{ijk} and β_{ijk}.

Step 7 Find the fuzzy optimal solution $\{\tilde{x}_{ijk}\}$ by putting the values of m_{ijk}, n_{ijk}, α_{ijk}, β_{ijk} into

$$\tilde{x}_{ijk} = (m_{ijk}, n_{ijk}, \alpha_{ijk}, \beta_{ijk})_{LR}.$$

Step 8 Find the minimum total fuzzy transportation cost by putting the values of $\tilde{x}_{ijk}$ into

$$\sum_i \sum_j \sum_k (\tilde{c}_{ijk} \otimes \tilde{x}_{ijk}),\ \forall\, i = 1, 2, \ldots, m + l + q + r;\ j = 1, 2, \ldots, t + q + r + l;\ k = 1, 2, \ldots, s.$$

We can formulate here the following important remark:

Remark 24 Let $\tilde{A} = (m_{ijk}, n_{ijk}, \alpha_{ijk}, \beta_{ijk})_{LR}$ be an LR flat fuzzy number with $L(x) = R(x) = \max\{0, 1 - x\}$. Then, we have

$$\eta_{ijk} = \frac{1}{2}((m_{ijk} + n_{ijk}) - \alpha_{ijk} \int_0^1 L^{-1}(\lambda)d\lambda + \beta_{ijk} \int_0^1 R^{-1}(\lambda)d\lambda)$$

$$= \frac{1}{4}(2m_{ijk} + 2n_{ijk} + \beta_{ijk} - \alpha_{ijk}),$$

$$\rho_{ijk} = \frac{1}{2}((m_{ijk} + n_{ijk}) - m_{ijk} \int_0^1 L^{-1}(\lambda)d\lambda + \beta_{ijk} \int_0^1 R^{-1}(\lambda)d\lambda)$$

$$= \frac{1}{4}(m_{ijk} + 2n_{ijk} + \beta_{ijk}),$$

$$\delta_{ijk} = \frac{1}{2}(n_{ijk} + \beta_{ijk} \int_0^1 R^{-1}(\lambda)d\lambda)$$

$$= \frac{1}{4}(2n_{ijk} + \beta_{ijk}),$$

and

$$\xi_{ijk} = \frac{1}{2}((n_{ijk} + \beta_{ijk}) \int_0^1 R^{-1}(\lambda)d\lambda)$$

$$= \frac{1}{4}(n_{ijk} + \beta_{ijk}).$$

Table 8.1 Tabular representation of the balanced fully fuzzy solid transportation problem obtained by adding a fuzzy buffer stock ($\tilde{P}$)

Conveyance	E_1			…	E_1			E_1			…	E_1		
		…		…		…			…		…		…	
			E_s	…			E_s			E_s	…			E_s
Destinations → Sources↓	N_{PD_1}			…	N_{PD_t}			N_{D_1}			…	N_{D_q}		
N_{PS_1}	$\tilde{c}_{111}$	…	$\tilde{c}_{11s}$	…	$\tilde{c}_{1t1}$	…	$\tilde{c}_{1ts}$	$\tilde{c}_{1(t+1)1}$	…	$\tilde{c}_{1(t+1)s}$	…	$\tilde{c}_{1(t+q)1}$	…	
⋮	⋮	⋮	⋮	⋮	⋮	⋮	⋮	⋮	⋮	⋮	⋮	⋮	⋮	
N_{PS_m}	$\tilde{c}_{m11}$	…	$\tilde{c}_{m1s}$	…	$\tilde{c}_{mt1}$	…	$\tilde{c}_{mts}$	$\tilde{c}_{m(t+1)1}$	…	$\tilde{c}_{m(t+1)s}$	…	$\tilde{c}_{m(t+q)1}$	…	$\tilde{c}_{m(t+q)s}$
N_{S_1}	$\tilde{c}_{(m+1)11}$	…	$\tilde{c}_{(m+1)1s}$	…	$\tilde{c}_{(m+1)t1}$	…	$\tilde{c}_{(m+1)ts}$	$\tilde{c}_{(m+1)(t+1)1}$	…	$\tilde{c}_{(m+1)(t+1)s}$	…	$\tilde{c}_{(m+1)(t+q)1}$	…	$\tilde{c}_{(m+1)(t+q)s}$
⋮	⋮	⋮	⋮	⋮	⋮	⋮	⋮	⋮	⋮	⋮	⋮	⋮	⋮	⋮
N_{S_l}	$\tilde{c}_{(m+l)11}$	…	$\tilde{c}_{(m+l)1s}$	…	$\tilde{c}_{(m+l)t1}$	…	$\tilde{c}_{(m+l)ts}$	$\tilde{c}_{(m+l)(t+1)1}$	…	$\tilde{c}_{(m+l)(t+1)s}$	…	$\tilde{c}_{(m+l)(t+q)1}$	…	$\tilde{c}_{(m+l)(t+q)s}$
N_{D_1}	$\tilde{c}_{(m+l+1)11}$	…	$\tilde{c}_{(m+l+1)1s}$	…	$\tilde{c}_{(m+l+1)t1}$	…	$\tilde{c}_{(m+l+1)ts}$	$\tilde{c}_{(m+l+1)(t+1)1}$	…	$\tilde{c}_{(m+l+1)(t+1)s}$	…	$\tilde{c}_{(m+l+1)(t+q)1}$	…	$\tilde{c}_{(m+l+1)(t+q)s}$
⋮	⋮	⋮	⋮	⋮	⋮	⋮	⋮	⋮	⋮	⋮	⋮	⋮	⋮	⋮
N_{D_q}	$\tilde{c}_{(m+l+q)11}$	…	$\tilde{c}_{(m+l+q)1s}$	…	$\tilde{c}_{(m+l+q)t1}$	…	$\tilde{c}_{(m+l+q)ts}$	$\tilde{c}_{(m+l+q)(t+1)1}$	…	$\tilde{c}_{(m+l+q)(t+1)s}$	…	$\tilde{c}_{(m+l+q)(t+q)1}$	…	$\tilde{c}_{(m+l+q)(t+q)s}$
N_{T_1}	$\tilde{c}_{(m+l+q+1)11}$	…	$\tilde{c}_{(m+l+q+1)1s}$	…	$\tilde{c}_{(m+l+q+1)t1}$	…	$\tilde{c}_{(m+l+q+1)ts}$	$\tilde{c}_{(m+l+q+1)(t+1)1}$	…	$\tilde{c}_{(m+l+q+1)(t+1)s}$	…	$\tilde{c}_{(m+l+q+1)(t+q)1}$	…	$\tilde{c}_{(m+l+q+1)(t+q)s}$
⋮	⋮	⋮	⋮	⋮	⋮	⋮	⋮	⋮	⋮	⋮	⋮	⋮	⋮	⋮
N_{T_r}	$\tilde{c}_{(m+l+q+r)11}$	…	$\tilde{c}_{(m+l+q+r)1s}$	…	$\tilde{c}_{(m+l+q+r)t1}$	…	$\tilde{c}_{(m+l+q+r)ts}$	$\tilde{c}_{(m+l+q+r)(t+1)1}$	…	$\tilde{c}_{(m+l+q+r)(t+1)s}$	…	$\tilde{c}_{(m+l+q+r)(t+q)1}$	…	$\tilde{c}_{(m+l+q+r)(t+q)s}$
Demand	$\tilde{b}_1$			…	$\tilde{b}_t$			$\tilde{b'}_{t+1}$			…	$\tilde{b'}_{t+q}$		

(continued)

Table 8.1 (continued)

															Capacity
Conveyance	E_1			...	E_1			E_1			...	$\tilde{E}_1$			$\tilde{e}_1$
		...		...		...			...		...		...		⋮
			E_s	...			E_s			E_s	...			E_s	$\tilde{e}_s$
Destinations → Sources↓	N_{T_1}			...	N_{T_r}			N_{S_1}			...	N_{S_l}			Availability
N_{PS_1}	$\tilde{c}_{1(t+q+1)1}$	...	$\tilde{c}_{1(t+q+1)s}$	...	$\tilde{c}_{1(t+q+r)1}$	...	$\tilde{c}_{1(t+q+r)s}$	$\tilde{c}_{1(t+q+r+1)1}$	...	$\tilde{c}_{1(t+q+r+1)s}$	...	$\tilde{c}_{1(t+q+r+l)1}$	...	$\tilde{c}_{1(t+q+r+l)s}$	$\tilde{a}_1$
⋮	⋮	⋮	⋮	⋮	⋮	⋮	⋮	⋮	⋮	⋮	⋮	⋮	⋮	⋮	⋮
N_{PS_m}	$\tilde{c}_{m(t+q+1)1}$	...	$\tilde{c}_{m(t+q+1)s}$	...	$\tilde{c}_{m(t+q+r)1}$	...	$\tilde{c}_{m(t+q+r)s}$	$\tilde{c}_{m(t+q+r+1)1}$	...	$\tilde{c}_{m(t+q+r+1)s}$	...	$\tilde{c}_{m(t+q+r+l)1}$	...	$\tilde{c}_{m(t+q+r+l)s}$	$\tilde{a}_m$
N_{S_1}	$\tilde{c}_{(m+1)(t+q+1)1}$	...	$\tilde{c}_{(m+1)(t+q+1)s}$	...	$\tilde{c}_{(m+1)(t+q+r)1}$	...	$\tilde{c}_{(m+1)(t+q+r)s}$	$\tilde{c}_{(m+1)(t+q+r+1)1}$	...	$\tilde{c}_{(m+1)(t+q+r+1)s}$	...	$\tilde{c}_{(m+1)(t+q+r+l)1}$	...	$\tilde{c}_{(m+1)(t+q+r+l)s}$	$\tilde{a}'_{m+1}$
⋮	⋮	⋮	⋮	⋮	⋮	⋮	⋮	⋮	⋮	⋮	⋮	⋮	⋮	⋮	⋮
N_{S_l}	$\tilde{c}_{(m+l)(t+q+1)1}$	...	$\tilde{c}_{(m+l)(t+q+1)s}$	...	$\tilde{c}_{(m+l)(t+q+r)1}$	...	$\tilde{c}_{(m+l)(t+q+r)s}$	$\tilde{c}_{(m+l)(t+q+r+1)1}$	...	$\tilde{c}_{(m+l)(t+q+r+1)s}$	...	$\tilde{c}_{(m+l)(t+q+r+l)1}$	...	$\tilde{c}_{(m+l)(t+q+r+l)s}$	$\tilde{a}'_{m+l}$
N_{D_1}	$\tilde{c}_{(m+l+1)(t+q+1)1}$	...	$\tilde{c}_{(m+l+1)(t+q+1)s}$	...	$\tilde{c}_{(m+l+1)(t+q+r)1}$	...	$\tilde{c}_{(m+l+1)(t+q+r)s}$	$\tilde{c}_{(m+l+1)(t+q+r+1)1}$	...	$\tilde{c}_{(m+l+1)(t+q+r+1)s}$	...	$\tilde{c}_{(m+l+1)(t+q+r+l)1}$	...	$\tilde{c}_{(m+l+1)(t+q+r+l)s}$	$\tilde{P}$
⋮	⋮	⋮	⋮	⋮	⋮	⋮	⋮	⋮	⋮	⋮	⋮	⋮	⋮	⋮	⋮
N_{D_q}	$\tilde{c}_{(m+l+q)(t+q+1)1}$	...	$\tilde{c}_{(m+l+q)(t+q+1)s}$	...	$\tilde{c}_{(m+l+q)(t+q+r)1}$	...	$\tilde{c}_{(m+l+q)(t+q+r)s}$	$\tilde{c}_{(m+l+q)(t+q+r+1)1}$	...	$\tilde{c}_{(m+l+q)(t+q+r+1)s}$	...	$\tilde{c}_{(m+l+q)(t+q+r+l)1}$	...	$\tilde{c}_{(m+l+q)(t+q+r+l)s}$	$\tilde{P}$
N_{T_1}	$\tilde{c}_{(m+l+q+1)(t+q+1)1}$	...	$\tilde{c}_{(m+l+q+1)(t+q+1)s}$	...	$\tilde{c}_{(m+l+q+1)(t+q+r)1}$	...	$\tilde{c}_{(m+l+q+1)(t+q+r)s}$	$\tilde{c}_{(m+l+q+1)(t+q+r+1)1}$	...	$\tilde{c}_{(m+l+q+1)(t+q+r+1)s}$	...	$\tilde{c}_{(m+l+q+1)(t+q+r+l)1}$	...	$\tilde{c}_{(m+l+q+1)(t+q+r+l)s}$	$\tilde{P}$
⋮	⋮	⋮	⋮	⋮	⋮	⋮	⋮	⋮	⋮	⋮	⋮	⋮	⋮	⋮	⋮
N_{T_r}	$\tilde{c}_{(m+l+q+r)(t+q+1)1}$	...	$\tilde{c}_{(m+l+q+r)(t+q+1)s}$	...	$\tilde{c}_{(m+l+q+r)(t+q+r)1}$	...	$\tilde{c}_{(m+l+q+r)(t+q+r)s}$	$\tilde{c}_{(m+l+q+r)(t+q+r+1)1}$	...	$\tilde{c}_{(m+l+q+r)(t+q+r+1)s}$	...	$\tilde{c}_{(m+l+q+r)(t+q+r+l)1}$	...	$\tilde{c}_{(m+l+q+r)(t+q+r+l)s}$	$\tilde{P}$
Demand	$\tilde{P}$			...	$\tilde{P}$			$\tilde{P}$			...	$\tilde{P}$			

$\tilde{c}_{ijk} = (m'_{ijk}, n'_{ijk}, \alpha'_{ijk}, \beta'_{ijk})LR =$

- $(0, 0, 0, 0)_{LR}$; If the product is supplied from some intermediate node to same intermediate node by a dummy conveyance
  or if the product is supplied from some dummy purely source node to any intermediate node or to any purely destination node by any conveyance
  or if the product is supplied from any purely source node to purely dummy destination node by any conveyance
  or if the product is supplied from any intermediate node to purely dummy destination node by any conveyance
  or if the product is supplied from some purely source node to any purely destination node or to any intermediate node by a dummy conveyance
  or if the product is supplied from any intermediate node to any purely destination node or to any intermediate node by a dummy conveyance
- $(M, M, 0, 0)_{LR}$; or if the product is supplied from some intermediate node to same intermediate node by such a conveyance which is not a dummy conveyance
  or if the product can not be directly supplied from the ith node to the jth node
  or if the product can not be supplied from the ith node to the jth node
- $(m''_{ijk}, n''_{ijk}, \alpha''_{ijk}, \beta''_{ijk})LR$; or if the product can be directly supplied from the ith node to the jth node

$\tilde{a}_i = (m_i, n_i, \alpha_i, \beta_i)LR$; $\tilde{a}'_{m+i} = (m''_{m+i}, n''_{m+i}, \alpha''_{m+i}, \beta''_{m+i})LR$; $\tilde{b}_j = (m'_j, n'_j, \alpha'_j, \beta'_j)LR$; $\tilde{b}'_{t+j} = (m''_{t+j}, n''_{t+j}, \alpha''_{t+j}, \beta''_{t+j})LR$ and $\tilde{P} = (P_1, P_2, \alpha_P, \beta_P)_{LR}$

Table 8.2 Tabular representation of the first crisp solid transportation problem

Conveyance	E_1			…	E_1			E_1			…	E_1		
		…		…		…			…		…		…	
			E_s	…			E_s			E_s	…			E_s
Destinations → Sources↓	N_{PD_1}			…	N_{PD_t}			N_{D_1}			…	N_{D_q}		
N_{PS_1}	η_{111}	…	η_{11s}	…	η_{1t1}	…	η_{1ts}	$\eta_{1(t+1)1}$	…	$\eta_{1(t+1)s}$	…	$\eta_{1(t+q)1}$	…	$\eta_{1(t+q)s}$
⋮	⋮	⋮	⋮	⋮	⋮	⋮	⋮	⋮	⋮	⋮	⋮	⋮	⋮	⋮
N_{PS_m}	η_{m11}	…	η_{m1s}	…	η_{mt1}	…	η_{mts}	$\eta_{m(t+1)1}$	…	$\eta_{m(t+1)s}$	…	$\eta_{m(t+q)1}$	…	$\eta_{m(t+q)s}$
N_{S_1}	$\eta_{(m+1)11}$	…	$\eta_{(m+1)1s}$	…	$\eta_{(m+1)t1}$	…	$\eta_{(m+1)ts}$	$\eta_{(m+1)(t+1)1}$	…	$\eta_{(m+1)(t+1)s}$	…	$\eta_{(m+1)(t+q)1}$	…	$\eta_{(m+1)(t+q)s}$
⋮	⋮	⋮	⋮	⋮	⋮	⋮	⋮	⋮	⋮	⋮	⋮	⋮	⋮	⋮
N_{S_l}	$\eta_{(m+l)11}$	…	$\eta_{(m+l)1s}$	…	$\eta_{(m+l)t1}$	…	$\eta_{(m+l)ts}$	$\eta_{(m+l)(t+1)1}$	…	$\eta_{(m+l)(t+1)s}$	…	$\eta_{(m+l)(t+q)1}$	…	$\eta_{(m+l)(t+q)s}$
N_{D_1}	$\eta_{(m+l+1)11}$	…	$\eta_{(m+l+1)1s}$	…	$\eta_{(m+l+1)t1}$	…	$\eta_{(m+l+1)ts}$	$\eta_{(m+l+1)(t+1)1}$	…	$\eta_{(m+l+1)(t+1)s}$	…	$\eta_{(m+l+1)(t+q)1}$	…	$\eta_{(m+l+1)(t+q)s}$
⋮	⋮	⋮	⋮	⋮	⋮	⋮	⋮	⋮	⋮	⋮	⋮	⋮	⋮	⋮
N_{D_q}	$\eta_{(m+l+q)11}$	…	$\eta_{(m+l+q)1s}$	…	$\eta_{(m+l+q)t1}$	…	$\eta_{(m+l+q)ts}$	$\eta_{(m+l+q)(t+1)1}$	…	$\eta_{(m+l+q)(t+1)s}$	…	$\eta_{(m+l+q)(t+q)1}$	…	$\eta_{(m+l+q)(t+q)s}$
N_{T_1}	$\eta_{(m+l+q+1)11}$	…	$\eta_{(m+l+q+1)1s}$	…	$\eta_{(m+l+q+1)t1}$	…	$\eta_{(m+l+q+1)ts}$	$\eta_{(m+l+q+1)(t+1)1}$	…	$\eta_{(m+l+q+1)(t+1)s}$	…	$\eta_{(m+l+q+1)(t+q)1}$	…	$\eta_{(m+l+q+1)(t+q)s}$
⋮	⋮	⋮	⋮	⋮	⋮	⋮	⋮	⋮	⋮	⋮	⋮	⋮	⋮	⋮
N_{T_r}	$\eta_{(m+l+q+r)11}$	…	$\eta_{(m+l+q+r)1s}$	…	$\eta_{(m+l+q+r)t1}$	…	$\eta_{(m+l+q+r)ts}$	$\eta_{(m+l+q+r)(t+1)1}$	…	$\eta_{(m+l+q+r)(t+1)s}$	…	$\eta_{(m+l+q+r)(t+q)1}$	…	$\eta_{(m+l+q+r)(t+q)s}$
Demand	$m'_1 - \alpha'_1$			…	$m'_t - \alpha'_t$			$m''_{t+1} - \alpha''_{t+1}$			…	$m''_{t+q} - \alpha''_{t+q}$		

(continued)

Table 8.2 (continued)

															Capacity
Conveyance	E_1			$\cdots$	E_1			E_1			$\cdots$	E_1			$m_1''' - \alpha_1'''$
		$\cdots$		$\cdots$		$\cdots$			$\cdots$		$\cdots$		$\cdots$		$\vdots$
			E_s	$\cdots$			E_s			E_s	$\cdots$			E_s	$m_s''' - \alpha_s'''$
Destinations→ Sources↓	N_{T_1}			$\cdots$	N_{T_r}			N_{S_1}			$\cdots$	N_{S_l}			Availability
N_{PS_1}	$\eta_{1(t+q+1)1}$	$\cdots$	$\eta_{1(t+q+1)s}$	$\cdots$	$\eta_{1(t+q+r)1}$	$\cdots$	$\eta_{1(t+q+r)s}$	$\eta_{1(t+q+r+1)1}$	$\cdots$	$\eta_{1(t+q+r+1)s}$	$\cdots$	$\eta_{1(t+q+r+l)1}$	$\cdots$	$\eta_{1(t+q+r+l)s}$	$m_1 - \alpha_1$
$\vdots$	$\vdots$	$\vdots$	$\vdots$	$\vdots$	$\vdots$	$\vdots$	$\vdots$	$\vdots$	$\vdots$	$\vdots$	$\vdots$	$\vdots$	$\vdots$	$\vdots$	$\vdots$
N_{PS_m}	$\eta_{m(t+q+1)1}$	$\cdots$	$\eta_{m(t+q+1)s}$	$\cdots$	$\eta_{m(t+q+r)1}$	$\cdots$	$\eta_{m(t+q+r)s}$	$\eta_{m(t+q+r+1)1}$	$\cdots$	$\eta_{m(t+q+r+1)s}$	$\cdots$	$\eta_{m(t+q+r+l)1}$	$\cdots$	$\eta_{m(t+q+r+l)s}$	$m_m - \alpha_m$
N_{S_1}	$\eta_{(m+1)(t+q+1)1}$	$\cdots$	$\eta_{(m+1)(t+q+1)s}$	$\cdots$	$\eta_{(m+1)(t+q+r)1}$	$\cdots$	$\eta_{(m+1)(t+q+r)s}$	$\eta_{(m+1)(t+q+r+1)1}$	$\cdots$	$\eta_{(m+1)(t+q+r+1)s}$	$\cdots$	$\eta_{(m+1)(t+q+r+l)1}$	$\cdots$	$\eta_{(m+1)(t+q+r+l)s}$	$m''_{m+1} - \alpha''_{m+1}$
$\vdots$	$\vdots$	$\vdots$	$\vdots$	$\vdots$	$\vdots$	$\vdots$	$\vdots$	$\vdots$	$\vdots$	$\vdots$	$\vdots$	$\vdots$	$\vdots$	$\vdots$	$\vdots$
N_{S_l}	$\eta_{(m+l)(t+q+1)1}$	$\cdots$	$\eta_{(m+l)(t+q+1)s}$	$\cdots$	$\eta_{(m+l)(t+q+r)1}$	$\cdots$	$\eta_{(m+l)(t+q+r)s}$	$\eta_{(m+l)(t+q+r+1)1}$	$\cdots$	$\eta_{(m+l)(t+q+r+1)s}$	$\cdots$	$\eta_{(m+l)(t+q+r+l)1}$	$\cdots$	$\eta_{(m+l)(t+q+r+l)s}$	$m''_{m+l} - \alpha''_{m+l}$
N_{D_1}	$\eta_{(m+l+1)(t+q+1)1}$	$\cdots$	$\eta_{(m+l+1)(t+q+1)s}$	$\cdots$	$\eta_{(m+l+1)(t+q+r)1}$	$\cdots$	$\eta_{(m+l+1)(t+q+r)s}$	$\eta_{(m+l+1)(t+q+r+1)1}$	$\cdots$	$\eta_{(m+l+1)(t+q+r+1)s}$	$\cdots$	$\eta_{(m+l+1)(t+q+r+l)1}$	$\cdots$	$\eta_{(m+l+1)(t+q+r+l)s}$	$P_1 - \alpha_p$
$\vdots$	$\vdots$	$\vdots$	$\vdots$	$\vdots$	$\vdots$	$\vdots$	$\vdots$	$\vdots$	$\vdots$	$\vdots$	$\vdots$	$\vdots$	$\vdots$	$\vdots$	$\vdots$
N_{D_q}	$\eta_{(m+l+q)(t+q+1)1}$	$\cdots$	$\eta_{(m+l+q)(t+q+1)s}$	$\cdots$	$\eta_{(m+l+q)(t+q+r)1}$	$\cdots$	$\eta_{(m+l+q)(t+q+r)s}$	$\eta_{(m+l+q)(t+q+r+1)1}$	$\cdots$	$\eta_{(m+l+q)(t+q+r+1)s}$	$\cdots$	$\eta_{(m+l+q)(t+q+r+l)1}$	$\cdots$	$\eta_{(m+l+q)(t+q+r+l)s}$	$P_1 - \alpha_p$
N_{T_1}	$\eta_{(m+l+q+1)(t+q+1)1}$	$\cdots$	$\eta_{(m+l+q+1)(t+q+1)s}$	$\cdots$	$\eta_{(m+l+q+1)(t+q+r)1}$	$\cdots$	$\eta_{(m+l+q+1)(t+q+r)s}$	$\eta_{(m+l+q+1)(t+q+r+1)1}$	$\cdots$	$\eta_{(m+l+q+1)(t+q+r+1)s}$	$\cdots$	$\eta_{(m+l+q+1)(t+q+r+l)1}$	$\cdots$	$\eta_{(m+l+q+1)(t+q+r+l)s}$	$P_1 - \alpha_p$
$\vdots$	$\vdots$	$\vdots$	$\vdots$	$\vdots$	$\vdots$	$\vdots$	$\vdots$	$\vdots$	$\vdots$	$\vdots$	$\vdots$	$\vdots$	$\vdots$	$\vdots$	$\vdots$
N_{T_r}	$\eta_{(m+l+q+r)(t+q+1)1}$	$\cdots$	$\eta_{(m+l+q+r)(t+q+1)s}$	$\cdots$	$\eta_{(m+l+q+r)(t+q+r)1}$	$\cdots$	$\eta_{(m+l+q+r)(t+q+r)s}$	$\eta_{(m+l+q+r)(t+q+r+1)1}$	$\cdots$	$\eta_{(m+l+q+r)(t+q+r+1)s}$	$\cdots$	$\eta_{(m+l+q+r)(t+q+r+l)1}$	$\cdots$	$\eta_{(m+l+q+r)(t+q+r+l)s}$	$P_1 - \alpha_p$
Demand	$P_1 - \alpha_p$			$\cdots$	$P_1 - \alpha_p$			$P_1 - \alpha_p$			$\cdots$	$P_1 - \alpha_p$			

Table 8.3 Tabular representation of the second crisp solid transportation problem

Conveyance	E_1			…	E_1			E_1			…	E_1		
		…		…		…			…		…		…	
			E_s	…			E_s			E_s	…			E_s
Destinations → Sources↓	N_{PD_1}			…	N_{PD_t}			N_{D_1}			…	N_{D_q}		
N_{PS_1}	ρ_{111}	…	ρ_{11s}	…	ρ_{1t1}	…	ρ_{1ts}	$\rho_{1(t+1)1}$	…	$\rho_{1(t+1)s}$	…	$\rho_{1(t+q)1}$	…	$\rho_{1(t+q)s}$
⋮	⋮	⋮	⋮	⋮	⋮	⋮	⋮	⋮	⋮	⋮	⋮	⋮	⋮	⋮
N_{PS_m}	ρ_{m11}	…	ρ_{m1s}	…	ρ_{mt1}	…	ρ_{mts}	$\rho_{m(t+1)1}$	…	$\rho_{m(t+1)s}$	…	$\rho_{m(t+q)1}$	…	$\rho_{m(t+q)s}$
N_{S_1}	$\rho_{(m+1)11}$	…	$\rho_{(m+1)1s}$	…	$\rho_{(m+1)t1}$	…	$\rho_{(m+1)ts}$	$\rho_{(m+1)(t+1)1}$	…	$\rho_{(m+1)(t+1)s}$	…	$\rho_{(m+1)(t+q)1}$	…	$\rho_{(m+1)(t+q)s}$
⋮	⋮	⋮	⋮	⋮	⋮	⋮	⋮	⋮	⋮	⋮	⋮	⋮	⋮	⋮
N_{S_l}	$\rho_{(m+l)11}$	…	$\rho_{(m+l)1s}$	…	$\rho_{(m+l)t1}$	…	$\rho_{(m+l)ts}$	$\rho_{(m+l)(t+1)1}$	…	$\rho_{(m+l)(t+1)s}$	…	$\rho_{(m+l)(t+q)1}$	…	$\rho_{(m+l)(t+q)s}$
N_{D_1}	$\rho_{(m+l+1)11}$	…	$\rho_{(m+l+1)1s}$	…	$\rho_{(m+l+1)t1}$	…	$\rho_{(m+l+1)ts}$	$\rho_{(m+l+1)(t+1)1}$	…	$\rho_{(m+l+1)(t+1)s}$	…	$\rho_{(m+l+1)(t+q)1}$	…	$\rho_{(m+l+1)(t+q)s}$
⋮	⋮	⋮	⋮	⋮	⋮	⋮	⋮	⋮	⋮	⋮	⋮	⋮	⋮	⋮
N_{D_q}	$\rho_{(m+l+q)11}$	…	$\rho_{(m+l+q)1s}$	…	$\rho_{(m+l+q)t1}$	…	$\rho_{(m+l+q)ts}$	$\rho_{(m+l+q)(t+1)1}$	…	$\rho_{(m+l+q)(t+1)s}$	…	$\rho_{(m+l+q)(t+q)1}$	…	$\rho_{(m+l+q)(t+q)s}$
N_{T_1}	$\rho_{(m+l+q+1)11}$	…	$\rho_{(m+l+q+1)1s}$	…	$\rho_{(m+l+q+1)t1}$	…	$\rho_{(m+l+q+1)ts}$	$\rho_{(m+l+q+1)(t+1)1}$	…	$\rho_{(m+l+q+1)(t+1)s}$	…	$\rho_{(m+l+q+1)(t+q)1}$	…	$\rho_{(m+l+q+1)(t+q)s}$
⋮	⋮	⋮	⋮	⋮	⋮	⋮	⋮	⋮	⋮	⋮	⋮	⋮	⋮	⋮
N_{T_r}	$\rho_{(m+l+q+r)11}$	…	$\rho_{(m+l+q+r)1s}$	…	$\rho_{(m+l+q+r)t1}$	…	$\rho_{(m+l+q+r)ts}$	$\rho_{(m+l+q+r)(t+1)1}$	…	$\rho_{(m+l+q+r)(t+1)s}$	…	$\rho_{(m+l+q+r)(t+q)1}$	…	$\rho_{(m+l+q+r)(t+q)s}$
Demand	α'_1			…	α'_t			α''_{t+1}			…	α''_{t+q}		

(continued)

Table 8.3 (continued)

															Capacity
Conveyance	E_1			⋯	E_1			E_1			⋯	E_1			α_1'''
		⋯		⋯		⋯			⋯		⋯		⋯		⋮
			E_s	⋯			E_s			E_s	⋯			E_s	α_s'''
Destinations → Sources↓	N_{T_1}			⋯	N_{T_r}			N_{S_1}			⋯	N_{S_l}			Availability
N_{PS_1}	$\rho_{1(t+q+1)1}$	⋯	$\rho_{1(t+q+1)s}$	⋯	$\rho_{1(t+q+r)1}$	⋯	$\rho_{1(t+q+r)s}$	$\rho_{1(t+q+r+1)1}$	⋯	$\rho_{1(t+q+r+1)s}$	⋯	$\rho_{1(t+q+r+l)1}$	⋯	$\rho_{1(t+q+r+l)s}$	α_1
⋮	⋮	⋮	⋮	⋮	⋮	⋮	⋮	⋮	⋮	⋮	⋮	⋮	⋮	⋮	⋮
N_{PS_m}	$\rho_{m(t+q+1)1}$	⋯	$\rho_{m(t+q+1)s}$	⋯	$\rho_{m(t+q+r)1}$	⋯	$\rho_{m(t+q+r)s}$	$\rho_{m(t+q+r+1)1}$	⋯	$\rho_{m(t+q+r+1)s}$	⋯	$\rho_{m(t+q+r+l)1}$	⋯	$\rho_{m(t+q+r+l)s}$	α_m
N_{S_1}	$\rho_{(m+1)(t+q+1)1}$	⋯	$\rho_{(m+1)(t+q+1)s}$	⋯	$\rho_{(m+1)(t+q+r)1}$	⋯	$\rho_{(m+1)(t+q+r)s}$	$\rho_{(m+1)(t+q+r+1)1}$	⋯	$\rho_{(m+1)(t+q+r+1)s}$	⋯	$\rho_{(m+1)(t+q+r+l)1}$	⋯	$\rho_{(m+1)(t+q+r+l)s}$	α''_{m+1}
⋮	⋮	⋮	⋮	⋮	⋮	⋮	⋮	⋮	⋮	⋮	⋮	⋮	⋮	⋮	⋮
N_{S_l}	$\rho_{(m+l)(t+q+1)1}$	⋯	$\rho_{(m+l)(t+q+1)s}$	⋯	$\rho_{(m+l)(t+q+r)1}$	⋯	$\rho_{(m+l)(t+q+r)s}$	$\rho_{(m+l)(t+q+r+1)1}$	⋯	$\rho_{(m+l)(t+q+r+1)s}$	⋯	$\rho_{(m+l)(t+q+r+l)1}$	⋯	$\rho_{(m+l)(t+q+r+l)s}$	α''_{m+l}
N_{D_1}	$\rho_{(m+l+1)(t+q+1)1}$	⋯	$\rho_{(m+l+1)(t+q+1)s}$	⋯	$\rho_{(m+l+1)(t+q+r)1}$	⋯	$\rho_{(m+l+1)(t+q+r)s}$	$\rho_{(m+l+1)(t+q+r+1)1}$	⋯	$\rho_{(m+l+1)(t+q+r+1)s}$	⋯	$\rho_{(m+l+1)(t+q+r+l)1}$	⋯	$\rho_{(m+l+1)(t+q+r+l)s}$	α_p
⋮	⋮	⋮	⋮	⋮	⋮	⋮	⋮	⋮	⋮	⋮	⋮	⋮	⋮	⋮	⋮
N_{D_q}	$\rho_{(m+l+q)(t+q+1)1}$	⋯	$\rho_{(m+l+q)(t+q+1)s}$	⋯	$\rho_{(m+l+q)(t+q+r)1}$	⋯	$\rho_{(m+l+q)(t+q+r)s}$	$\rho_{(m+l+q)(t+q+r+1)1}$	⋯	$\rho_{(m+l+q)(t+q+r+1)s}$	⋯	$\rho_{(m+l+q)(t+q+r+l)1}$	⋯	$\rho_{(m+l+q)(t+q+r+l)s}$	α_p
N_{T_1}	$\rho_{(m+l+q+1)(t+q+1)1}$	⋯	$\rho_{(m+l+q+1)(t+q+1)s}$	⋯	$\rho_{(m+l+q+1)(t+q+r)1}$	⋯	$\rho_{(m+l+q+1)(t+q+r)s}$	$\rho_{(m+l+q+1)(t+q+r+1)1}$	⋯	$\rho_{(m+l+q+1)(t+q+r+1)s}$	⋯	$\rho_{(m+l+q+1)(t+q+r+l)1}$	⋯	$\rho_{(m+l+q+1)(t+q+r+l)s}$	α_p
⋮	⋮	⋮	⋮	⋮	⋮	⋮	⋮	⋮	⋮	⋮	⋮	⋮	⋮	⋮	⋮
N_{T_r}	$\rho_{(m+l+q+r)(t+q+1)1}$	⋯	$\rho_{(m+l+q+r)(t+q+1)s}$	⋯	$\rho_{(m+l+q+r)(t+q+r)1}$	⋯	$\rho_{(m+l+q+r)(t+q+r)s}$	$\rho_{(m+l+q+r)(t+q+r+1)1}$	⋯	$\rho_{(m+l+q+r)(t+q+r+1)s}$	⋯	$\rho_{(m+l+q+r)(t+q+r+l)1}$	⋯	$\rho_{(m+l+q+r)(t+q+r+l)s}$	α_p
Demand	α_p			⋯	α_p			α_p			⋯	α_p			

Table 8.4 Tabular representation of the third crisp solid transportation problem

Conveyance	E_1			…	E_1			E_1			…	E_1		
		…		…		…			…		…		…	
			E_s	…			E_s			E_s	…			E_s
Destinations → Sources↓	N_{PD_1}			…	N_{PD_t}			N_{D_1}			…	N_{D_q}		
N_{PS_1}	δ_{111}	…	δ_{11s}	…	δ_{1t1}	…	δ_{1ts}	$\delta_{1(t+1)1}$	…	$\delta_{1(t+1)s}$	…	$\delta_{1(t+q)1}$	…	$\delta_{1(t+q)s}$
⋮	⋮	⋮	⋮	⋮	⋮	⋮	⋮	⋮	⋮	⋮	⋮	⋮	⋮	⋮
N_{PS_m}	δ_{m11}	…	δ_{m1s}	…	δ_{mt1}	…	δ_{mts}	$\delta_{m(t+1)1}$	…	$\delta_{m(t+1)s}$	…	$\delta_{m(t+q)1}$	…	$\delta_{m(t+q)s}$
N_{S_1}	$\delta_{(m+1)11}$	…	$\delta_{(m+1)1s}$	…	$\delta_{(m+1)t1}$	…	$\delta_{(m+1)ts}$	$\delta_{(m+1)(t+1)1}$	…	$\delta_{(m+1)(t+1)s}$	…	$\delta_{(m+1)(t+q)1}$	…	$\delta_{(m+1)(t+q)s}$
⋮	⋮	⋮	⋮	⋮	⋮	⋮	⋮	⋮	⋮	⋮	⋮	⋮	⋮	⋮
N_{S_l}	$\delta_{(m+l)11}$	…	$\delta_{(m+l)1s}$	…	$\delta_{(m+l)t1}$	…	$\delta_{(m+l)ts}$	$\delta_{(m+l)(t+1)1}$	…	$\delta_{(m+l)(t+1)s}$	…	$\delta_{(m+l)(t+q)1}$	…	$\delta_{(m+l)(t+q)s}$
N_{D_1}	$\delta_{(m+l+1)11}$	…	$\delta_{(m+l+1)1s}$	…	$\delta_{(m+l+1)t1}$	…	$\delta_{(m+l+1)ts}$	$\delta_{(m+l+1)(t+1)1}$	…	$\delta_{(m+l+1)(t+1)s}$	…	$\delta_{(m+l+1)(t+q)1}$	…	$\delta_{(m+l+1)(t+q)s}$
⋮	⋮	⋮	⋮	⋮	⋮	⋮	⋮	⋮	⋮	⋮	⋮	⋮	⋮	⋮
N_{D_q}	$\delta_{(m+l+q)11}$	…	$\delta_{(m+l+q)1s}$	…	$\delta_{(m+l+q)t1}$	…	$\delta_{(m+l+q)ts}$	$\delta_{(m+l+q)(t+1)1}$	…	$\delta_{(m+l+q)(t+1)s}$	…	$\delta_{(m+l+q)(t+q)1}$	…	$\delta_{(m+l+q)(t+q)s}$
N_{T_1}	$\delta_{(m+l+q+1)11}$	…	$\delta_{(m+l+q+1)1s}$	…	$\delta_{(m+l+q+1)t1}$	…	$\delta_{(m+l+q+1)ts}$	$\delta_{(m+l+q+1)(t+1)1}$	…	$\delta_{(m+l+q+1)(t+1)s}$	…	$\delta_{(m+l+q+1)(t+q)1}$	…	$\delta_{(m+l+q+1)(t+q)s}$
⋮	⋮	⋮	⋮	⋮	⋮	⋮	⋮	⋮	⋮	⋮	⋮	⋮	⋮	⋮
N_{T_r}	$\delta_{(m+l+q+r)11}$	…	$\delta_{(m+l+q+r)1s}$	…	$\delta_{(m+l+q+r)t1}$	…	$\delta_{(m+l+q+r)ts}$	$\delta_{(m+l+q+r)(t+1)1}$	…	$\delta_{(m+l+q+r)(t+1)s}$	…	$\delta_{(m+l+q+r)(t+q)1}$	…	$\delta_{(m+l+q+r)(t+q)s}$
Demand	$n'_1 - m'_1$			…	$n'_t - m'_t$			$n''_{t+1} - m''_{t+1}$			…	$n''_{t+q} - m''_{t+q}$		

(continued)

Table 8.4 (continued)

															Capacity
Conveyance	E_1			$\cdots$	E_1			E_1			$\cdots$	E_1			$n_1''' - m_1'''$
		$\cdots$		$\cdots$		$\cdots$			$\cdots$		$\cdots$		$\cdots$		$\vdots$
			E_s	$\cdots$			E_s			E_s	$\cdots$			E_s	$n_s''' - m_s'''$
Destinations → Sources↓	N_{T_1}			$\cdots$	N_{T_r}			N_{S_1}			$\cdots$	N_{S_t}			Availability
N_{PS_1}	$\delta_{1(t+q+1)1}$	$\cdots$	$\delta_{1(t+q+1)s}$	$\cdots$	$\delta_{1(t+q+r)1}$	$\cdots$	$\delta_{1(t+q+r)s}$	$\delta_{1(t+q+r+1)1}$	$\cdots$	$\delta_{1(t+q+r+1)s}$	$\cdots$	$\delta_{1(t+q+r+l)1}$	$\cdots$	$\delta_{1(t+q+r+l)s}$	$n_1 - m_1$
$\vdots$	$\vdots$	$\vdots$	$\vdots$	$\vdots$	$\vdots$	$\vdots$	$\vdots$	$\vdots$	$\vdots$	$\vdots$	$\vdots$	$\vdots$	$\vdots$	$\vdots$	$\vdots$
N_{PS_m}	$\delta_{m(t+q+1)1}$	$\cdots$	$\delta_{m(t+q+1)s}$	$\cdots$	$\delta_{m(t+q+r)1}$	$\cdots$	$\delta_{m(t+q+r)s}$	$\delta_{m(t+q+r+1)1}$	$\cdots$	$\delta_{m(t+q+r+1)s}$	$\cdots$	$\delta_{m(t+q+r+l)1}$	$\cdots$	$\delta_{m(t+q+r+l)s}$	$n_m - m_m$
N_{S_1}	$\delta_{(m+1)(t+q+1)1}$	$\cdots$	$\delta_{(m+1)(t+q+1)s}$	$\cdots$	$\delta_{(m+1)(t+q+r)1}$	$\cdots$	$\delta_{(m+1)(t+q+r)s}$	$\delta_{(m+1)(t+q+r+1)1}$	$\cdots$	$\delta_{(m+1)(t+q+r+1)s}$	$\cdots$	$\delta_{(m+1)(t+q+r+l)1}$	$\cdots$	$\delta_{(m+1)(t+q+r+l)s}$	$n''_{m+1} - m''_{m+1}$
$\vdots$	$\vdots$	$\vdots$	$\vdots$	$\vdots$	$\vdots$	$\vdots$	$\vdots$	$\vdots$	$\vdots$	$\vdots$	$\vdots$	$\vdots$	$\vdots$	$\vdots$	$\vdots$
N_{S_t}	$\delta_{(m+l)(t+q+1)1}$	$\cdots$	$\delta_{(m+l)(t+q+1)s}$	$\cdots$	$\delta_{(m+l)(t+q+r)1}$	$\cdots$	$\delta_{(m+l)(t+q+r)s}$	$\delta_{(m+l)(t+q+r+1)1}$	$\cdots$	$\delta_{(m+l)(t+q+r+1)s}$	$\cdots$	$\delta_{(m+l)(t+q+r+l)1}$	$\cdots$	$\delta_{(m+l)(t+q+r+l)s}$	$n''_{m+l} - m''_{m+l}$
N_{D_1}	$\delta_{(m+l+1)(t+q+1)1}$	$\cdots$	$\delta_{(m+l+1)(t+q+1)s}$	$\cdots$	$\delta_{(m+l+1)(t+q+r)1}$	$\cdots$	$\delta_{(m+l+1)(t+q+r)s}$	$\delta_{(m+l+1)(t+q+r+1)1}$	$\cdots$	$\delta_{(m+l+1)(t+q+r+1)s}$	$\cdots$	$\delta_{(m+l+1)(t+q+r+l)1}$	$\cdots$	$\delta_{(m+l+1)(t+q+r+l)s}$	$P_2 - P_1$
$\vdots$	$\vdots$	$\vdots$	$\vdots$	$\vdots$	$\vdots$	$\vdots$	$\vdots$	$\vdots$	$\vdots$	$\vdots$	$\vdots$	$\vdots$	$\vdots$	$\vdots$	$\vdots$
N_{D_q}	$\delta_{(m+l+q)(t+q+1)1}$	$\cdots$	$\delta_{(m+l+q)(t+q+1)s}$	$\cdots$	$\delta_{(m+l+q)(t+q+r)1}$	$\cdots$	$\delta_{(m+l+q)(t+q+r)s}$	$\delta_{(m+l+q)(t+q+r+1)1}$	$\cdots$	$\delta_{(m+l+q)(t+q+r+1)s}$	$\cdots$	$\delta_{(m+l+q)(t+q+r+l)1}$	$\cdots$	$\delta_{(m+l+q)(t+q+r+l)s}$	$P_2 - P_1$
N_{T_1}	$\delta_{(m+l+q+1)(t+q+1)1}$	$\cdots$	$\delta_{(m+l+q+1)(t+q+1)s}$	$\cdots$	$\delta_{(m+l+q+1)(t+q+r)1}$	$\cdots$	$\delta_{(m+l+q+1)(t+q+r)s}$	$\delta_{(m+l+q+1)(t+q+r+1)1}$	$\cdots$	$\delta_{(m+l+q+1)(t+q+r+1)s}$	$\cdots$	$\delta_{(m+l+q+1)(t+q+r+l)1}$	$\cdots$	$\delta_{(m+l+q+1)(t+q+r+l)s}$	$P_2 - P_1$
$\vdots$	$\vdots$	$\vdots$	$\vdots$	$\vdots$	$\vdots$	$\vdots$	$\vdots$	$\vdots$	$\vdots$	$\vdots$	$\vdots$	$\vdots$	$\vdots$	$\vdots$	$\vdots$
N_{T_r}	$\delta_{(m+l+q+r)(t+q+1)1}$	$\cdots$	$\delta_{(m+l+q+r)(t+q+1)s}$	$\cdots$	$\delta_{(m+l+q+r)(t+q+r)1}$	$\cdots$	$\delta_{(m+l+q+r)(t+q+r)s}$	$\delta_{(m+l+q+r)(t+q+r+1)1}$	$\cdots$	$\delta_{(m+l+q+r)(t+q+r+1)s}$	$\cdots$	$\delta_{(m+l+q+r)(t+q+r+l)1}$	$\cdots$	$\delta_{(m+l+q+r)(t+q+r+l)s}$	$P_2 - P_1$
Demand	$P_2 - P_1$			$\cdots$	$P_2 - P_1$			$P_2 - P_1$			$\cdots$	$P_2 - P_1$			

Table 8.5 Tabular representation of the fourth crisp solid transportation problem

Conveyance	E_1			⋯	E_1			E_1			⋯	E_1		
		⋯		⋯		⋯			⋯		⋯		⋯	
			E_s	⋯			E_s			E_s	⋯			E_s
Destinations → Sources↓	N_{PD_1}			⋯	N_{PD_t}			N_{D_1}			⋯	N_{D_q}		
N_{PS_1}	ξ_{111}	⋯	ξ_{11s}	⋯	ξ_{1t1}	⋯	ξ_{1ts}	$\xi_{1(t+1)1}$	⋯	$\xi_{1(t+1)s}$	⋯	$\xi_{1(t+q)1}$	⋯	$\xi_{1(t+q)s}$
⋮	⋮	⋮	⋮	⋮	⋮	⋮	⋮	⋮	⋮	⋮	⋮	⋮	⋮	⋮
N_{PS_m}	ξ_{m11}	⋯	ξ_{m1s}	⋯	ξ_{mt1}	⋯	ξ_{mts}	$\xi_{m(t+1)1}$	⋯	$\xi_{m(t+1)s}$	⋯	$\xi_{m(t+q)1}$	⋯	$\xi_{m(t+q)s}$
N_{S_1}	$\xi_{(m+1)11}$	⋯	$\xi_{(m+1)1s}$	⋯	$\xi_{(m+1)t1}$	⋯	$\xi_{(m+1)ts}$	$\xi_{(m+1)(t+1)1}$	⋯	$\xi_{(m+1)(t+1)s}$	⋯	$\xi_{(m+1)(t+q)1}$	⋯	$\xi_{(m+1)(t+q)s}$
⋮	⋮	⋮	⋮	⋮	⋮	⋮	⋮	⋮	⋮	⋮	⋮	⋮	⋮	⋮
N_{S_l}	$\xi_{(m+l)11}$	⋯	$\xi_{(m+l)1s}$	⋯	$\xi_{(m+l)t1}$	⋯	$\xi_{(m+l)ts}$	$\xi_{(m+l)(t+1)1}$	⋯	$\xi_{(m+l)(t+1)s}$	⋯	$\xi_{(m+l)(t+q)1}$	⋯	$\xi_{(m+l)(t+q)s}$
N_{D_1}	$\xi_{(m+l+1)11}$	⋯	$\xi_{(m+l+1)1s}$	⋯	$\xi_{(m+l+1)t1}$	⋯	$\xi_{(m+l+1)ts}$	$\xi_{(m+l+1)(t+1)1}$	⋯	$\xi_{(m+l+1)(t+1)s}$	⋯	$\xi_{(m+l+1)(t+q)1}$	⋯	$\xi_{(m+l+1)(t+q)s}$
⋮	⋮	⋮	⋮	⋮	⋮	⋮	⋮	⋮	⋮	⋮	⋮	⋮	⋮	⋮
N_{D_q}	$\xi_{(m+l+q)11}$	⋯	$\xi_{(m+l+q)1s}$	⋯	$\xi_{(m+l+q)t1}$	⋯	$\xi_{(m+l+q)ts}$	$\xi_{(m+l+q)(t+1)1}$	⋯	$\xi_{(m+l+q)(t+1)s}$	⋯	$\xi_{(m+l+q)(t+q)1}$	⋯	$\xi_{(m+l+q)(t+q)s}$
N_{T_1}	$\xi_{(m+l+q+1)11}$	⋯	$\xi_{(m+l+q+1)1s}$	⋯	$\xi_{(m+l+q+1)t1}$	⋯	$\xi_{(m+l+q+1)ts}$	$\xi_{(m+l+q+1)(t+1)1}$	⋯	$\xi_{(m+l+q+1)(t+1)s}$	⋯	$\xi_{(m+l+q+1)(t+q)1}$	⋯	$\xi_{(m+l+q+1)(t+q)s}$
⋮	⋮	⋮	⋮	⋮	⋮	⋮	⋮	⋮	⋮	⋮	⋮	⋮	⋮	⋮
N_{T_r}	$\xi_{(m+l+q+r)11}$	⋯	$\xi_{(m+l+q+r)1s}$	⋯	$\xi_{(m+l+q+r)t1}$	⋯	$\xi_{(m+l+q+r)ts}$	$\xi_{(m+l+q+r)(t+1)1}$	⋯	$\xi_{(m+l+q+r)(t+1)s}$	⋯	$\xi_{(m+l+q+r)(t+q)1}$	⋯	$\xi_{(m+l+q+r)(t+q)s}$
Demand	β'_1			⋯	β'_t			β''_{t+1}			⋯	β''_{t+q}		

(continued)

Table 8.5 (continued)

															Capacity
Conveyance	E_1			$\cdots$	E_1			E_1			$\cdots$	E_1			β_1'''
		$\cdots$		$\cdots$		$\cdots$			$\cdots$		$\cdots$		$\cdots$		$\vdots$
			E_s	$\cdots$			E_s			E_s	$\cdots$			E_s	β_s'''
Destinations → Sources↓	N_{T_1}			$\cdots$	N_{T_r}			N_{S_1}			$\cdots$	N_{S_l}			Availability
N_{PS_1}	$\xi_{1(t+q+1)1}$	$\cdots$	$\xi_{1(t+q+1)s}$	$\cdots$	$\xi_{1(t+q+r)1}$	$\cdots$	$\xi_{1(t+q+r)s}$	$\xi_{1(t+q+r+1)1}$	$\cdots$	$\xi_{1(t+q+r+1)s}$	$\cdots$	$\xi_{1(t+q+r+l)1}$	$\cdots$	$\xi_{1(t+q+r+l)s}$	β_1
$\vdots$	$\vdots$	$\vdots$	$\vdots$	$\vdots$	$\vdots$	$\vdots$	$\vdots$	$\vdots$	$\vdots$	$\vdots$	$\vdots$	$\vdots$	$\vdots$	$\vdots$	$\vdots$
N_{PS_m}	$\xi_{m(t+q+1)1}$	$\cdots$	$\xi_{m(t+q+1)s}$	$\cdots$	$\xi_{m(t+q+r)1}$	$\cdots$	$\xi_{m(t+q+r)s}$	$\xi_{m(t+q+r+1)1}$	$\cdots$	$\xi_{m(t+q+r+1)s}$	$\cdots$	$\xi_{m(t+q+r+l)1}$	$\cdots$	$\xi_{m(t+q+r+l)s}$	β_m
N_{S_1}	$\xi_{(m+1)(t+q+1)1}$	$\cdots$	$\xi_{(m+1)(t+q+1)s}$	$\cdots$	$\xi_{(m+1)(t+q+r)1}$	$\cdots$	$\xi_{(m+1)(t+q+r)s}$	$\xi_{(m+1)(t+q+r+1)1}$	$\cdots$	$\xi_{(m+1)(t+q+r+1)s}$	$\cdots$	$\xi_{(m+1)(t+q+r+l)1}$	$\cdots$	$\xi_{(m+1)(t+q+r+l)s}$	β''_{m+1}
$\vdots$	$\vdots$	$\vdots$	$\vdots$	$\vdots$	$\vdots$	$\vdots$	$\vdots$	$\vdots$	$\vdots$	$\vdots$	$\vdots$	$\vdots$	$\vdots$	$\vdots$	$\vdots$
N_{S_l}	$\xi_{(m+l)(t+q+1)1}$	$\cdots$	$\xi_{(m+l)(t+q+1)s}$	$\cdots$	$\xi_{(m+l)(t+q+r)1}$	$\cdots$	$\xi_{(m+l)(t+q+r)s}$	$\xi_{(m+l)(t+q+r+1)1}$	$\cdots$	$\xi_{(m+l)(t+q+r+1)s}$	$\cdots$	$\xi_{(m+l)(t+q+r+l)1}$	$\cdots$	$\xi_{(m+l)(t+q+r+l)s}$	β''_{m+l}
N_{D_1}	$\xi_{(m+l+1)(t+q+1)1}$	$\cdots$	$\xi_{(m+l+1)(t+q+1)s}$	$\cdots$	$\xi_{(m+l+1)(t+q+r)1}$	$\cdots$	$\xi_{(m+l+1)(t+q+r)s}$	$\xi_{(m+l+1)(t+q+r+1)1}$	$\cdots$	$\xi_{(m+l+1)(t+q+r+1)s}$	$\cdots$	$\xi_{(m+l+1)(t+q+r+l)1}$	$\cdots$	$\xi_{(m+l+1)(t+q+r+l)s}$	β_p
$\vdots$	$\vdots$	$\vdots$	$\vdots$	$\vdots$	$\vdots$	$\vdots$	$\vdots$	$\vdots$	$\vdots$	$\vdots$	$\vdots$	$\vdots$	$\vdots$	$\vdots$	$\vdots$
N_{D_q}	$\xi_{(m+l+q)(t+q+1)1}$	$\cdots$	$\xi_{(m+l+q)(t+q+1)s}$	$\cdots$	$\xi_{(m+l+q)(t+q+r)1}$	$\cdots$	$\xi_{(m+l+q)(t+q+r)s}$	$\xi_{(m+l+q)(t+q+r+1)1}$	$\cdots$	$\xi_{(m+l+q)(t+q+r+1)s}$	$\cdots$	$\xi_{(m+l+q)(t+q+r+l)1}$	$\cdots$	$\xi_{(m+l+q)(t+q+r+l)s}$	β_p
N_{T_1}	$\xi_{(m+l+q+1)(t+q+1)1}$	$\cdots$	$\xi_{(m+l+q+1)(t+q+1)s}$	$\cdots$	$\xi_{(m+l+q+1)(t+q+r)1}$	$\cdots$	$\xi_{(m+l+q+1)(t+q+r)s}$	$\xi_{(m+l+q+1)(t+q+r+1)1}$	$\cdots$	$\xi_{(m+l+q+1)(t+q+r+1)s}$	$\cdots$	$\xi_{(m+l+q+1)(t+q+r+l)1}$	$\cdots$	$\xi_{(m+l+q+1)(t+q+r+l)s}$	β_p
$\vdots$	$\vdots$	$\vdots$	$\vdots$	$\vdots$	$\vdots$	$\vdots$	$\vdots$	$\vdots$	$\vdots$	$\vdots$	$\vdots$	$\vdots$	$\vdots$	$\vdots$	$\vdots$
N_{T_r}	$\xi_{(m+l+q+r)(t+q+1)1}$	$\cdots$	$\xi_{(m+l+q+r)(t+q+1)s}$	$\cdots$	$\xi_{(m+l+q+r)(t+q+r)1}$	$\cdots$	$\xi_{(m+l+q+r)(t+q+r)s}$	$\xi_{(m+l+q+r)(t+q+r+1)1}$	$\cdots$	$\xi_{(m+l+q+r)(t+q+r+1)s}$	$\cdots$	$\xi_{(m+l+q+r)(t+q+r+l)1}$	$\cdots$	$\xi_{(m+l+q+r)(t+q+r+l)s}$	β_p
Demand	β_p			$\cdots$	β_p			β_p			$\cdots$	β_p			

8.3.3 *Advantages of the New Methods*

The advantages of the methods proposed in Chaps. 6 and 7 over the methods proposed in Chaps. 4, 5 and over the known existing methods proposed by Ghatee and Hashemi [1] and Liu [2] have already been discussed in Chaps. 6 and 7, respectively. Therefore, in this section we will discuss the advantages of the methods proposed in this chapter over the methods proposed in Chaps. 6 and 7.

These advantages can be briefly summarized as follows:

1. The new methods proposed in Chap. 6 can be used to find the fuzzy optimal solution of the fully fuzzy transportation problems and the fully fuzzy transshipment problems but these methods can neither be used for solving the fully fuzzy solid transportation problems nor for solving the fully fuzzy solid transshipment problems. Since the fully fuzzy transportation problems, the fully fuzzy transshipment problems and the fully fuzzy solid transportation problems are special types of the fully fuzzy solid transshipment problems, then the methods proposed in this chapter, can be used for finding the fuzzy optimal solution of all those problems.
2. The new methods proposed in Chap. 7 can be used to find the fuzzy optimal solution of the fully fuzzy transportation problems and the fully fuzzy solid transportation problems but these methods can neither be used for solving the fully fuzzy transshipment problems nor for solving the fully fuzzy solid transshipment problems. Since the fully fuzzy transportation problems, the fully fuzzy transshipment problems and the fully fuzzy solid transportation problems are special types of the fully fuzzy solid transshipment problems, then the methods proposed in this chapter can be used for finding the fuzzy optimal solution of all those problems.

8.4 Illustrative Example

In this section the fuzzy optimal solution of the fully fuzzy solid transshipment problem shown in Example 21, is obtained by using the proposed methods.

8.4.1 Determination of the Fuzzy Optimal Solution of the Fully Fuzzy Solid Transshipment Problem Using the Method Based on the Fuzzy Linear Programming Formulation

By using the proposed method, based on the fuzzy linear programming formulation, the fuzzy optimal solution of the fully fuzzy solid transshipment problem considered in Example 21 can be obtained as follows:

Step 1 We have:

- the total fuzzy availability

$$\sum_{i\in N_{PS}} \tilde{a}_i \oplus \sum_{i\in N_S} \tilde{a}'_i = (110, 150, 40, 40)_{LR},$$

- the total fuzzy demand

$$\sum_{j\in N_D} \tilde{b}_j \oplus \sum_{j\in N_{PD}} \tilde{b}_j = (50, 80, 30, 50)_{LR}$$

- and the total fuzzy capacity

$$\sum_{k\in S_C} \tilde{e}_k = (110, 150, 40, 50)_{LR}.$$

Since

$$\sum_{i\in N_{PS}} \tilde{a}_i \oplus \sum_{i\in N_S} \tilde{a}'_i \neq \sum_{j\in N_D} \tilde{b}_j \oplus \sum_{j\in N_{PD}} \tilde{b}_j \neq \sum_{k\in S_C} \tilde{e}_k$$

so this is an unbalanced fully fuzzy solid transshipment problem.

Step 2 By comparing

$$\sum_{i\in N_{PS}} \tilde{a}_i \oplus \sum_{i\in N_S} \tilde{a}'_i = (110, 150, 40, 40)_{LR}$$

with

$$\sum_{i\in N_{PS}} \tilde{a}_i \oplus \sum_{i\in N_S} \tilde{a}'_i = (m, n, \alpha, \beta)_{LR}$$

and

$$\sum_{j \in N_{PD}} \tilde{b}_j \oplus \sum_{j \in N_D} \tilde{b'}_j = (50, 80, 30, 50)_{LR}$$

with

$$\sum_{j \in N_{PD}} \tilde{b}_j \oplus \sum_{j \in N_D} \tilde{b}'_j = (m', n', \alpha', \beta')_{LR}$$

we obtain the values of $m, n, \alpha, \beta, m', n', \alpha'$ and β' as 110, 150, 40, 40, 50, 80, 30 and 50, respectively.

Since

$$\sum_{i \in N_{PS}} \tilde{a}_i \oplus \sum_{i \in N_S} \tilde{a}'_i \neq \sum_{j \in N_{PD}} \tilde{b}_j \oplus \sum_{j \in N_D} \tilde{b'}_j$$

and neither the conditions

$$m - \alpha \leq m' - \alpha', \alpha \leq \alpha', n - m \leq n' - m', \beta \leq \beta'$$

nor the conditions

$$m - \alpha \geq m' - \alpha', \alpha \geq \alpha', n - m \geq n' - m', \beta \geq \beta'$$

are satisfied, then—as described in Step 2 (Case (2c)) of the proposed method—there is a need to introduce:

- a dummy purely source node 4 with the fuzzy availability $\tilde{a}_4 = (0, 0, 0, 10)_{LR}$, and
- a dummy purely destination node 5 with fuzzy demand $\tilde{b}_5 = (60, 70, 10, 0)_{LR}$,

so that

$$\sum_{i \in N_{PS}} \tilde{a}_i \oplus \sum_{i \in N_S} \tilde{a}'_i = \sum_{j \in N_D} \tilde{b}_j \oplus \sum_{j \in N_{PD}} \tilde{b}_j.$$

Step 3 Since

$$\sum_{i \in N_{PS}} \tilde{a}_i \oplus \sum_{i \in N_S} \tilde{a}'_i = \sum_{j \in N_D} \tilde{b}_j \oplus \sum_{j \in N_{PD}} \tilde{b}_j = (110, 150, 40, 50)_{LR} = \sum_{k \in S_C} \tilde{e}_k,$$

then the fully fuzzy solid transshipment problem, obtained in Step 2, is a balanced fully fuzzy solid transshipment problem.

Step 4 Since a dummy purely source node (4) and a dummy purely destination node (5) are introduced, then—as described in Step 4 of the proposed method—by assuming

$$\tilde{c}_{4jk} = \tilde{c}_{i5k} = (0, 0, 0, 0)_{LR};\ \forall i = 1, 2, 4;\ j = 3, 5;\ k = 1, 2;$$

the fuzzy linear programming formulation of the balanced fully fuzzy solid transshipment problem, obtained in Step 3, can be written as:

$$\begin{cases} \min((8, 10, 2, 2)_{LR} \otimes \tilde{x}_{131} \oplus (4, 8, 3, 2)_{LR} \otimes \tilde{x}_{132} \oplus (8, 10, 4, 4)_{LR} \otimes \tilde{x}_{211} \\ \oplus(6, 8, 4, 4)_{LR} \otimes \tilde{x}_{212} \oplus (9, 12, 6, 3)_{LR} \otimes \tilde{x}_{231} \oplus (3, 6, 2, 3)_{LR} \otimes \tilde{x}_{232} \oplus \\ (0, 0, 0, 0)_{LR} \otimes \tilde{x}_{411} \oplus (0, 0, 0, 0)_{LR} \otimes \tilde{x}_{412} \oplus (0, 0, 0, 0)_{LR} \otimes \tilde{x}_{431} \oplus \\ (0, 0, 0, 0)_{LR} \otimes \tilde{x}_{432} \oplus (0, 0, 0, 0)_{LR} \otimes \tilde{x}_{251} \oplus (0, 0, 0, 0)_{LR} \otimes \tilde{x}_{252} \oplus \\ (0, 0, 0, 0)_{LR} \otimes \tilde{x}_{451} \oplus (0, 0, 0, 0)_{LR} \otimes \tilde{x}_{452} \oplus (0, 0, 0, 0)_{LR} \otimes \tilde{x}_{151} \oplus \\ (0, 0, 0, 0)_{LR} \otimes \tilde{x}_{152}) \\ \text{subject to:} \\ \sum_{k=1}^{2} (\tilde{x}_{13k} \oplus \tilde{x}_{15k}) \ominus_H \sum_{k=1}^{2} (\tilde{x}_{41k} \oplus \tilde{x}_{21k}) = (60, 80, 20, 20)_{LR} \\ \sum_{k=1}^{2} (\tilde{x}_{21k} \oplus \tilde{x}_{23k} \oplus \tilde{x}_{25k}) = (50, 70, 20, 20)_{LR} \\ \sum_{k=1}^{2} (\tilde{x}_{13k} \oplus \tilde{x}_{23k} \oplus \tilde{x}_{43k}) = (50, 80, 30, 50)_{LR} \\ \sum_{k=1}^{2} (\tilde{x}_{41k} \oplus \tilde{x}_{43k} \oplus \tilde{x}_{45k}) = (0, 0, 0, 10)_{LR} \\ \sum_{k=1}^{2} (\tilde{x}_{15k} \oplus \tilde{x}_{25k} \oplus \tilde{x}_{45k}) = (60, 70, 10, 0)_{LR} \\ \tilde{x}_{131} \oplus \tilde{x}_{151} \oplus \tilde{x}_{211} \oplus \tilde{x}_{231} \oplus \tilde{x}_{251} \oplus \tilde{x}_{411} \oplus \tilde{x}_{431} \oplus \tilde{x}_{451} = (60, 80, 20, 10)_{LR} \\ \tilde{x}_{132} \oplus \tilde{x}_{152} \oplus \tilde{x}_{212} \oplus \tilde{x}_{232} \oplus \tilde{x}_{252} \oplus \tilde{x}_{412} \oplus \tilde{x}_{432} \oplus \tilde{x}_{452} = (50, 70, 20, 40)_{LR} \\ \tilde{x}_{ijk} \text{ are non-negative } LR \text{ flat fuzzy numbers };\ \forall i = 1, 2, 4;\ j = 3, 5;\ k = 1, 2. \end{cases} \tag{8.8}$$

Step 5 Using Step 6 to Step 9 of the method proposed in Sect. 8.3.1, the fuzzy linear programming problem obtained in Step 5 can be converted into the following crisp linear programming problem:

$$
\left\{
\begin{array}{l}
\min(\frac{1}{4}(14m_{131} - 6\alpha_{131} + 22n_{131} + 12\beta_{131} + 5m_{132} - \alpha_{132} + 18n_{132} + 10\beta_{132} + \\
+ 12m_{211} - 4\alpha_{211} + 24n_{211} + 14\beta_{211} + 8m_{212} - 2\alpha_{212} + 20n_{212} + 12\beta_{212} + 12m_{231} \\
-3\alpha_{231} + 27n_{231} + 15\beta_{231} + 4m_{232} - 1\alpha_{232} + 15n_{232} + 9\beta_{232})) \\
\text{subject to:} \\
\sum_{k=1}^{2}(m_{13k} + m_{15k}) - \sum_{k=1}^{2}(m_{41k} + m_{21k}) = 60 \\
\sum_{k=1}^{2}(n_{13k} + n_{15k}) - \sum_{k=1}^{2}(n_{41k} + n_{21k}) = 80 \\
\sum_{k=1}^{2}(\alpha_{13k} + \alpha_{15k}) - \sum_{k=1}^{2}(\alpha_{41k} + \alpha_{21k}) = 20 \\
\sum_{k=1}^{2}(\beta_{13k} + \beta_{15k}) - \sum_{k=1}^{2}(\beta_{41k} + \beta_{21k}) = 20 \\
\sum_{k=1}^{2}(m_{21k} + m_{23k} + m_{25k}) = 50 \\
\sum_{k=1}^{2}(n_{21k} + n_{23k} + n_{25k}) = 70 \\
\sum_{k=1}^{2}(\alpha_{21k} + \alpha_{23k} + \alpha_{25k}) = 20 \\
\sum_{k=1}^{2}(\beta_{21k} + \beta_{23k} + \beta_{25k}) = 20 \\
\sum_{k=1}^{2}(m_{13k} + m_{23k} + m_{43k}) = 50 \\
\sum_{k=1}^{2}(n_{13k} + n_{23k} + n_{43k}) = 80 \\
\sum_{k=1}^{2}(\alpha_{13k} + \alpha_{23k} + \alpha_{43k}) = 30 \\
\sum_{k=1}^{2}(\beta_{13k} + \beta_{23k} + \beta_{43k}) = 50 \\
\sum_{k=1}^{2}(m_{41k} + m_{43k} + m_{45k}) = 0 \\
\sum_{k=1}^{2}(n_{41k} + n_{43k} + n_{45k}) = 0 \\
\sum_{k=1}^{2}(\alpha_{41k} + \alpha_{43k} + \alpha_{45k}) = 0 \\
\sum_{k=1}^{2}(\beta_{41k} + \beta_{43k} + \beta_{45k}) = 10 \\
\sum_{k=1}^{2}(m_{15k} + m_{25k} + m_{45k}) = 60 \\
\sum_{k=1}^{2}(n_{15k} + n_{25k} + n_{45k}) = 70 \\
\sum_{k=1}^{2}(\alpha_{15k} + \alpha_{25k} + \alpha_{45k}) = 10 \\
\sum_{k=1}^{2}(\beta_{15k} + \beta_{25k} + \beta_{45k}) = 0 \\
m_{131} + m_{151} + m_{211} + m_{231} + m_{251} + m_{411} + m_{431} + m_{451} = 60 \\
n_{131} + n_{151} + n_{211} + n_{231} + n_{251} + n_{411} + n_{431} + n_{451} = 80 \\
\alpha_{131} + \alpha_{151} + \alpha_{211} + \alpha_{231} + \alpha_{251} + \alpha_{411} + \alpha_{431} + \alpha_{451} = 20 \\
\beta_{131} + \beta_{151} + \beta_{211} + \beta_{231} + \beta_{251} + \beta_{411} + \beta_{431} + \beta_{451} = 10 \\
m_{132} + m_{152} + m_{212} + m_{232} + m_{252} + m_{412} + m_{432} + m_{452} = 50 \\
n_{132} + n_{152} + n_{212} + n_{232} + n_{252} + n_{412} + n_{432} + n_{452} = 70 \\
\alpha_{132} + \alpha_{152} + \alpha_{212} + \alpha_{232} + \alpha_{252} + \alpha_{412} + \alpha_{432} + \alpha_{452} = 20 \\
\beta_{132} + \beta_{152} + \beta_{212} + \beta_{232} + \beta_{252} + \beta_{412} + \beta_{432} + \beta_{452} = 40 \\
m_{ijk} - \alpha_{ijk}, n_{ijk} - m_{ijk}, \alpha_{ijk}, \beta_{ijk} \geq 0; \forall i = 1, 2, 4;\ j = 3, 5;\ k = 1, 2.
\end{array}
\right.
\tag{8.9}
$$

Step 6 The optimal solution of the crisp linear programming problem presented above is:

$$\begin{cases} m_{131} = 10, \; n_{131} = 20, \; \alpha_{131} = 10, \; \beta_{131} = 0, \\ m_{132} = 0, \; n_{132} = 0, \; \alpha_{132} = 0, \; \beta_{132} = 20, \\ m_{232} = 40, \; n_{232} = 60, \; \alpha_{232} = 20, \; \beta_{232} = 20, \\ m_{151} = 40, \; n_{151} = 50, \; \alpha_{151} = 10, \; \beta_{151} = 0, \\ m_{152} = 10, \; n_{152} = 10, \; \alpha_{152} = 0, \; \beta_{152} = 0, \\ m_{251} = 10, \; n_{251} = 10, \; \alpha_{251} = 0, \; \beta_{251} = 0, \\ m_{431} = 0, \; n_{431} = 0, \; \alpha_{431} = 0, \; \beta_{431} = 10, \end{cases}$$

and the remaining values of $m_{ijk}, n_{ijk}, \alpha_{ijk}$, and β_{ijk} are zero.

Step 7 Putting the values of $m_{ijk}, n_{ijk}, \alpha_{ijk}$ and β_{ijk} into

$$\tilde{x}_{ij} = (m_{ijk}, n_{ijk}, \alpha_{ijk}, \beta_{ijk})_{LR},$$

for all i, j, k, the fuzzy optimal solution is:

$$\begin{cases} \tilde{x}_{131} = (10, 20, 10, 0)_{LR}, \\ \tilde{x}_{132} = (0, 0, 0, 20)_{LR}, \\ \tilde{x}_{232} = (40, 60, 20, 20)_{LR}, \\ \tilde{x}_{151} = (40, 50, 10, 0)_{LR}, \\ \tilde{x}_{251} = (10, 10, 0, 0)_{LR}, \\ \tilde{x}_{431} = (0, 0, 0, 10)_{LR}, \\ \tilde{x}_{152} = (10, 10, 0, 0)_{LR} \end{cases}$$

and the remaining values of $\tilde{x}_{ijk} = (0, 0, 0, 0)_{LR}$, for all i, j, k.

Step 8 Putting the values of

$$\begin{cases} \tilde{x}_{131}, \tilde{x}_{131}, \\ \tilde{x}_{211}, \tilde{x}_{212}, \\ \tilde{x}_{231}, \tilde{x}_{232}, \\ \tilde{x}_{411}, \tilde{x}_{412}, \\ \tilde{x}_{431}, \tilde{x}_{432}, \\ \tilde{x}_{451}, \tilde{x}_{452}, \\ \tilde{x}_{251}, \tilde{x}_{252}, \\ \tilde{x}_{151}, \tilde{x}_{152}, \end{cases}$$

into

$$((8, 10, 2, 2)_{LR} \otimes \tilde{x}_{131} \oplus (4, 8, 3, 2)_{LR} \otimes \tilde{x}_{132} \oplus (8, 10, 4, 4)_{LR} \otimes \tilde{x}_{211} \oplus (6, 8, 4, 4)_{LR} \otimes \tilde{x}_{212} \oplus (9, 12, 6, 3)_{LR} \otimes \tilde{x}_{231} \oplus (3, 6, 2, 3)_{LR} \otimes \tilde{x}_{232} \oplus (0, 0, 0, 0)_{LR} \otimes \tilde{x}_{411} \oplus (0, 0, 0, 0)_{LR} \otimes \tilde{x}_{412} \oplus (0, 0, 0, 0)_{LR} \otimes \tilde{x}_{431} \oplus (0, 0, 0, 0)_{LR} \otimes \tilde{x}_{432} \oplus (0, 0, 0, 0)_{LR} \otimes \tilde{x}_{251} \oplus (0, 0, 0, 0)_{LR} \otimes \tilde{x}_{252} \oplus (0, 0, 0, 0)_{LR} \otimes \tilde{x}_{451} \oplus (0, 0, 0, 0)_{LR} \otimes \tilde{x}_{452} \oplus (0, 0, 0, 0)_{LR} \otimes \tilde{x}_{151} \oplus (0, 0, 0, 0)_{LR} \otimes \tilde{x}_{152}),$$

the minimum total fuzzy transportation cost is $(200, 560, 180, 600)_{LR}$.

8.4.2 Determination of the Fuzzy Optimal Solution of the Fully Fuzzy Solid Transshipment Problem Using the Method Based on the Tabular Representation

Using the proposed method for the solution of the fully fuzzy solid transshipment problem based on tabular representation, the fuzzy optimal solution of problem considered in Example 21, can be obtained by following the below steps:

Step 1 The balanced fully fuzzy solid transshipment problem, obtained from Step 1 to Step 3 of the solution process shown in Sect. 8.4.1, can be represented by Table 8.6.

Step 2 Due to Step 2 of the method, proposed in Sect. 8.3.2, add an amount of the fuzzy buffer stock

$$\tilde{P} = \sum_{i \in N_{PS}} \tilde{a}_i \oplus \sum_{i \in N_S} \tilde{a}_i' = \sum_{j \in N_{PD}} \tilde{b}_j \oplus \sum_{j \in N_D} \tilde{b}_j' = \sum_{k \in S_C} \tilde{e}_k = (110, 150, 40, 50)_{LR}$$

into the fuzzy availability and fuzzy demand corresponding to each intermediate node, and introduce a dummy conveyance E_3 with the fuzzy capacity $\tilde{e}_3 = (110, 150, 40, 50)_{LR}$.

After adding the fuzzy buffer stock $\tilde{P}$ and introducing the dummy conveyance E_3, Table 8.6 is converted into Table 8.7.

Step 3 Following Step 4 of the method proposed in Sect. 8.3.2, Table 8.7 can be split into four crisp solid transportation tables, i.e., Tables 8.8, 8.9, 8.10 and 8.11.

Step 4 The optimal solution of crisp solid transportation problems represented by Tables 8.8, 8.9, 8.10 and 8.11, respectively, are:

Table 8.6 Tabular representation of balanced fully fuzzy solid transshipment problem

							Capacity
	E_1		E_1		E_1		$(60, 80, 20, 10)_{LR}$
		E_2		E_2		E_2	$(50, 70, 20, 40)_{LR}$
	1		3		5		Availability
1	$(M, M, 0, 0)_{LR}$	$(M, M, 0, 0)_{LR}$	$(8, 10, 2, 2)_{LR}$	$(4, 8, 3, 2)_{LR}$	$(0, 0, 0, 0)_{LR}$	$(0, 0, 0, 0)_{LR}$	$(60, 80, 20, 20)_{LR}$
2	$(8, 10, 4, 4)_{LR}$	$(6, 8, 4, 4)_{LR}$	$(9, 12, 6, 3)_{LR}$	$(3, 6, 2, 3)_{LR}$	$(0, 0, 0, 0)_{LR}$	$(0, 0, 0, 0)_{LR}$	$(50, 70, 20, 20)_{LR}$
4	$(0, 0, 0, 0)_{LR}$	$(0, 0, 0, 0)_{LR}$	$(0, 0, 0, 0)_{LR}$	$(0, 0, 0, 0)_{LR}$	$(0, 0, 0, 0)_{LR}$	$(0, 0, 0, 0)_{LR}$	$(0, 0, 0, 10)_{LR}$
	–		$(50, 80, 30, 50)_{LR}$		$(60, 70, 10, 0)_{LR}$		

Table 8.7 Tabular representation of balanced fuzzy solid transportation problem after adding a fuzzy buffer stock

										Capacity
	E_1			E_1			E_1			$(60, 80, 20, 10)_{LR}$
		E_2			E_2			E_2		$(50, 70, 20, 40)_{LR}$
			E_3			E_3			E_3	$(110, 150, 40, 50)_{LR}$
	1			3			5			Availability
1	$(M, M, 0, 0)_{LR}$	$(M, M, 0, 0)_{LR}$	$(0, 0, 0, 0)_{LR}$	$(8, 10, 2, 2)_{LR}$	$(4, 8, 3, 2)_{LR}$	$(0, 0, 0, 0)_{LR}$	$(0, 0, 0, 0)_{LR}$	$(0, 0, 0, 0)_{LR}$	$(0, 0, 0, 0)_{LR}$	$(170, 230, 60, 70)_{LR}$
2	$(8, 10, 4, 4)_{LR}$	$(6, 8, 4, 4)_{LR}$	$(0, 0, 0, 0)_{LR}$	$(9, 12, 6, 3)_{LR}$	$(3, 6, 2, 3)_{LR}$	$(0, 0, 0, 0)_{LR}$	$(0, 0, 0, 0)_{LR}$	$(0, 0, 0, 0)_{LR}$	$(0, 0, 0, 0)_{LR}$	$(50, 70, 20, 20)_{LR}$
4	$(0, 0, 0, 0)_{LR}$	$(0, 0, 0, 0)_{LR}$	$(0, 0, 0, 0)_{LR}$	$(0, 0, 0, 0)_{LR}$	$(0, 0, 0, 0)_{LR}$	$(0, 0, 0, 0)_{LR}$	$(0, 0, 0, 0)_{LR}$	$(0, 0, 0, 0)_{LR}$	$(0, 0, 0, 0)_{LR}$	$(0, 0, 0, 10)_{LR}$
	$(110, 150, 40, 50)_{LR}$			$(50, 80, 30, 50)_{LR}$			$(60, 70, 10, 0)_{LR}$			

Table 8.8 Tabular representation of first crisp transportation problem

										Capacity
	E_1			E_1			E_1			30
		E_2			E_2			E_2		30
			E_3			E_3			E_3	70
	1			3			5			Availability
1	M	M	0	9	5.75	0	0	0	0	110
2	9	7	0	9.75	4.75	0	0	0	0	30
4	0	0	0	0	0	0	0	0	0	0
	70			20			50			

Table 8.9 Tabular representation of second crisp transportation problem

										Capacity
	E_1			E_1			E_1			30
		E_2			E_2			E_2		30
			E_3			E_3			E_3	70
	1			3			5			Availability
1	M	M	0	9	5.75	0	0	0	0	110
2	9	7	0	9.75	4.75	0	0	0	0	30
4	0	0	0	0	0	0	0	0	0	0
	70			20			50			

Table 8.10 Tabular representation of third crisp transportation problem

										Capacity
	E_1			E_1			E_1			20
		E_2			E_2			E_2		20
			E_3			E_3			E_3	40
	1			3			5			Availability
1	$\frac{M}{2}$	$\frac{M}{2}$	0	5.5	4.5	0	0	0	0	60
2	6	5	0	6.75	3.75	0	0	0	0	20
4	0	0	0	0	0	0	0	0	0	0
	40			30			10			

Table 8.11 Tabular representation of fourth crisp transportation problem

										Capacity
	E_1			E_1			E_1			10
		E_2			E_2			E_2		40
			E_3			E_3			E_3	50
	1			3			5			Availability
1	$\frac{M}{4}$	$\frac{M}{4}$	0	3	2.5	0	0	0	0	70
2	3.5	3	0	3.75	2.25	0	0	0	0	20
4	0	0	0	0	0	0	0	0	0	10
	50			50			0			

$$\begin{cases} m_{113} - \alpha_{113} = 70, \\ m_{151} - \alpha_{151} = 40, \\ m_{232} - \alpha_{232} = 20, \\ m_{252} - \alpha_{252} = 10, \\ \alpha_{113} = 40, \\ \alpha_{131} = 10, \\ \alpha_{151} = 10, \\ \alpha_{232} = 20, \\ n_{113} - m_{113} = 40, \\ n_{131} - m_{131} = 10, \\ n_{151} - m_{151} = 10, \\ n_{232} - m_{232} = 20, \\ \beta_{113} = 40, \\ \beta_{131} = 10, \\ \beta_{151} = 10, \\ \beta_{232} = 20, \\ \beta_{411} = 10, \end{cases}$$

respectively.

Step 5 By solving the equations obtained in Step 4, the obtained values of m_{ijk}, n_{ijk}, α_{ijk} and β_{ijk} are:

$$\begin{cases} m_{113} = 110, & n_{113} = 150, & \alpha_{113} = 40, & \beta_{113} = 40, \\ m_{151} = 50, & n_{151} = 60, & \alpha_{151} = 10, & \beta_{151} = 0, \\ m_{232} = 40, & n_{232} = 60, & \alpha_{232} = 20, & \beta_{232} = 20, \\ m_{252} = 10, & n_{252} = 10, & \alpha_{252} = 0, & \beta_{252} = 0, \\ m_{131} = 10, & n_{131} = 20, & \alpha_{131} = 10, & \beta_{131} = 0, \\ m_{132} = 0, & n_{132} = 0, & \alpha_{132} = 0, & \beta_{132} = 20, \\ m_{133} = 0, & n_{133} = 0, & \alpha_{133} = 0, & \beta_{133} = 10, \\ m_{411} = 0, & n_{411} = 0, & \alpha_{411} = 0, & \beta_{411} = 10 \end{cases}$$

and the remaining values of $m_{ijk}, n_{ijk}, \alpha_{ijk}, \beta_{ijk}$ are zero.

Step 6 Putting the values of $m_{ijk}, n_{ijk}, \alpha_{ijk}$ and β_{ijk} into $\tilde{x}_{ijk} = (m_{ijk}, n_{ijk}, \alpha_{ijk}, \beta_{ijk})_{LR}$, the fuzzy optimal solution becomes

$$\begin{cases} \tilde{x}_{113} = (110, 150, 40, 40)_{LR}, \\ \tilde{x}_{151} = (50, 60, 10, 0)_{LR}, \\ \tilde{x}_{232} = (40, 60, 20, 20)_{LR}, \\ \tilde{x}_{252} = (10, 10, 0, 0)_{LR}, \\ \tilde{x}_{131} = (10, 20, 10, 0)_{LR}, \\ \tilde{x}_{132} = (0, 0, 0, 20)_{LR}, \\ \tilde{x}_{133} = (0, 0, 0, 10)_{LR}, \\ \tilde{x}_{411} = (0, 0, 0, 10)_{LR}, \end{cases}$$

and the remaining values of $\tilde{x}_{ijk}$ are zero.

Step 7 Putting the values of

$$\begin{cases} \tilde{x}_{111}, \tilde{x}_{112}, \tilde{x}_{113}, \\ \tilde{x}_{131}, \tilde{x}_{132}, \tilde{x}_{133}, \\ \tilde{x}_{151}, \tilde{x}_{152}, \tilde{x}_{153}, \\ \tilde{x}_{211}, \tilde{x}_{212}, \tilde{x}_{213}, \\ \tilde{x}_{231}, \tilde{x}_{232}, \tilde{x}_{233}, \\ \tilde{x}_{251}, \tilde{x}_{252}, \tilde{x}_{253}, \\ \tilde{x}_{411}, \tilde{x}_{412}, \tilde{x}_{413}, \\ \tilde{x}_{431}, \tilde{x}_{432}, \tilde{x}_{433}, \\ \tilde{x}_{451}, \tilde{x}_{452}, \tilde{x}_{453} \end{cases}$$

into

$$\begin{aligned} &((M, M, 0, 0)_{LR} \otimes \tilde{x}_{111} \oplus (M, M, 0, 0)_{LR} \otimes \tilde{x}_{112} \oplus (0, 0, 0, 0)_{LR} \otimes \tilde{x}_{113} \oplus (8, 10, 2, 2)_{LR} \otimes \\ &\tilde{x}_{131} \oplus (4, 8, 3, 2)_{LR} \otimes \tilde{x}_{132} \oplus (0, 0, 0, 0)_{LR} \otimes \tilde{x}_{133} \oplus (0, 0, 0, 0)_{LR} \otimes \tilde{x}_{151} \oplus (0, 0, 0, 0)_{LR} \\ &\otimes \tilde{x}_{152} \oplus (0, 0, 0, 0)_{LR} \otimes \tilde{x}_{153} \oplus (8, 10, 4, 4)_{LR} \otimes \tilde{x}_{211} \oplus (6, 8, 4, 4)_{LR} \otimes \tilde{x}_{212} \oplus (0, 0, 0, 0)_{LR} \\ &\otimes \tilde{x}_{213} \oplus (0, 0, 0, 0)_{LR} \otimes \tilde{x}_{251} \oplus (0, 0, 0, 0)_{LR} \otimes \tilde{x}_{252} \oplus (0, 0, 0, 0)_{LR} \otimes \tilde{x}_{253} \oplus (9, 12, 6, 3)_{LR} \\ &\otimes \tilde{x}_{231} \oplus (3, 6, 2, 3)_{LR} \otimes \tilde{x}_{232} \oplus (0, 0, 0, 0)_{LR} \otimes \tilde{x}_{411} \oplus (0, 0, 0, 0)_{LR} \otimes \tilde{x}_{412} \oplus (0, 0, 0, 0)_{LR} \\ &\otimes \tilde{x}_{413} \oplus (0, 0, 0, 0)_{LR} \otimes \tilde{x}_{431} \oplus (0, 0, 0, 0)_{LR} \otimes \tilde{x}_{432} \oplus (0, 0, 0, 0)_{LR} \otimes \tilde{x}_{433} \oplus (0, 0, 0, 0)_{LR} \\ &\otimes \tilde{x}_{451} \oplus (0, 0, 0, 0)_{LR} \otimes \tilde{x}_{452} \oplus (0, 0, 0, 0)_{LR} \otimes \tilde{x}_{453}), \end{aligned}$$

the minimum total fuzzy transportation cost is $(200, 560, 180, 600)_{LR}$.

8.4.3 *Interpretation of Results*

By using the proposed method for the solution of the fully fuzzy solid transshipment problem, the minimum total fuzzy transportation cost is obtained as $(200, 560, 180, 600)_{LR}$ which can be interpreted as follows:

1. the least amount of the minimum total transportation cost is 20 units,

2. the most possible amount of the minimum total transportation cost lies between 200 units and 560 units,
3. the greatest amount of the minimum total transportation cost is 1160 units, i.e., the minimum total transportation cost will always be greater than 20 units and less than 1160 units, and the maximum chances are that the minimum total transportation cost will lie between 200 units and 560 units.

8.5 A Comparison of Results Obtained

The results obtained by the existing method by Ghatee and Hashemi [1] and by the methods proposed in this chapter and in the previous chapters, are briefly compared in Table 8.12.

The results shown in Table 8.12 can be explained as follows:

1. The well known existing method by Ghatee and Hashemi [1] is proposed for solving the balanced fully fuzzy transshipment problems. Since the balanced fully fuzzy transportation problems are special types of the balanced fully fuzzy transshipment problems, then this method can also be used for solving these type of problems. Since the existing fully fuzzy transshipment problem (cf. Ghatee and Hashemi [3], and Example 17, and the fully fuzzy transportation problem illustrated on Example 18, are the balanced problems so that they can be solved by using the known Ghatee and Hashemi's [1] method [1]. However, the fully fuzzy transportation problems illustrated on Examples 15, 17, and the fully fuzzy transshipment problem illustrated on Example 19, are the unbalanced problems, then they can not be solved by using the known existing method by Ghatee and Hashemi [1].
   There is however no link between the fully fuzzy transshipment problems and the fully fuzzy solid transportation problems. Moreover, the fully fuzzy solid transshipment problems are generalizations of the fully fuzzy transshipment problems. Therefore, neither the fully fuzzy solid transportation problem illustrated on Example 20, nor the fully fuzzy solid transshipment problem illustrated on Example 21, can be solved by the known existing method by Ghatee and Hashemi [1].
2. The new methods proposed in Chap. 4 can only be used for solving such fully fuzzy transportation problems in which all the parameters are either represented by triangular fuzzy numbers or by trapezoidal fuzzy numbers. Similarly, the methods proposed in Chap. 5 can be used for solving such fully fuzzy transportation problems in which all the parameters are represented by the LR flat fuzzy numbers. Since, in the fully fuzzy transportation problem illustrated on Example 17, all the parameters are represented by the LR flat fuzzy numbers so that this problem can not be solved by the methods proposed in Chap. 4 but it can be solved by the methods proposed in Chap. 5.

Table 8.12 Results obtained by using the existing known method by Ghatee and Hashemi [1] and the proposed methods

Exam	Minimum total fuzzy transportation cost					
	Existing method	Methods proposed	Methods proposed	Methods proposed	Methods proposed	Methods proposed
	Ghatee and Hashemi's [1]	Proposed in Chap. 4	Proposed in Chap. 5	Proposed in Chap. 6	Proposed in Chap. 7	Proposed in Chap. 8
2.2	Not applicable	(2100, 2900, 3500, 4200)	(2100, 2900, 3500, 4200)	(2100, 2900, 3500, 4200)	(2100, 2900, 3500, 4200)	(2100, 2900, 3500, 4200)
3.1	Not applicable	Not applicable	$(5800, 8400, 2800, 2900)_{LR}$	$(5800, 8400, 2800, 2900)_{LR}$	$(5800, 8400, 2800, 2900)_{LR}$	$(5800, 8400, 2800, 2900)_{LR}$
3.5 [3]	$(1924000, 1903300, 7299800)_{LR}$	Not applicable	Not applicable	$(1924000, 1903300, 7299800)_{LR}$	Not applicable	$(1924000, 1903300, 7299800)_{LR}$
4.1	$(4100, 6600, 2000, 2600)_{LR}$	(2100, 4100, 6600, 9200)	$(4100, 6600, 2000, 2600)_{LR}$	$(4100, 6600, 2000, 2600)_{LR}$	$(4100, 6600, 2000, 2600)_{LR}$	$(4100, 6600, 2000, 2600)_{LR}$
4.2	Not applicable	Not applicable	Not applicable	$(360, 560, 270, 350)_{LR}$	Not applicable	$(360, 560, 270, 350)_{LR}$
5.1	Not applicable	Not applicable	Not applicable	Not applicable	$(1900, 1900, 100, 900)_{LR}$	$(1900, 1900, 100, 900)_{LR}$
6.1	Not applicable	Not applicable	Not applicable	Not applicable	Not applicable	$(200, 560, 180, 600)_{LR}$

Since, the fully fuzzy transshipment problems, the fully fuzzy solid transportation problems and the fully fuzzy solid transshipment problems are generalizations of the fully fuzzy transportation problems. Therefore, the existing fully fuzzy transshipment problem (cf. Ghatee and Hashemi [3], and Example 17), and the fully fuzzy transshipment problem illustrated on Example 19, the fully fuzzy solid transportation problem illustrated on Example 20, and the fully fuzzy solid transshipment problem illustrated on Example 21, can not be solved by the methods proposed in Chaps. 4 and 5.

3. Since the methods, proposed in Chap. 6, can be used for solving such balanced and unbalanced fully fuzzy transshipment problems in which all the parameters are represented by the LR flat fuzzy numbers and the fully fuzzy transportation problems are also a special type of the fully fuzzy transhipment problems, then the fully fuzzy transportation problems illustrated on Examples 15, and 17, the existing fully fuzzy transshipment problem proposed by Ghatee and Hashemi [3] illustrated on Example 17, and the fully fuzzy transshipment problem illustrated on Example 19, can be solved by the methods proposed in Chap. 6.
   There is however no link between the fully fuzzy transshipment problems and the fully fuzzy solid transportation problems. Moreover, the fully fuzzy solid transshipment problems are generalizations of the fully fuzzy transshipment problems. Therefore, neither the fully fuzzy solid transportation problem illustrated on Example 20, nor the fully fuzzy solid transshipment problem illustrated on Example 21, can be solved by the methods proposed in Chap. 6.
4. Since, the methods proposed in Chap. 7, can be used for solving such balanced and unbalanced fully fuzzy solid transportation problems in which all the parameters are represented by the LR flat fuzzy numbers and the fully fuzzy transportation problems are special types of the fully fuzzy solid transportation problems, then the fully fuzzy transportation problems illustrated on Example 15, Examples 17, 18, and the fully fuzzy solid transportation problem illustrated on Example 20, can be solved by the new methods proposed in Chap. 7.
   There is however no link between the fully fuzzy solid transportation problems and the fully fuzzy transshipment problems. Moreover, the fully fuzzy solid transshipment problems are generalizations of the fully fuzzy solid transportation problems. Therefore, the known existing fully fuzzy transshipment problem proposed by Ghatee and Hashemi [3], illustrated on Example 17), the fully fuzzy transshipment problem illustrated on Example 19, and the fully fuzzy solid transshipment problem illustrated on Example 21, can not be solved by the method proposed in Chap. 7.
5. Since the fully fuzzy transportation problems, the fully fuzzy transshipment problems and the fully fuzzy solid transportation problems are special types of

the fully fuzzy solid transshipment problems, then the new methods proposed in this chapter for solving the fully fuzzy solid transshipment problems can be used for solving all these problems. For the same reason the method proposed in this chapter can be used to find the fuzzy optimal solution of all the problems mentioned above.

8.6 Concluding Remarks

By comparing the results obtained it can be concluded that all the problems which can be solved by using the known existing method by Ghatee and Hashemi [1] and the methods proposed in the previous chapters can also be solved by the new method proposed in this chapter. However, there exist some problems which can be solved by the methods proposed in this chapter but can neither be solved by using any of the existing method nor by the methods proposed in the previous chapters. Hence, it is better to use the methods proposed in this chapter as compared to the known existing method by Ghatee and Hashemi [1] and the methods proposed in the previous chapters.

References

1. M. Ghatee, S.M. Hashemi, Ranking function-based solutions of fully fuzzified minimal cost flow problem. Inf. Sci. **177**, 4271–4294 (2007)
2. S.T. Liu, Fuzzy total transportation cost measures for fuzzy solid transportation problem. Appl. Math. Comput. **174**, 927–941 (2006)
3. M. Ghatee, S.M. Hashemi, Generalized minimal cost flow problem in fuzzy nature: an application in bus network planning problem. Appl. Math. Model. **32**, 2490–2508 (2008)

Chapter 9
Conclusions and Future Research Directions

The results obtained in this book suggest the following conclusions:

1. It is better to use the methods proposed in Chap. 4 as compared to the known existing methods for solving such fully fuzzy transportation problems in which parameters are either represented by the triangular fuzzy numbers or trapezoidal fuzzy numbers.
2. Neither the methods proposed in Chap. 4 nor any existing method can be used to find the optimal solution of such fully fuzzy transportation problems in which the parameters are represented by the LR flat fuzzy numbers. However, the fuzzy optimal solution of all similar type of problems, the problems which can be solved by the existing methods and the methods proposed in Chap. 4, can be obtained by using the method proposed in Chap. 5.
3. The known existing method by Ghatee and Hashemi [1] can be used to find the fuzzy optimal solution of the balanced fully fuzzy transportation problems and the balanced fully fuzzy transshipment problems. However, this known existing method can not be used to find the fuzzy optimal solution of similar types of the unbalanced problems. The fuzzy optimal solution of all these problems can be obtained by using the methods proposed in Chap. 6.
4. It is better to use the methods proposed in Chap. 7 than the existing method by Liu [2] for solving the fully fuzzy solid transportation problems.
5. Neither the existing well known methods nor the methods proposed in Chaps. 4, 5, 6 and 7 can be used to find the fuzzy optimal solution of fully fuzzy solid transshipment, problems. However, the fuzzy optimal solution of these problems can be obtained by using the methods proposed in Chap. 8. These methods can be used to find the fuzzy optimal solution of the single objective fully fuzzy transportation problems, the transshipment problems, the solid transportation problems and the solid transshipment problems. However, these methods can not be used to find the fuzzy optimal solution of similar types of the multi-objective problems. In future, it may expedient be try to extend the methods proposed in Chap. 8 for solving the multi-objective problems. Moreover, it may also be

A. Kaur et al., *Fuzzy Transportation and Transshipment Problems*, Studies in Fuzziness and Soft Computing 385,
https://doi.org/10.1007/978-3-030-26676-9_9

expedient to extend the known existing method by Yang and Liu [3], which is used to find the crisp optimal solution of the single-objective fuzzy solid fixed charge transportation problems, to find the fuzzy optimal solution of the single-objective fuzzy solid fixed charge transportation problems.

Acknowledgements Partial contributions of the Project 691249,RUC-APS: Enhancing and implementing Knowledge based ICT solutions within high Risk and Uncertain Conditions for Agriculture Production Systems (www.rucaps.eu), funded by the European Union under their funding scheme H2020-MSCARISE-2015 and the Shota Rustaveli National Science Foundation of Georgia (SRNFG) Project FR-18-466 is acknowledged by Janusz Kacprzyk.

References

1. M. Ghatee, S.M. Hashemi, Ranking function-based solutions of fully fuzzified minimal cost flow problem. Inf. Sci. **177**, 4271–4294 (2007)
2. S.T. Liu, Fuzzy total transportation cost measures for fuzzy solid transportation problem. Appl. Math. Comput. **174**, 927–941 (2006)
3. L. Yang, L. Liu, Fuzzy fixed charge solid transportation problem and algorithm. Appl. Soft Comput. **7**, 879–889 (2007)

Printed in the United States
By Bookmasters